Soil Science & Management

4th Edition

Soil Science & Management

4th Edition

Edward J. Plaster

THOMSON
DELMAR LEARNING

Australia Canada Mexico Singapore Spain United Kingdom United States

Soil Science & Management, 4th edition
Edward J. Plaster

Business Unit Director:
Susan L. Simpfenderfer

Acquisitions Editor:
Zina M. Lawrence

Developmental Editor:
Andrea Edwards

Executive Production Manager:
Wendy A. Troeger

Production Manager:
Carolyn Miller

Production Editor:
Kathryn B. Kucharek

Technology Project Manager:
Joseph Saba

Executive Marketing Manager:
Donna J. Lewis

Channel Manager:
Nigar Hale

Cover Image:
Eye Wire

Printed in the United States
2 3 4 5 XXX 06 05 04 03

For more information contact
Delmar Learning, 5 Maxwell Drive,
Clifton Park, NY 12065-2919.

Or you can visit our internet site at
http://www.delmarlearning.com

Library of Congress
Cataloging-in-Publication Data
Plaster, Edward J.
Soil science & management /
Edward J. Plaster.—4th ed.
p. cm.
ISBN 0-7668-3935-4
1. Soil science. 2. Soil management. I. Title: Soil science and management. II. Title.
S591.P513 2002
631.4—dc21
2002074051

NOTICE TO THE READER

Contents

Preface

This fourth edition of *Soil Science and Management* continues the primary objectives of earlier editions. First, to acquaint the reader with the soil and water resources of the United States. To fully appreciate the importance of these resources, we must know what they are. Second, to present soil science theory tied to the practice of those who use the soil. Third, to stress the sustainable use of soil and water resources by devoting some detail to such subjects as soil and water conservation, conservation tillage, nutrient management, Best Management Practices, and sustainable agriculture.

The most obvious change in this edition is the reordering of chapters. The chapter "Soil Classification and Survey" has become chapter 3, just after "Soil Origin and Development." This move places the related topics of soil pedology and classification together. "Physical Properties of Soils" moved to chapter 4, followed by the chapters on soil life and organic matter. Other chapters follow in their prior order.

Of course I have updated information such as the addition of a twelfth soil order and data on national soil usage, erosion rates, and other matters. The national soil map, appendix 3, has been replaced with a new map that includes the new soil orders added since the first edition of this text.

In this fourth edition I have added some insights of ecosystem ecology, the study of the flow of energy and matter within and between ecosystems. This changed the description of the process of organic matter decay somewhat and led me to add protista and methane bacteria to the chapter on life. An important addition in this light is the interaction of soil and global climate, a topic of great current importance. Also a discussion of human alteration of the global nitrogen cycle and its consequences, such as ocean hypoxic zones, has been added.

Most chapters have been modified to some degree; details can be found in the instructor's guide. For example, "Soil Origin and Development" expands the history of soil study, and "Soil Classification and Survey" covers the lower levels of the soil taxonomy in greater detail than prior editions. The chapter "Life in the Soil" has been greatly altered.

"Organic Amendments" has also been rewritten, including more information on the health and ecological consequences of inadequate nutrient management. Important new information on "stem-girdling roots" of landscape trees appears in "Horticultural Uses of Soil." "Urban Soil" has been expanded, with new information on modified and structured soils as well as other matters.

As in earlier editions, chapters are arranged to enhance learning, with introductory objectives and terms, plentiful graphics, chapter summary, and review questions at the end. Two major changes will be immediately obvious. First, review questions have been rewritten to require short essay answers. Some of these take the form of case studies. This change answers requests from reviewers, but also reflects results of a survey conducted by the author's college, which revealed that critical thinking and writing were important nontechnical skills desired by employers.

Second, the enrichment exercises for each chapter now include numerous Internet activities. The author recognizes that Web sites tend to be transitory and hopes that most remain stable for the life of this edition.

Most are government or university extension sites, and pages that are part of an instructor's course, even if available to the public, have been avoided. Some activities involve Internet searches rather than specific sites. Instructors who wish to assign any of these activities may want to check them regularly. The author and Delmar affirm that the Web site URLs referenced herein were accurate at the time of printing. However, due to the fluid nature of the Internet, we cannot guarantee their accuracy for the life of the edition.

A final note about the sources of photographs in this text: if there is no attribution, they were taken by the author, and if no location is noted, they are probably Minnesota scenes.

Author's Acknowledgments

I would like to thank Drs. Margaret Davis and Sarah Hobbe for their instruction in the science of ecology and for useful information on global climate change. Katy Deshotels-Moore and Jo Anna Hebberger contributed photographs to this edition that filled a couple of holes in the graphics program. As usual, I am greatly obligated to the United States Department of Agriculture for the many photographs used in this text and the volumes of information they make available. I also greatly appreciate the input of my own students, as well as that of reviewers and instructors who use this text. I thank you all.

Delmar and the author wish to thank the following individuals who devoted their time and professional experience reviewing this manuscript:

Philip Gibson
Gwinnett Technical College
Lawrenceville, GA

Galen Zumbach
Creston High School
Creston, IA

Charles Speck
Warren Central High School
Bowling Green, KY

Sherry Wiggs
Grand Saline High School
Grand Saline, TX

CHAPTER 1

The Importance of Soil

OBJECTIVES

After completing this chapter, you should be able to:

- summarize the ecological functions of soil and its role in recycling resources needed for plant growth
- describe four ways plants use soil
- explain how soil is a three-phase system
- list and explain some agricultural and engineering uses of soil
- discuss the concept of soil quality

TERMS TO KNOW

anchorage
Best Management Practices
cropland
hardpans
hydroponic crops
load-bearing capacity
nutrients
photosynthesis
pore space
prime farmland
respiration
shrink-swell potential
soil aeration
soil air
soil degradation
soil matrix
soil quality
soil solution
waterlogged soil

Soil Shapes Human History

Life and human civilization depend, at their base, on the planet's limited soil and water resources. We have ample historical proof. Most early civilizations in the Old World grew and thrived on the rich soil and waters of major river floodplains. Early Chinese culture, for instance, began to develop six thousand to seven thousand years ago on the floodplains of the Yellow River, where periodic flooding deposited fresh soil for agriculture and canals could carry river water to fields for irrigation. Similarly, farming began to flourish some ten thousand years ago in the Mideast along major rivers.

The famous Greek historian Herodotus passed on to us his notes about the dependence of Egypt on the soil supplied by the flooding of the Nile. In 340 B.C.E., in *The History of Herodotus,* he observed that the black, crumbly, silty deposits of the river were easy to work and productive—that these soils could produce plentiful food with less toil. These soils, he noted, were the foundation for Egyptian civilization.

North America too depends on its soil and water, and its history was influenced by its soils. Part of its success stands on rich soil and water resources; those of the United States are described later in this chapter and throughout the text. But we have often not been wise in their use. The Dust Bowl of the 1930s, caused by drought, soil misuse, and widespread wind erosion, drove farmers out of several of the middle plains states. At its peak, severe wind erosion damaged some 150,000 square miles of prairie farmland (figure 1-1). Photographs from that era create striking and poignant images of people driven off their land by the destruction of the Dust Bowl and scattered to new lives elsewhere (figure 1-2).

Soil and water resource problems also cross national borders, sometimes in ways hard to imagine. For instance, North and South America and Europe receive dust blown over the Atlantic Ocean from the desert lands of Africa. As soils of Africa degrade, that movement of dust has been increasing, and it contains not only soil particles, but also spores of plant diseases, chemicals like arsenic and pesticides, and even insects. More obviously, shortages of soil and water resources cause conflicts and migration of refugees.

In the future, soil will become even more crucial. World population doubles every forty years, yet only about 7 percent of the earth's surface is suitable for agriculture. Of that land, some is being lost to degradation

FIGURE 1-1 A Dust Bowl "roller" moves across a Colorado landscape, carrying off the soil and damaging everything in its path. *(Courtesy of USDA NRCS)*

FIGURE 1-2 A famous Dorothea Lange photograph of a migrant worker in California displaced by the Dust Bowl. Many poignant images of the misery caused by the Dust Bowl remain. *(Courtesy of USDA NRCS)*

and urbanization. It cannot easily be replaced: soil is a nonrenewable resource within the time frame of a human generation.

Many experts have noted that part of the rhythm of human history is the rise and fall of cultures founded on the use, abuse, and final exhaustion of soil and water resources. But this is not simply a problem of the past, but of today. Human society, indeed, most life is possible only because the earth's crust is dusted with a bit of soil where we can grow food. This book is dedicated to describing those soil and water resources and how we can use them wisely. We begin here by looking at soil's ecological functions in supporting life on the planet, its role as a medium for plant growth, and its use by humans.

Soil Is a Life-Supporting Layer of Material

Although we often take soil for granted, it is a very thin and often fragile layer of life-supporting material. The earth, as shown in figure 1-3, consists of a solid part (core, mantle, and crust) and the atmosphere surrounding it. The continental crust, made of rock, is about 50 miles thick, and the atmosphere is about 170 miles deep. The soil forms a very thin interface between the two.

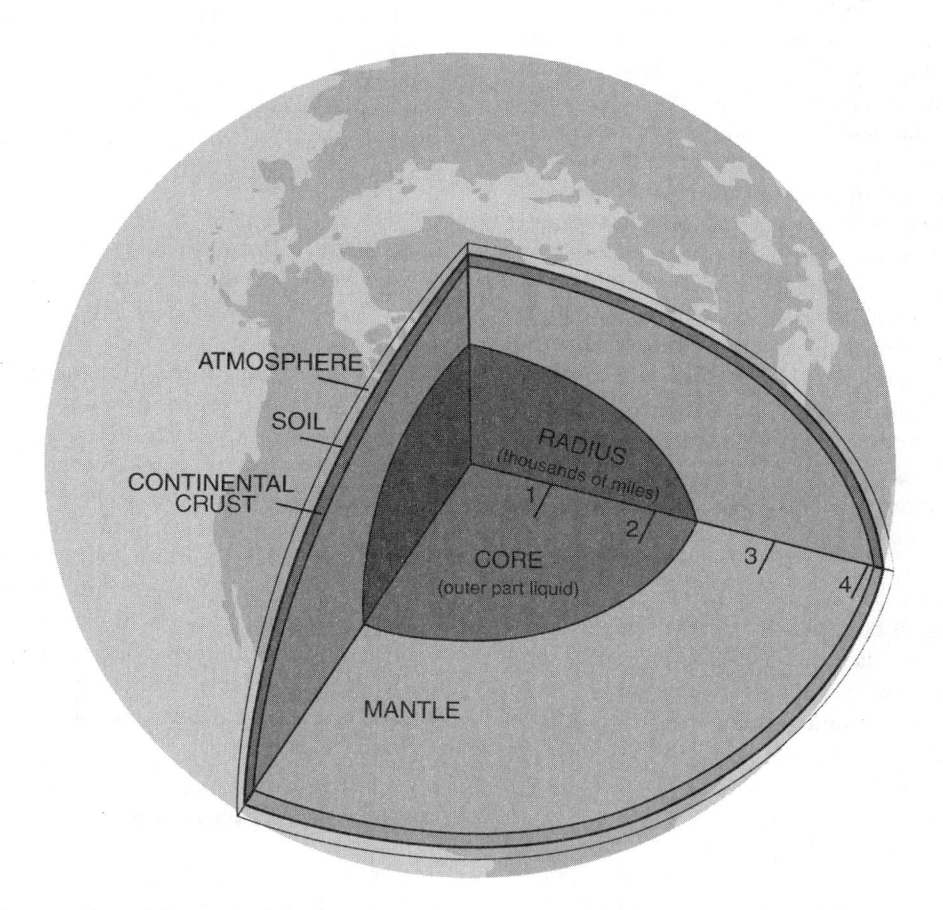

FIGURE 1-3 Cross section of the earth through a continent. The soil is a very thin layer between the atmosphere and the earth's crust. If the soil were drawn to scale, it would be 1/2,000,000 of an inch thick.

The atmosphere, crust, and soil interact to provide plants and animals with the resources they need. Living things need proper temperature, oxygen, water, carbon (the basic element of all living bodies), and other nutrients. These factors are exchanged in the soil, as shown in figure 1-4, usually in cycles that allow elements to be recycled and stored rather than lost. Although later chapters will discuss the cycles in greater detail, this chapter will preview them briefly.

Temperature. Plant roots grow best in certain soil temperature ranges. For instance, most plant roots in temperate climates grow when the soil temperature is above 40°F to 50°F. The cool-season grasses grown in northern states cease root growth at soil temperatures about 85°F. Seed germination also depends on soil temperature. Wheat seed, for example, germinates between 40°F and 50°F, while sorghum needs temperatures above 80°F. Soil temperature, and to some degree the air above the soil, is controlled by a heat-exchange mechanism.

Pedestrians standing on a tar road on a hot summer day sense how heat is exchanged as the road both gains and loses heat. Feet in contact with the pavement get hot, because the road is absorbing energy from the sun. Looking down the road, one sees heat waves rising from the tar, an effect of the road losing excess heat to the air. Energy is also radiating away as light in a wavelength that humans cannot see. In the same way, soil maintains temperatures for growing plants. On a larger scale, this heat exchange influences air temperature, weather, and even global climate.

Gases. Plant roots and other soil organisms need oxygen and give off carbon dioxide as they respire. Some important soil bacteria need nitrogen gas as well. Figure 1-4 shows that these gases pass into and out of the soil to maintain proper amounts of each. In this process of air exchange, the soil also acts to filter and purify the earth's atmosphere.

Water. Water seldom stays in one place long, always on its way to the next stage of its cycle. Water evaporates from the land, lakes, and oceans and forms clouds in the atmosphere. Rain falls from the clouds, moistens the soil, and fills streams and lakes. Much of the water finally reaches the oceans. More evaporation begins the cycle again. Some water seeps deep into the ground where it is held as groundwater. When moisture falls on the soil, however, some water is temporarily stored for plant use. In the process, soil can also purify the water that lands on the earth's surface.

Carbon. Plant leaves collect sunlight to use the suns energy in the process known as **photosynthesis,** which involves converting atmospheric carbon (carbon dioxide) to biological carbon (simple sugars). In the process, light energy is converted to chemical energy usable by plants and creatures that eat plants. Some of the carbon is recycled directly back to the atmosphere by plant and animal **respiration,** while other carbon is recycled by organic matter decay in the soil. In this process, some carbon is retained in the soil as organic matter. Soil acts as a vast carbon reserve keeping carbon dioxide out of the atmosphere, where it would contribute to the greenhouse effect. This actually influences global climate (see chapter 6).

Nutrients. Plant **nutrients** (chemicals a plant needs to grow) also cycle through the soil. Two kinds of nutrient cycles are shown in figure 1-4: the nitrogen cycle and other mineral cycles.

Nitrogen comes entirely from the atmosphere, where it occurs as a gas that plants cannot use. Soil organisms change gaseous nitrogen to forms that plants can use. Some nitrogen recycles as once-living material decays in the soil, while water carries some nitrogen deeper into the ground. Some nitrogen returns to the air when other microbes change it back to its original form.

Other nutrients are released from rocks in the earth's crust when the rocks are broken down by weather, plants, and other factors. These nutrients are continuously reused by plants until some return deep into the ground by leaching, get washed into the ocean, or are removed by cropping.

Soil Is a Medium for Plant Growth

In the broad view, soil has important ecological functions in recycling resources needed for plant growth. In the narrow view, an individual plant depends on soil to supply four needs: **anchorage,** water, oxygen, and nutrients for roots. Let's look at these four needs next.

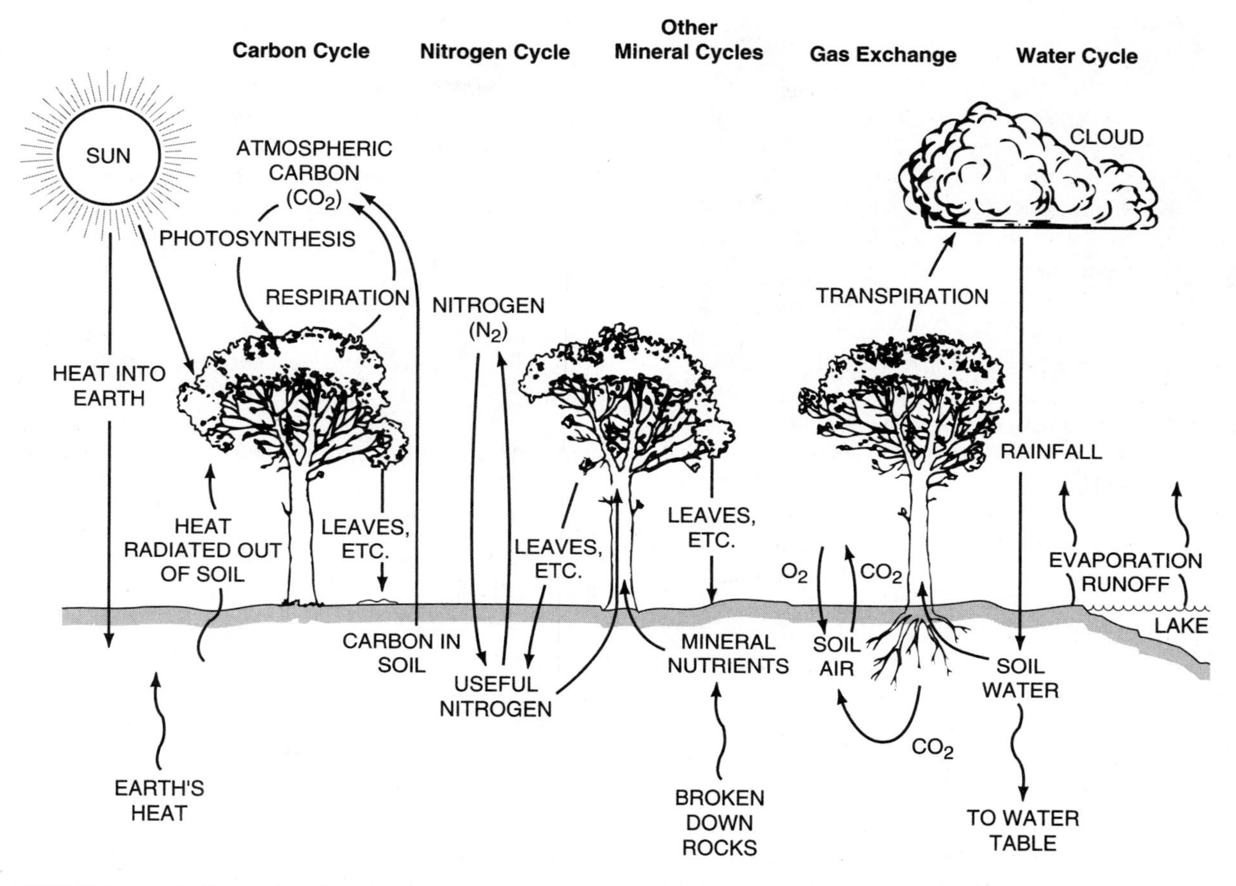

FIGURE 1-4 Cycling and exchange between atmosphere, crust, and soil. The soil temporarily stores resources needed for plant growth.

Anchorage. In deep soil, where roots grow freely, plants are firmly supported, or anchored, so they can grow to reach for sunlight. When people grow plants in ways that deprive plants of soil support, artificial support is often required. Growers of **hydroponic crops** (roots growing not in soil but in fertilizer solutions) often support plants with a wire framework. Landscapers may stake or "guy" a newly planted tree until the tree is firmly rooted, though staking weakens or even damages the trunk and is no longer recommended except in special cases. Poorly anchored trees can even cause serious safety or economic issues. Figure 1-5 shows a tree toppled by a windstorm because rebuilding a sidewalk severed support roots.

Water. Because roots are a plant's best water-absorbing body, soil supplies nearly all the water a plant uses. For each pound of dry matter produced by growth, plants obtain between 200 and 1,000 pounds of water from the soil for photosynthesis, sap flow, and other uses. It is obvious that the water-holding capacity of a soil is important in its agricultural use.

Oxygen. Except for some microscopic organisms, all living creatures, including plants, need oxygen. Plants release oxygen during photosynthesis but consume it during respiration. The parts of a plant above ground, suspended in an atmosphere that is 21 percent oxygen

FIGURE 1-5 A tree upended by high winds in Minneapolis. Support roots were cut during curb construction, and the tree was no longer well-anchored.

(figure 1-6), have all the oxygen they need. Underground, plant roots and soil organisms use up the oxygen and give off carbon dioxide. As a result, **soil air** has less oxygen and more carbon dioxide than the atmosphere.

In the absence of factors that limit it, the process known as **soil aeration** exchanges soil and atmospheric air to maintain adequate oxygen for plant roots. Aeration varies according to soil condition. Saturated, or **waterlogged soil,** which is completely soaked with water, is an example of a soil with poor aeration. The oxygen content near the surface of a well-aerated soil rarely drops below 20 percent but may approach zero in a saturated soil.

Nutrients. Of seventeen nutrients usually considered to be needed by most plants, plants obtain fourteen

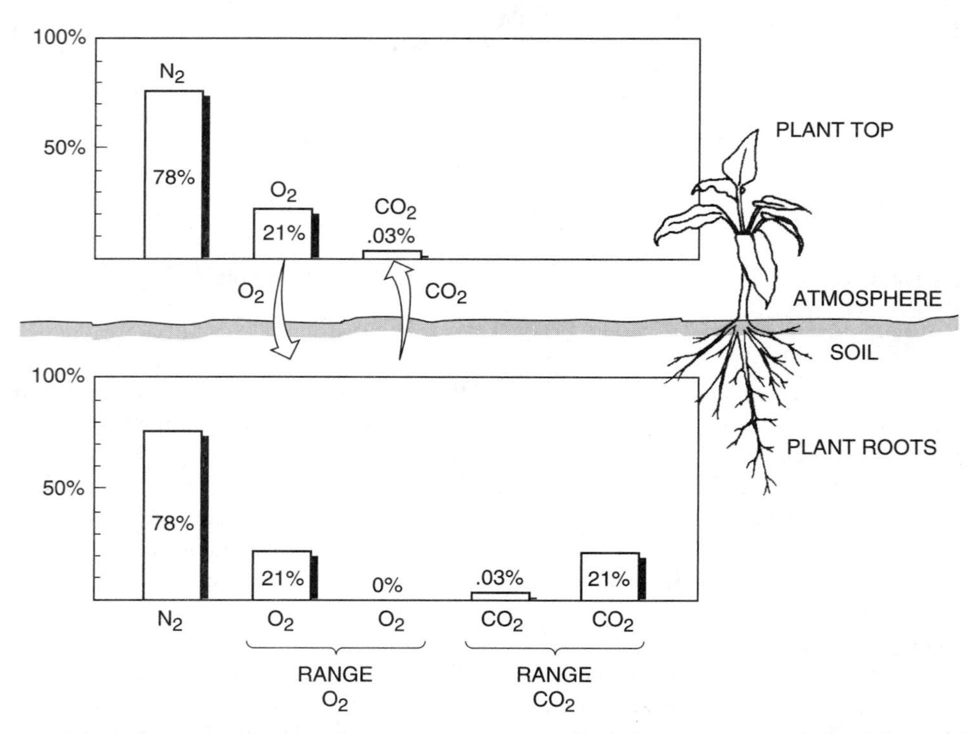

FIGURE 1-6 Soil air and aeration. Most of the gas in air and soil is nitrogen. Above the soil, air is about 21 percent oxygen. In the soil, respiration of living things replaces oxygen with carbon dioxide. Aeration is the process by which carbon dioxide and oxygen are exchanged.

from soil. Carbon, oxygen, and hydrogen come from air and water; the rest are stored in the soil. While leaves are able to absorb some nutrients, roots are specialized for the purpose. Root hairs absorb plant nutrients dissolved in soil water (called the **soil solution**) by an active process that moves nutrients into root cells. The energy that powers this process is produced by respiration in the roots.

Soil: A Three-Phase System

How does soil fulfill the four functions described? Any grower knows that soil is made of solid particles. In most soils, these solid particles largely consist of mineral matter with about 1 percent to 10 percent organic material. Between these solid particles are open spaces, or voids, which we call **pore spaces.** This arrangement of solid particles and pore spaces is called the **soil matrix.** Commonly, about half the volume of soil is solid material and half is pore space.

The pore space is always filled with some combination of air and water. Chemists call solids, liquids, and gases phases of matter, so we can describe soil as a three-phase system. A combination of about half air and half water provides an excellent supply of both for roots, but the actual ratio varies dramatically over time. Right after a heavy rain, pores may be filled almost entirely with water and have almost no air. As the soil dries, air replaces water in the pore spaces, and pores in a very dry soil may contain almost entirely air and very little water. These soils are too wet or too dry for good plant growth.

The amount and composition of soil air varies not only over time, but in space as well. That is, it varies greatly from spot to spot within the soil matrix. A large pore may be rich in oxygen, while a small pore a fraction of an inch away contains water and is devoid of oxygen.

Root Growth. From a root's point of view, it is the pores that matter. As roots grow into the soil, they follow continuous pore spaces between solid particles, absorbing water and nutrients from the soil solution. Root tips easily penetrate pores larger than themselves. Behind the root tip, the diameter of the root increases, pushing aside or deforming the soil to make room for itself. Root tips may also enter smaller pores if the root can exert enough pressure to deform the soil ahead of the tip. If pores are too small or discontinuous, or if soil particles are too difficult to push aside, root growth suffers. Such conditions can result if soil is severely compressed or compacted.

Water reaches roots in two ways: either water flows toward the root, or the root grows into moist soil. Roots spread to contact as much soil as possible. Tree roots, for instance, typically cover an area 60 percent to 100 percent beyond the spread of the tree canopy (figure 1-7). One authority estimated that a mature oak tree has about one million live root tips. Alfalfa roots grow to a depth of five or six feet and may go much deeper in loose soils. However, roots do not grow below the depth of aeration, unless they are specially adapted to do so, so most roots do not extend below five or six feet.

While roots try to grow widely or deeply, they grow best and most thickly where air, water, or nutrients are present in optimal amounts. Roots are able to sense soil gradients of the resources they need and grow towards concentrations of those resources. Where roots find rich supplies of resources, be it air, water, nutrients, or warmth, they will branch and fill that area with numerous roots. For instance, a city tree growing in a small pit in concrete may grow a few rope-like roots under the sidewalk, where the soil is dry and low in oxygen, to the lawn on the other side, where it will develop lush fans of absorbing roots. This pattern of root growth allows plants to most efficiently exploit soil resources.

Oxygen levels especially determine where roots grow. Most roots—even those of large trees—occupy the upper twelve inches of soil, where the greatest amount of oxygen is found. An example of the importance of oxygen levels in setting root distribution is readily seen by removing plastic mulch from an old shrub bed in a landscape planting (figure 1-8). Because plastic is a barrier to air exchange, numerous roots grow between the plastic and soil surface where oxygen is present. For this and other reasons, plastic sheeting has been largely replaced by porous landscape fabric.

Different soil types can affect how roots grow. For instance, some soils have hard layers under the surface, called **hardpans,** which prevent roots from growing deeply. As a result, the plant has less soil to draw on for water and nutrients. The same effect results from bedrock or waterlogged soil near the surface. Numerous other conditions inhibit root growth as well, including low oxygen, nutrient deficiencies, high soil salts, pH extremes,

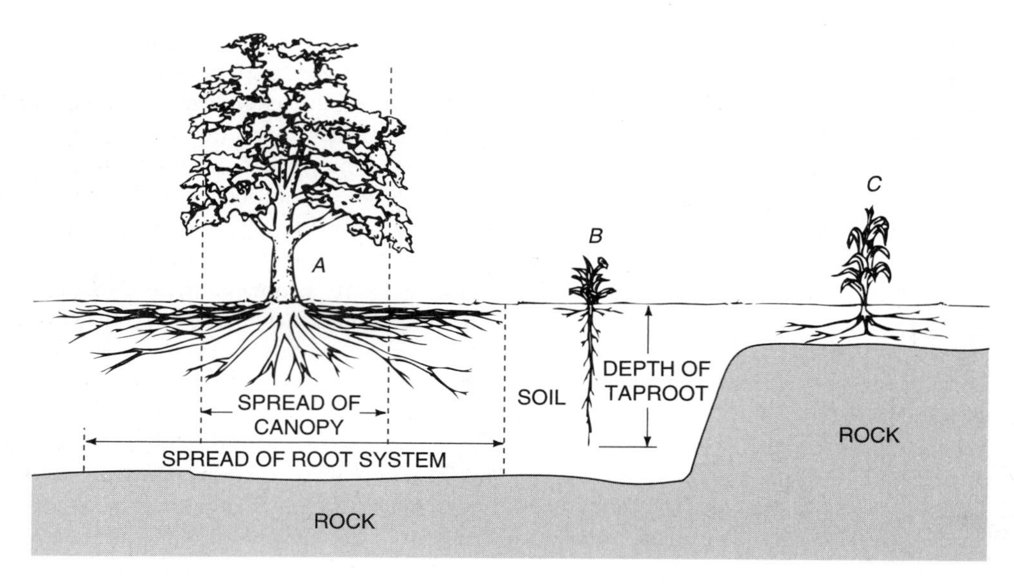

FIGURE 1-7 Plant roots spread far from the plant seeking water and nutrients. *(A)* Most roots spread laterally near the soil surface where the oxygen content is highest. *(B)* A few plants, like the taprooted alfalfa, send roots deep into the soil. *(C)* Conditions that limit the spread of roots, like solid rock near the surface, also limit plant growth.

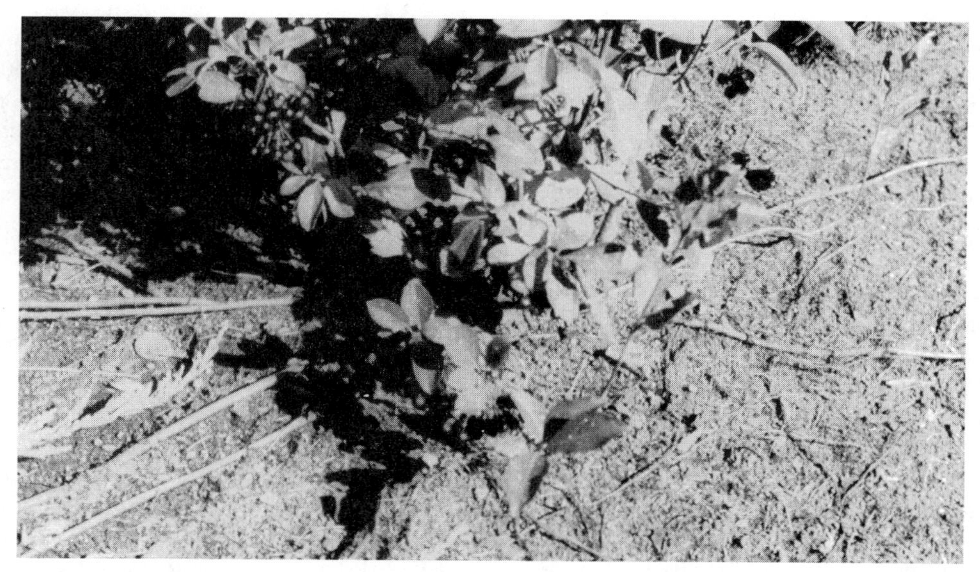

FIGURE 1-8 Shrub roots that grew under plastic mulch, now removed. The plastic acted as a barrier to oxygen, so the roots grew just below the plastic where oxygen was most concentrated.

toxic materials, compaction, or temperature extremes. All of these are covered in later chapters.

The health of the entire plant depends on the health of its root system. Healthy turfgrass roots, for example, lead to turf that resists damage from golf or baseball shoes, diseases, or drought, and will host fewer weeds. Poorly rooted turf is difficult and expensive to maintain, and requires much greater use of chemicals like pesticides.

To summarize, only where a soil has the proper proportion of solid, liquid, and gas can roots grow actively to obtain good anchorage and sufficient water and nutrients.

Agricultural Uses of Soil

Human societies depend on soil to grow food, fiber, timber, and ornamental plants. Different agricultural uses require different soil management practices, as described in the following paragraphs.

Cropland. **Cropland** is land on which soil is worked and crops are planted, cared for, and harvested. Worldwide, the greatest acreage of cropland is devoted to annual crops—those planted and harvested within one growing season. Annual crops include agronomic products such as corn (figure 1-9) and soybeans, fiber plants such as cotton, and horticultural crops like most vegetables. Annuals require yearly soil preparation. This activity gives growers a chance each year to control weeds and to work fertilizer and organic matter into the soil. Because the soil surface is bare much of the time, growers must be careful to keep soil from washing away.

Perennial forages, such as alfalfa, are in the ground for a few years. They may be harvested as hay to feed animals, or be used for grazing. These crops cover the soil completely and so keep the soil from washing away. Because the soil is not worked each year, fertilization is different than for annual crops. Perennial crops also tend to build up and improve the soil.

Perennial horticultural crops include fruits, nuts, and nursery stock (figure 1-10). Crops stay in the ground for three to as many as twenty years. Many crops are clean-cultivated to keep the ground bare and weed-free. Challenges to the grower of horticultural crops are controlling weeds, reducing erosion, preventing soil compaction, and keeping the level of organic matter stable.

FIGURE 1-9 Agronomic crops occupy most of the world's cropland. *(Courtesy of USDA NRCS)*

Grazing Land. Much land in the United States is grazed by cattle and sheep. In the eastern half of the country, pasture is planted to perennial forage. In the western half of the country, which has a drier climate, most grazing is on rangeland (figure 1-11). Range consists largely of native grasses and shrubs, with some nonnative grasses planted through the existing vegetation. Partly because of the size of much rangeland, it is usually loosely managed.

Forest. Foresters probably disturb soil the least, but soil management is still a concern. When trees are harvested after many years' growth, logging equipment tears up the vegetative cover and compacts the soil. Increased erosion results, and the soil is a less desirable medium for growth of newly planted seedlings. Other concerns of forestry include choosing the best trees for each soil type and ensuring good conditions for newly planted seedlings.

Other Uses. Some crops such as flowers, houseplants, and some nursery stock are grown in pots. Plants growing in the tiny root zone of a pot require great care. Growing media for containers may be highly modified to improve their properties or be entirely soilless mixtures of peat, perlite, and other materials.

Landscapers plant and maintain ornamental plants to beautify our surroundings. Their activities also conserve

FIGURE 1-10 A nursery of spruce trees in Minnesota. An important use of soil is to grow trees for nurseries, landscapes, farm shelterbelts, and forests. *(Courtesy of USDA NRCS)*

FIGURE 1-11 Rangeland in Montana. *(Courtesy of USDA NRCS)*

soil in urban areas. Soil knowledge is needed to decide what plants to put where and how to manage the plantings. All too often, landscapers must plant in soil heavily disturbed by construction or in rocky fill. Landscapes stay in place for decades, during which time soil can be subjected to injury from deicing salt, and foot and equipment traffic.

Nonagricultural Uses of Soil

Other human activities—in addition to growing plants—require soils. At its most basic, soil is a surface that people inhabit. Specific nonfarming soil uses include recreation and engineering projects such as building foundations and waste disposal. Let's look at a few of these uses.

Recreation. Recreational uses of the soil surface are important. Sit in an urban park and you will see children in the playground, softball teams on the field, and runners on jogging paths. Golf courses (figure 1-12),

FIGURE 1-12 The design, building, and maintenance of golf courses and parks, like farming, requires knowledge of soil.

FIGURE 1-13 Foundation of a building under construction. The soil must have a good load-bearing capacity and low shrink-swell potential for the foundation to be sound.

parks, and campgrounds are examples of large areas used for recreation. The design of recreational facilities is a specialized skill that requires knowledge of soil properties.

Sports playing fields are probably the most demanding of all soil uses. To grow turf that withstands the punishment of football cleats or soccer shoes challenges even the best of managers.

Soils in the best playing fields are highly engineered mixes of loam, specific sizes of sand, and other ingredients. They may even include a plastic mesh to hold the soil together. The fields generally have several soil layers, are carefully graded and drained, and are well-maintained.

Those who manage playing fields must worry about sideways pressure, or shear, from shoes tearing the soil surface. Playing fields are designed to have good shear resistance. Fields must be of a certain hardness to provide a proper playing surface and reduce injuries. They must dry quickly after a rain yet hold enough water to grow healthy turfgrass. These and other considerations require a knowledge of soil science.

Foundations. Before constructing a home, the builder tests the soil to a depth of several feet. People know that the structural soundness of a building depends not only on the builders skill but also on the soil under the house. Building foundations, for instance, will crack if the soil settles under the building. Even stricter requirements apply to soils for larger structures such as office buildings (figure 1-13). In some towns, landscapers require an engineer's service in designing retaining walls to ensure they will hold firmly in the soil. Civil engineers also need firm soils that settle little for the roadbeds of highways and foundations of bridges.

Examples of important engineering properties include **shrink-swell potential** and **load-bearing capacity.** Many soils swell when wet and shrink as they dry, cracking walls, destroying foundations, and breaking buried pipes. Soils high in clay or organic matter have low load-bearing capacity. Foundations of buildings constructed on such soils may shift and crack. Roads and other structures built on such soils may also have structural problems. In 1989, San Francisco shook to a major earthquake that brought down many buildings—most located on a loose "fill" soil that could not support structures when the earth began to shake.

Waste Disposal. Newspaper headlines about hazardous waste disposal focus attention on the difficulties of safely handling wastes generated by society. Soil has long been used for waste disposal, sometimes with unfortunate results.

Treatment of human sanitary waste often relies on soil because it filters out some of the material, while microorganisms break down organic portions into less dangerous compounds. The common home septic drain field is an example.

One way for sewage treatment plants to handle their end products is to spread them on soil. Sewage sludge may be useful to farmers as a source of nutrients and organic matter, as long as possible harmful materials in the sludge are taken into account. To avoid problems from sludge, its use is regulated by government agencies and may not be legal in some localities.

Sanitary or especially hazardous waste landfills require soils that will not allow hazardous materials to leach into the water table or run into neighboring streams or lakes. The search for landfill sites often arouses conflict in a community. Many people feel landfills cannot be entirely safe, and even those who agree landfills are necessary do not want them nearby.

Building Materials. Before long-distance shipping of building materials became practical, people built their homes of locally available materials, including soil. Early settlers in the Great Plains built huts out of sod, a thick carpet of grass, its roots, and soil. Adobe, a sun-baked mixture of three parts sandy soil to one part clay soil, has been used as a building material for thousands of years and continues to be used in the American Southwest. The walls of Jericho were adobe.

Modern applications of soils are being developed in the search for energy-efficient housing. Buildings can be built underground, into hillsides, or even with soil piled over them. These earth-sheltered buildings are warm in winter and cool in summer, lowering both heating and cooling costs. A few homes have been built of packed earthen walls, constructed by tamping earth into erected forms.

Land Use in the United States

Figure 1-14 shows how nonfederal land was used in 1997, but land use does not remain static. In any given year some forest is cleared for cropland, while somewhere else cropland returns to forest. Market forces such as land or grain prices, technological change such as irrigation, or government programs spur changes in land use.

Figure 1-14 displays data gathered during a USDA inventory of nonfederal land in the United States (excluding Alaska) called the 1997 National Resource Inventory (NRI). NRIs were also conducted in 1982, 1987, and 1992; together they show trends in land use in our nation. The 1997 NRI shows about 80 percent of nonfederal land is evenly divided between crop, forest, and rangelands as the major uses of our soils. Because much of the 402 million acres of federal land is forest or range, they cover a larger proportion of the total United States land surface than shown here.

During the 1982–1997 period, nonfederal rangeland declined by about 3 million acres, cropland decreased by 45 million acres, pastureland shrank by 18 million acres, and forest grew slightly. About 34 million acres of what had been crop and rangeland were enrolled in the Conservation Reserve Program (chapter 20). Some portion of CRP land will later be returned to its prior use.

One land use continues to grow at the expense of other uses—urbanization (figure 1-15). This includes the building of cities, towns, factories, and roads. Comparisons with the 1982 NRI show that during the years 1982–1997 25 million acres of rural land were developed for urban uses. About 7 million acres of **prime farmland,** that best suited for agriculture, were converted to suburban development.

Soil Quality

Soil quality, also called soil health, is the capacity of a specific soil to provide the needed functions for human or natural ecosystems over the long term. That is, it can sustain plant and animal growth and productivity, main-

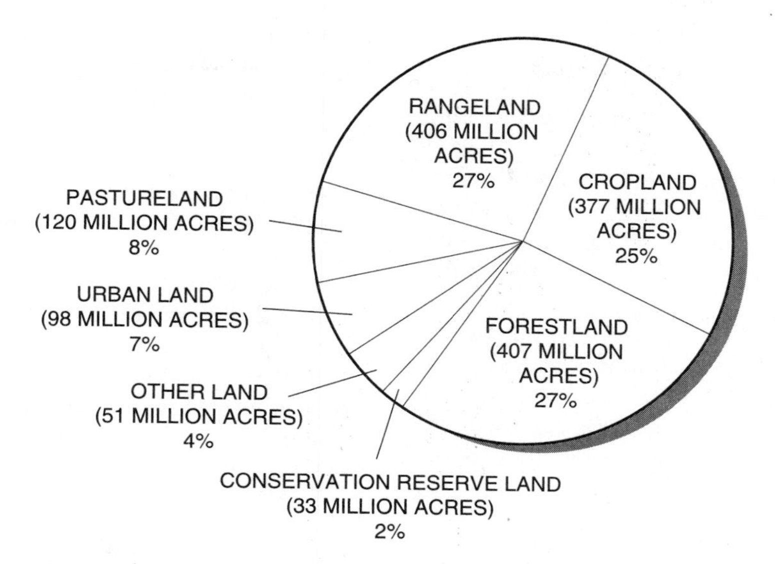

FIGURE 1-14 Use of nonfederal land in the United States (excluding Alaska) in 1997. Urban land includes built-up and rural transportation. *(Source: USDA, 1997 National Resource Inventory)*

FIGURE 1-15 One land use continues to grow at the expense of others—urbanization. Here we see new subdivisions outside Las Vegas, Nevada. *(Courtesy of USDA)*

tain air and water quality, and support human health. Quality soil helps keep a forest healthy and grows excellent crops or attractive landscapes; in short, it performs the functions described in this chapter.

Soil degradation is the loss of soil quality. A United Nations report, *Global Environment Outlook 2000,* estimates degradation of some 4.7 billion acres (1.9 billion hectares) of land worldwide. Examples of soil degradation include:

- erosion of soil from the land
- pollution by industrial chemicals, oil spills, and many others
- conversion of dry grasslands to desert, called desertification
- changes in soil chemistry, like severe changes in soil acidity
- increases in soil salt levels, or salination
- loss of soil organic matter

A major goal of this text is to present ways to prevent such problems and to conserve soil quality. While preserving soil quality involves understanding basic soil processes and management, it also includes specific practices that preserve our soil and water resources while being practical and profitable for a soil user. We call these practices **Best Management Practices,** or BMPs. Use of these BMPs is part of being a good citizen for those who use soil, and the population of every nation has the right to expect its soil users to understand the soil, to follow Best Management Practices, and to stay abreast of new methods for soil quality preservation.

SUMMARY

The significance of soil is best explained by describing its function in three ways. First, soil serves ecological functions that support life on earth, including supporting plant growth, recycling and storing carbon and nutrients, and purifying air and water. These functions are often served through interactions between the earth's crust, soil, and atmosphere.

Second, soil supplies anchorage, water, and nutrients to the plant and oxygen to roots. Soil answers these plant needs because it is a three-phase matrix of solid particles with water and air in the pores between the particles. When the soil is healthy, roots can explore these pores to find the water and nutrients needed by the plant.

Third, people inhabit the soil surface and have both agricultural and nonagricultural uses for soil. Agricultural uses include the production of food, fiber, timber, and ornamental plants. Engineering or nonagricultural uses include recreation, building or foundations and roadbeds, and waste disposal. Soil also provides a source of building material. Properties suitable for engineering uses, such as a low shrink-swell potential, often differ from those needed for agriculture.

This chapter stressed how important soil is and noted that the world's soil base is shrinking because of expanding urban areas. Other threats to the soil, noted later in the text, are erosion, shrinking supplies of water, and salt buildup. These problems emphasize the need to use and conserve soil correctly.

REVIEW

1. Is the need for maintaining good soil quality likely to increase or decrease in the years to come? Explain your answer.
2. In older homes, tile sewer lines have openings that are often invaded by tree roots, plugging the line. Explain why this happens.
3. In building a home, an old tree has its roots covered with an additional foot of soil to make a flat, level area for a patio; the tree slowly dies. Explain why the tree died using information from this chapter.
4. In many city neighborhoods, boulevard trees are planted on a narrow strip of land with the street on one side and a sidewalk on the other. Beyond the sidewalk lies a lawn. Explain why tree roots may damage the sidewalk and the possible consequences of cutting the roots while repairing the damage.
5. Which of the land uses of the United States is permanent—that is, once being used for this purpose, it can no longer be easily converted to another use? Is this use growing? Do you consider this a problem? What are some possible solutions?
6. Considering the four plant needs supplied by soil, speculate how these needs are supplied when plants are grown hydroponically.
7. A Case Study: The symbol of a town near the author's college was an old tree called The Lone Oak. When the nearby highway was completely remodeled and rebuilt, the health of the Lone Oak declined over several years and finally died. What factors might have led to its demise?

ENRICHMENT ACTIVITIES

1. Anybody whose career is in agriculture, horticulture, natural resources, or engineering needs a basic knowledge of soils. Some jobs we might consider soil science careers include a soil surveyor or soil conservation specialist. The Internet contains many sites about soil science careers; check out these examples:
   - Canadian Soil Science Society career page at <http://www.csss.ca>. Go to "Students," then "Careers."
   - Professional Soil Scientists Association of Texas career page at <http://www.io.org/PSSAT>. Go to "Careers."

- Soil Science Society of America has a brochure you can order from its Web site <http://www.soils.org>. Also check out job listings there.

2. Part of being a professional is participating in professional organizations. They are also good sources of information. Check out the home pages of these groups:
   - Soil Science Society of America (SSSA) at <http://www.soils.org>.
   - Association of Women Soil Scientists at <http://awss.homestead.com>.
   - Soil and Water Conservation Society at <http://www.swcs.org>.
3. Internships are an important part of learning about soil use. For information on internships in soil science, simply type "internships in soils" in your favorite Web browser and look for details of programs offered by various institutions.
4. For further information on some of the topics discussed in this chapter, try these Web sites.
   - The Soil Quality Institute for more information on soil quality, at <http://www.statlab.iastate.edu/survey/SQI>.
   - The SSSA statement about carbon storage in soils at <http://www.soils.org/carbseq.html>.
   - The History of Herodotus at <http://classics.mit.edu/Herodotus/history.2.ii.html>.

CHAPTER 2

Soil Origin and Development

OBJECTIVES

After completing this chapter, you should be able to:

- define a soil body
- list examples of the five soil-forming factors
- describe how soils develop
- describe the horizons of the soil profile

TERMS TO KNOW

alluvial fan
alluvial soil
chemical weathering
colluvium
delta
eluviation
eolian deposit
floodplains
frost wedging
glacial drift
glacial outwash
glacial till
hydrolysis
igneous rock
illuviation
lacustrine
leaching
levee
loess
marine sediment
master horizon
metamorphic rock
mineral soil
organic soil
oxidation-reduction
parent material
pedology
pedon
physical weathering
plow layer
polypedon
residual soil
river terrace
root wedging
sedimentary rock
soil genesis
soil horizon
soil profile
solum
solution
subsoil
talus
topsoil
transported soil
weathering

Soil is a very slowly renewable resource. Although many of our soils originated long ago, the forces that created them continue to operate. Some soils are probably hundreds of thousands of years old, while others may have just begun to make their appearance yesterday. Soils grow, change, and develop. This chapter describes the soil-formation processes.

Pedology is the study of soil formation, also known as **soil genesis,** and soil classification and mapping. We cover the latter subjects in the next chapter. Modern pedology dates to the eighteenth and nineteenth centuries in Germany, the United States, and especially Russia. These early researchers developed concepts of the soil as an evolving body arising from weathered rocks of the crust under a variety of influences. V. V. Dokuchaev (1846–1903), a Russian often credited with laying the foundation of modern pedology, published a careful study of Russian soils in 1883 that applied these concepts. He identified soil-forming factors, developed an early soil classification system, and began naming the soil horizons we use today. Dokuchaev's and other Russian publications remained unknown in the United States until into the twentieth century.

In the United States, Hans Jenny's 1941 publication *Factors of Soil Formation* further developed the five factors of soil formation detailed in this text. Jenny continued to develop and quantify these factors throughout his career and connected them with ecological principle. Much of the information in this chapter comes from the work of Jenny and other soil scientists practicing in the United States in the twentieth century.

Before describing how a body of soil forms, let's define what we mean by a soil body.

The Soil Body

The soil is a collection of natural bodies of the earth's surface containing living matter that is able to support the growth of plants. It ends at the top where the atmosphere or shallow water begins. It ends at the bottom at the farthest reach of the deepest rooted plants. The soil varies across the landscape: in one area it may be mostly made of decayed plant parts, in another place it may be mostly sand.

It is not possible to learn everything about a soil just by standing on the surface. One must dig a hole to see what it looks like below the surface. Because a soil scientist cannot dig up acres of ground to study a whole body of soil, the soil is broken up into small parts that can be easily studied. This small body is called the **pedon** (figure 2-1). A pedon is a section of soil, extending from the surface to the depth of root penetration of the deepest rooted plants, but generally examined to a depth of five feet. Generally a pedon has dimensions of about one meter by one meter, and about one and a half meters deep (about 3 ft. × 3 ft. × 5 ft.). Soil scientists use the pedon as a unit of soil easily studied by digging a pit in the ground (figure 2-2).

The traits of a pedon are set by the combination of factors that formed it. In the landscape near the pedon being studied are other pedons that are probably very similar. As one moves across the landscape, however, one will reach a pedon that is different, because the combination of factors that formed it were different. A collection of pedons that are much the same is called a **polypedon.** Later in this text we will learn how these polypedons are mapped into units called a soil series.

How does a soil pedon form? Picture a section of bare rock that will someday become a soil pedon. In the process of soil formation, this rock is changed into a layer of small, broken rock particles with some organic matter mixed in. Weather and plants are the major agents responsible for forming soil from rock, and the process is called **weathering.**

Physical weathering is the disintegration of rock by temperature, water, wind, and other factors. For instance, in cold climates **frost wedging** occurs when water freezes and expands in rocks or in cracks in the rock, causing it to break apart. The alternate expansion and contraction of rock caused by heating and cooling cycles also stresses the fabric of rock. Both cause rock to fracture or outer layers to peel away. Rain, running water, and windblown dust also wear away at rock surfaces.

Chemical weathering changes the chemical makeup of rock and breaks it down. The simplest process is **solution.** Rainwater is mildly acidic and can slowly dissolve many soil minerals, as in the solution of lime, calcium carbonate:

$$CaCO_3 + 2H^+ \rightarrow H_2CO_3 + Ca^{+2}$$

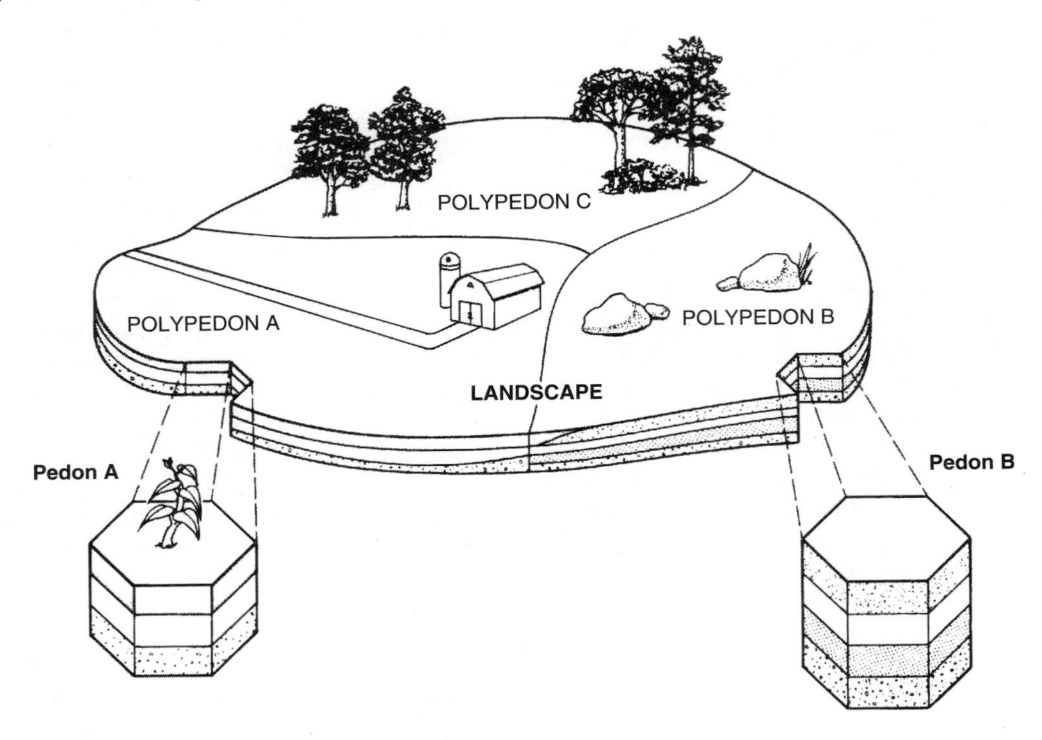

FIGURE 2-1 Two soil pedons show how each relates to the polypedons and the total landscape.

FIGURE 2-2 Digging a soil pit is required to fully study a soil body. *(Courtesy of USDA NRCS)*

In this reaction, lime dissolves in acidic water by reacting with hydrogen ions to form carbonic acid and calcium ions. In **hydrolysis,** water reacts with minerals to produce new, softer compounds, as in the hydrolysis of the feldspar mineral orthoclase to a softer feldspar mineral:

$$NaAlSi_3O_8 + H_2O \rightarrow HAlSi_3O_8 + Na^+ + OH^-$$

Oxidation-reduction and other reactions are also important in chemical weathering. Refer to appendix 1 if you need help understanding these reactions.

Plants also play an important role in rock crumbling. Roots can exert up to 150 pounds per square inch of pressure when growing into a crack in rock. **Root wedging** from the pressure pries apart stone.

Lichens growing on bare rock (figure 2-3) form mild acids that slowly dissolve rock. When lichen dies, its dry matter is added to the slowly growing mixture of mineral particles and organic matter. When a small bit of soil forms in a rock crevice, plants begin to grow

FIGURE 2-3 Plants help break down rocks to form soil. Here, patches of lichens grow on bare rocks of Lake Superior's North Shore, and plants grow in soil forming in the cracks.

from seed that has blown into the crevice, continuing the cycle.

Soil formation does not stop when a layer of young soil covers the surface. The new soil continues to slowly age and develop over thousands of years. Soil scientists state that five factors operate during the process of soil formation and development: parent material, climate, life, topography, and time. One could say that over time, climate and living things act on parent materials with a certain topography to create soil. Some have suggested that human activities might be named a sixth factor, because most soils have been modified to some degree by humans.

Soil formation begins with rock, which supplies the parent materials for most soils. Before studying the five factors, let's look at rocks of earth's crust.

Rocks and Minerals

The original source of most soils is rock—the solid, unweathered material of the earth's crust. Solid rock breaks into smaller particles, which are the **parent materials** of soil. Rock is a mixture of minerals that, when broken down, supply plant nutrients. Geologists classify rock into three broad types: igneous, sedimentary, and metamorphic. Refer to figure 2-4 for help in understanding the following paragraphs.

Igneous Rock. The basic material of the earth's crust is **igneous rock,** created by the cooling and solidification of molten materials from deep in the earth (figure 1-3). Igneous rocks, such as granite, contain minerals that supply fourteen of the seventeen required plant nutrients (listed in chapter 10 of this text). Granite, which is mined for monuments and building material, is a hard, coarse-grained rock made of feldspar, quartz, and other minerals. Feldspar, a fairly soft mineral containing potassium and calcium, weathers easily to clay. Quartz, a very hard and resistant mineral, weathers slowly to sand. Figure 2-5 lists the nutrient content of two sample igneous rocks: a granite and a basalt. Granite tends to weather slowly to create acidic parent materials high in sand, while basalt, a softer, finer-grained rock, weathers more quickly to less acidic materials low in sand.

Sedimentary Rock. Igneous rock comprises only about one-quarter of the earth's actual surface, even if most of the crust is igneous. This is because **sedimentary rock** overlays about three-quarters of the igneous crust. Sedimentary rock forms when loose materials like mud or sand are deposited by water, wind, or other agents, slowly cemented by chemicals and/or pressure into rock. Much of the sedimentary rock covering North America was deposited in prehistoric seas.

The parent materials of many American soils derive from sandstone and limestone. Sandstone, which consists of cemented quartz grains, weathers to sandy soils. Generally, these soils are infertile and droughty. Limestone is high in calcium and weathers easily to soils high in pH, calcium, and magnesium. Figure 2-5 lists contents of a typical sandstone and limestone.

Metamorphic Rock. If igneous and sedimentary rocks are subjected to great heat and pressure, they change to form **metamorphic rock.** For instance, limestone is a fairly soft, gritty rock. When subjected to heat and pressure, it changes to marble, which is harder and

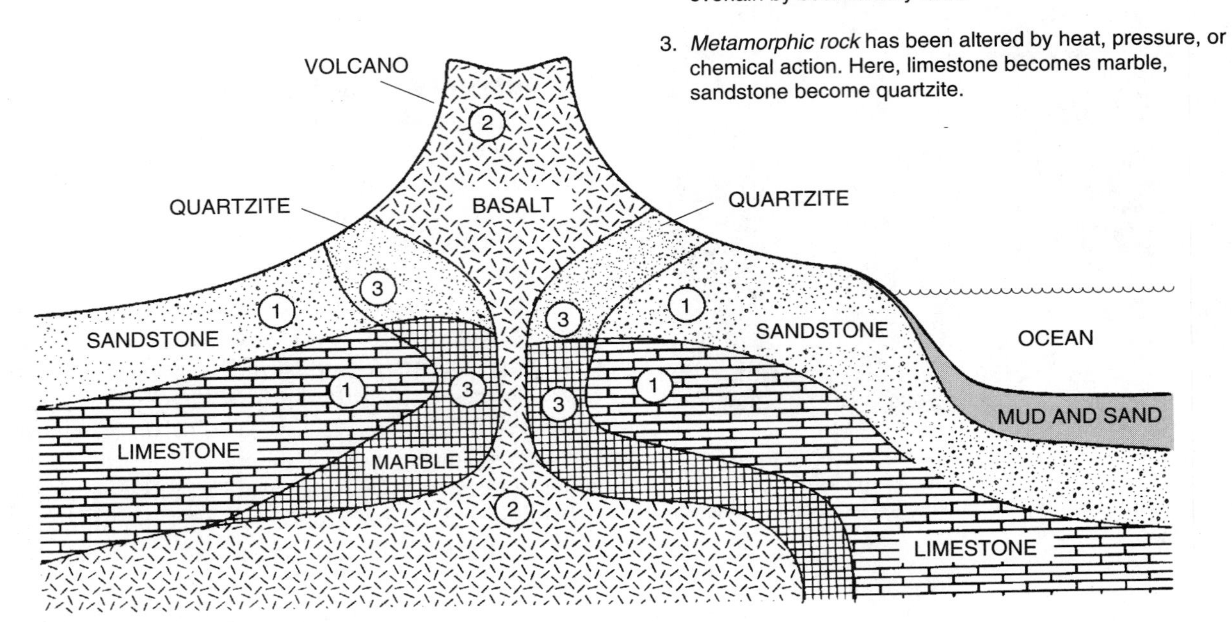
1. *Sedimentary rock*, like limestone or sandstone, formed by deposition of materials in water or by wind. Note fresh mud and sand that will someday be sedimentary rock.
2. *Igneous rock*, like basalt, formed from molten rock, as in this volcano. Most of the earth's crust is igneous rock overlain by sedimentary rock.
3. *Metamorphic rock* has been altered by heat, pressure, or chemical action. Here, limestone becomes marble, sandstone become quartzite.
VOLCANO
QUARTZITE
BASALT
QUARTZITE
SANDSTONE
SANDSTONE
OCEAN
LIMESTONE
MARBLE
MUD AND SAND
LIMESTONE

FIGURE 2-4 Three types of rock: sedimentary, igneous, and metamorphic.

MINERALS	Gray Granite	Basalt	Hinckley Sandstone	Platteville Limestone
% quartz	64	49	94	7.5
% feldspars, others	20	20	2	—
% calcite, dolomite	7	15	5	90
ELEMENTS		**Pounds per Ton of Rock**		
Calcium	69	150	—	704
Potassium	66	17	—	—
Magnesium	36	66	—	18
Iron	23	35	15*	
Phosphorus	5	5	—	—
Manganese	—	3	—	—

FIGURE 2-5 Composition of several igneous and sedimentary rocks of Minnesota, according to the Minnesota Geological Survey. Dashes mean only trace amounts, and the starred number was estimated by the author.

can be cut and polished. Soils arising from metamorphic parent materials resemble soils from the original sedimentary or igneous rock. Figure 2-6 shows relationships between common sedimentary, igneous, and metamorphic rock.

Parent Material

The description of soil origin at the beginning of this chapter was of soil formed directly from bedrock. These **residual soils,** as they are called, are actually less common than soils of parent materials carried from elsewhere by wind, water, ice, or gravity. Residual soils (figure 2-7) form very slowly, as solid rock must be weathered first. **Transported soils,** however, developed from already weathered material, form more quickly. Figure 2-8 shows parent materials of the United States.

Glacial Ice. Glacial ice carried parent materials over the northern part of North America (figure 2-9) during numerous glacial periods over the past two million years. The last four left the most evidence, and the most recent glacier, that of the Wisconsin period, reached its peak expanse about 18,000 years ago and melted back out the United States about 10,000 to 12,000 years ago. Most of the glacial deposits that cover the northern states come from the Wisconsin ice sheet. Glaciers expanded out of several centers in Canada carving and grinding the earth, picking up and transporting soil, gravel, rocks, and other debris. As the glaciers melted and shrunk between glacial periods, transported material remained in deposits called **glacial drift.** In the process, they left behind a very distinctive landscape over much of the northern United States and Canada.

Glaciers deposited materials in many ways, so there are several kinds of glacial drift. During the melting process, some debris simply dropped in place to form deposits called **glacial till.** Because there was no sorting action in the deposition, glacial till is extremely variable, and so are the soils derived from it. Till soils often contain pebbles, stones, and even boulders (figure 2-10).

Other materials carried by the glacier washed away in meltwater to form sediments in streams and lakes. During the process, the materials were sorted by size.

FIGURE 2-7 A residual soil. Four to eight inches of soil have formed atop basaltic bedrock. *(Courtesy of USDA NRCS)*

Rocks			
Sedimentary	*Igneous*	*Metamorphic*	*Main Components*
Sandstone		Quartzite	Quartz sand
Limestone		Marble	Calcite
Shale		Slate	Feldspar clays
	Granite	Gneiss	Quartz, mica, feldspar
	Basalt	Schist	Feldspar, mica, olivine

FIGURE 2-6 Relationships of some common sedimentary, igneous, and metamorphic rock.

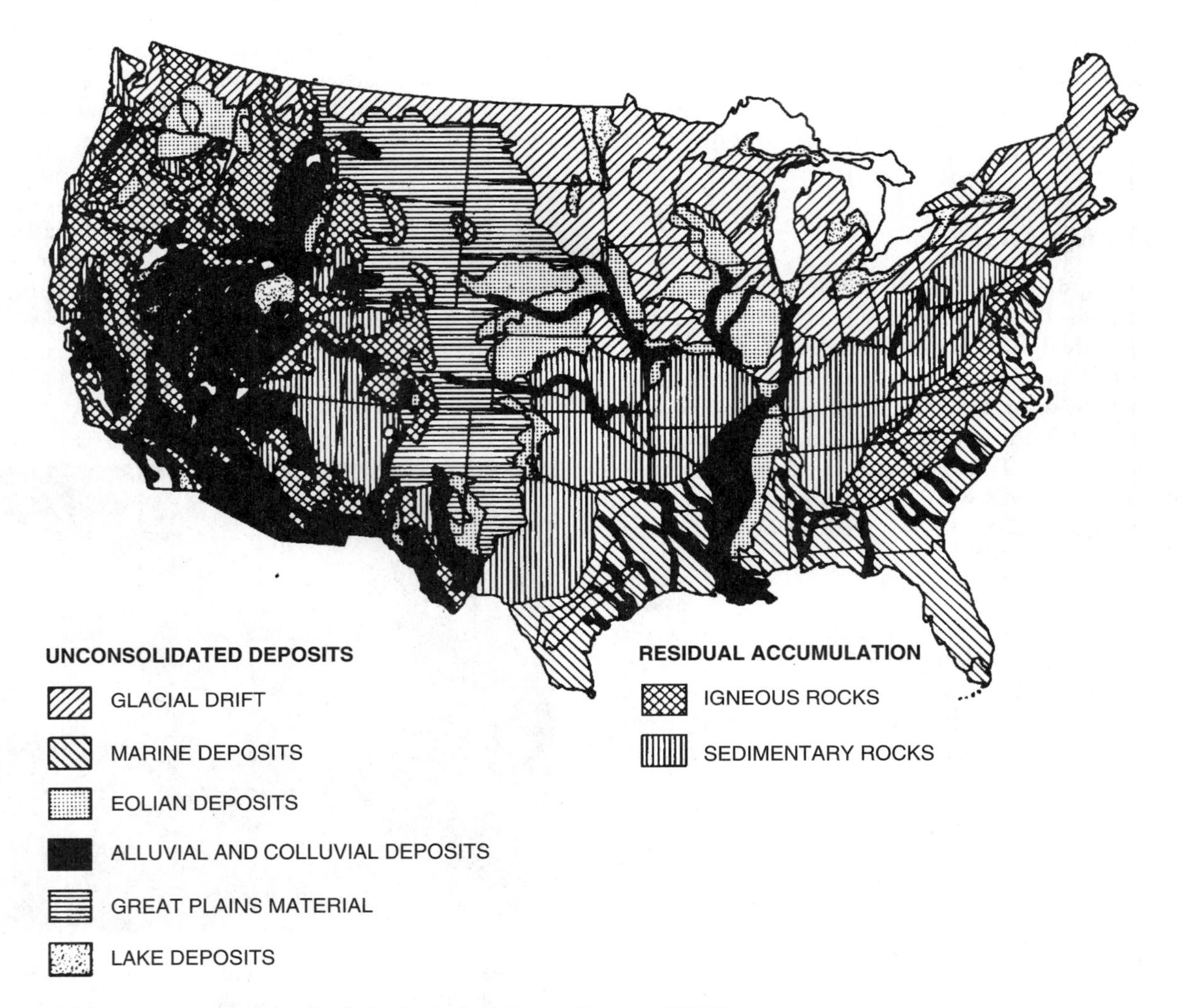

FIGURE 2-8 Parent materials of soils in the United States. *(Source: USDA)*

Coarser material, being larger and heavier, was deposited near the glacier and in streams and rivers to form **glacial outwash.** Outwash soils tend to be sandy. Smaller particles reached glacial lakes to form **lacustrine** deposits on the lake bottoms.

Wind. Some parent materials were carried by wind, leaving **eolian deposits.** For example, some soils in Nebraska formed from sand dunes, deposits of sand carried by rolling in the wind. Most eolian soils in the United States are actually a result of the last glacial period.

After the last glaciers melted and meltwaters subsided, large expanses of land were exposed to a dry climate with strong westerly winds. Winds picked up silt-size (medium) particles and deposited them in the Mississippi and Missouri River valleys and elsewhere. These **loess** soils—wind-deposited silt—are important agricultural soils in much of Iowa, Illinois, and neighboring states.

Water. **Alluvial soils** are soils whose parent materials were carried and deposited in moving fresh water to

FIGURE 2-9 Glaciers have covered much of North America several times.

FIGURE 2-10 Glacial till. Till soils often contain rocks and gravel. *(Courtesy of Howard Hobbs, Minnesota Geological Survey)*

form sediments (figure 2-11). Alluvial materials can be deposited in several ways. **Alluvial fans** form below hills and mountain ranges where streams flowing down the slope deposit material in a fan shape at the base. As water speed slows abruptly at the foot of the slope, large particles drop out first. As a result, alluvial fans are generally sandy or gravelly. Finer materials are carried away in rivers.

Flooding rivers also leave deposits behind. Often coarser materials are deposited in low ridges, or **levees,** along the river bank. Away from the river, floodwaters spread over large flat areas called **floodplains.** Here the water will be shallow and slow moving; fine particles will settle out (figure 2-12). Floodplains tend to be fertile because new soil is added at each flood, but the soils tend to stay wet. Levees, being coarser and elevated, dry more quickly. Floodplain soils are especially important along the Mississippi and its tributaries and along rivers that flow into the ocean on the East and Gulf coasts. Many important soils of California are from river alluvium.

Sometimes a river will cut deeply into its floodplain to flow at a lower elevation. This establishes a new riverbed and floodplain, while the old floodplain is left higher as a **river terrace.** An example of river terrace soils is some soil of the San Joaquin Valley of California.

Lacustrine deposits form under still, fresh water. Most of our lacustrine soils remain from giant glacial lakes that have since dried up. Examples include Glacial Lake Agassiz of northern Minnesota, North Dakota, and Canada, and Glacial Lake Bonneville of Utah. When glacial runoff water ran into the lake, the heaviest materials were left near the shore, while the smallest particles were carried to the center of the lake. Thus, lacustrine soils are sandy near the old shoreline and grade to soils with smaller particles toward the old lake center.

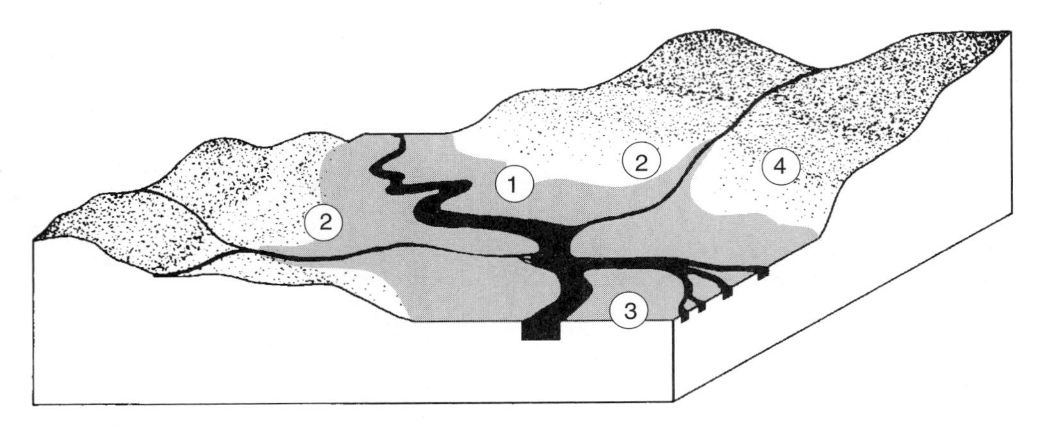

FIGURE 2-11 Water and marine deposited soils. *(1)* Floodplains form along rivers from materials deposited during flooding. *(2)* Alluvial fans are deposited at the base of slopes by running water. *(3)* Deltas form when smaller particles drop out as a river enters an ocean. *(4)* River terraces are old flood plains left above a new river level.

FIGURE 2-12 Mississippi mud. This alluvial deposit remained after floodwaters receded.

Marine sediments form in the ocean. Many scattered soils of the Great Plains and the Imperial Valley of California are beaches of prehistoric seas that once covered the United States. Other beach soils are common along the Atlantic coastline and the Gulf of Mexico. These all tend to be sandy soils. **Deltas,** in contrast, have very small particles and tend to be wet. Deltas form when rivers flowing into an ocean deposit sediments at the mouth of the river. The Mississippi River Delta of Louisiana is a prime example, as is the Rio Grande Valley of Texas and Mexico.

Gravity. Some parent materials move simply by sliding or rolling down a slope. This material, called **colluvium,** is scattered in hilly or mountainous areas. An example of a colluvial material is a **talus**—sand and rocks that collect at the foot of a slope. Avalanches, mudslides, and landslides are other examples.

Volcanic Deposits. The ash blown out of a volcano and deposited nearby or carried some distance by wind forms a chemically distinct, dark, and lightweight parent material. The Pacific Northwest, Hawaii, and Alaska are areas of the United States where such deposits are common.

Organic Deposits. Characteristics of the soils formed from parent materials described so far are set by mineral particles in the soil. **Mineral soils** contain less than 20 percent organic matter, except for a surface layer of plant debris. **Organic soils,** containing 20 percent or more organic matter, form under water as aquatic plants die. Low oxygen conditions under water retard decay of these dead plants, so they tend to pile up on the lake bottom. Eventually the lake fills in and is replaced by an organic soil. Organic soils are extensive in Minnesota, Wisconsin, and Florida.

Climate

Climate first affects soils by causing physical and chemical weathering of rock. However, climate continues to affect soil development long beyond this initial stage. The main effects are due to temperature and rainfall.

Temperature affects the speed of chemical reactions in the soil—the higher the temperature, the faster a reaction. Chemical weathering in soils occurs mostly when the soil is warmer than 60°F. Thus, in cold areas, like tundra, soils develop slowly. In warm areas, like the tropics, soils develop more rapidly.

Another result of temperature is its effect on organic matter. Warmth promotes greater vegetation, so more organic matter is added to the soil. However, warm temperatures also speed the decay and loss of organic matter. Thus, soils of warm climates tend to be low in organic matter.

Rainfall affects soil development mainly by leaching. **Leaching** moves materials deeper into the soil via water moving downward through the soil. Leached materials include lime, clay, plant nutrients, and other chemicals. These materials are then deposited in lower parts of the soil.

High rainfall areas also tend to grow more vegetation, so the soils of humid areas tend to have more organic matter than soils of drier regions. To summarize, rainfall tends to cause leaching and the accumulation of organic matter.

The United States is a good example of the effects of climate on soil (figure 2-13). The climate of the United States cools from south to north. This is reflected in an increase in average organic matter content from south to north. Also, the most weathered soils in the United States are in the South. The average rainfall of the United States increases from west to east. As a result, the organic matter content of the United States' soils also tends to increase from west to east.

Soil color also follows north-south trends. Because organic matter is black, soils tend to appear darker as one moves from warmer to cooler climates. Because of changes in chemical reactions involving iron, soils tend to appear redder as one moves from cooler to warmer climates.

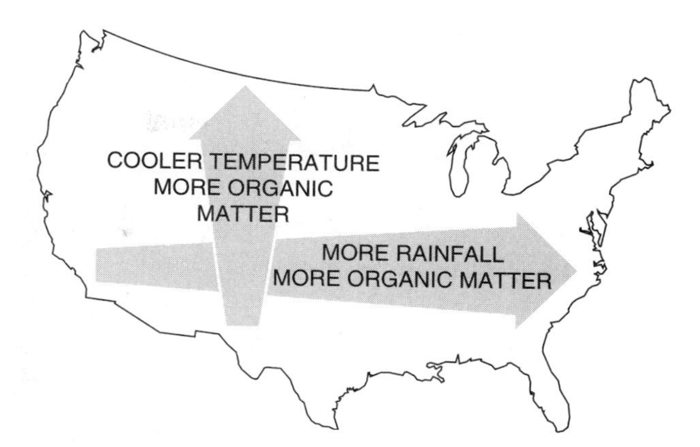

FIGURE 2-13 One effect of climate on United States soils. The average organic matter of soils increases to the east and north because of cooler temperatures and higher rainfall. Other factors also affect organic matter, like vegetation, which alters these general trends regionally.

Organisms

Organisms that live in soil—like plants, insects, and microbes—actively affect soil formation. The actual properties of a developing soil are influenced especially by the type of plants growing on it. Figure 2-14 shows the parent vegetation of soils of the United States.

Mineral soils having the highest organic matter content form under grasslands. Grasses usually have a dense mat of fibrous roots, some of which die each year. This keeps the organic matter content high and the soil color dark. In a forest, much of the organic material is above ground in the trees. When the leaves fall or the tree dies, the material falls to the soil where it creates a surface layer of organic matter that does not mix with deeper layers. As a result, forest soils have less organic matter than prairie soils and are lighter in color. The type of trees also influences the soil. Compared to hardwoods (deciduous trees), softwood (conifer) foliage is acidic and resistant to decay, therefore their soils tend to be thinner, lower in organic matter, and more acidic. Deserts, with very sparse vegetation, have the least organic matter (figure 2-15).

Vegetation also affects the location of nutrients and other ions in the soil. Plants absorb ions in the roots and carry them to the tops, where they are returned to the soil surface when leaves drop. This recycles ions from

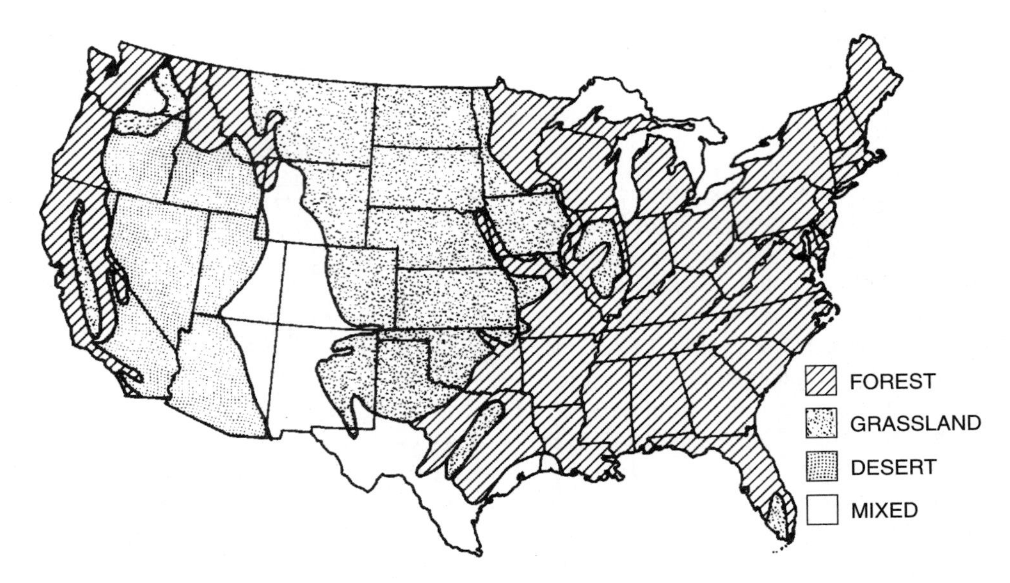

FIGURE 2-14 Native vegetation of the United States. *(Source: USDA)*

FIGURE 2-15 The soils of this piece of the Sonoran Desert outside Phoenix are influenced by sparse vegetation and arid climate.

deeper in the soil to the surface and helps reduce their loss from leaching. Deep tree roots, for instance, extract ions from deep in the soil, leaving the surface horizon of forest soils enriched in ions.

We tend to stress vegetation as the main living factor in soil formation, but other life impacts soil as well, such as burrowing animals that bring subsoil to the surface, earthworms that create large, deep pores and speed organic matter decay, or nitrogen-fixing bacteria. These other organisms are covered in greater detail in chapter 5.

Topography

Topography, or the soil's position in the landscape, influences soil development mainly by affecting water movement. Water runs off slopes, making them drier, and collects in low areas, making them more moist. This, in turn, affects leaching, chemical reactions, and types of vegetation. Slope effects vary according to a number of characteristics: steep or south-facing slopes are drier than gentle or north-facing slopes, and the top portion of a slope is drier than the bottom portion of one.

If enough water runs off a slope, it may carry away soil as fast as it is formed. Thus, soil may be thin on a slope and thick at its base. The effect of topography is most obvious in rolling fields, where sloped areas are light brown from topsoil loss while lower areas are black from accumulating topsoil and organic matter (figure 2-16).

Because running water tends to carry off smaller particles, soils in lower areas may be finer than those of higher areas. Depressions may also intersect the water table at least part of the year, keeping them wet for long periods.

Time

Soils change over time, undergoing an aging process. Initially, a thin layer of soil forms on the parent material. Such a young, immature soil takes as little as a hundred

FIGURE 2-16 One effect of topography on soils. Topsoil has eroded off the knolls and has been deposited in lower spots. *(Courtesy of USDA NRCS)*

years to form from well-weathered parent materials under warm, humid conditions. Under other conditions, it may take hundreds of years.

Weathering of the young soil continues, and many generations of plants live and die, so the young soil becomes deeper and higher in organic matter. If there is enough rainfall, leaching begins to carry some material deeper into the soil, creating the soil profile described later in this chapter.

As soils age, biological processes tend to increase the nitrogen content, while leaching tends to reduce phosphorus. Thus, young soils tend to be low in nitrogen but high in phosphorus, while older soils are the opposite. Mature soils are generally productive, but as soils continue to age, they become more severely weathered, more highly leached, and often less productive. In general, as soil ages it becomes deeper, develops distinct layers, and becomes more acidic and leached.

However, the aging process is not static. Time zero for a soil usually begins when some dramatic event like landslides or glaciers changes everything and resets the clock. Such events can happen at any time. A soil might age through the years until it reaches some steady state and remains unchanged thereafter, but this is rare. Soils can erode away, be buried, or even become the parent material for a new soil. If soil factors change, the direction of soil development can be deflected into a new path. For instance, if forest invades prairie, the soil embarks on a new path towards a forest-type soil.

Humans

Humans may be considered just another living entity that modifies soil, but their action can be so rapid, dramatic, and different from other life that they might be considered a separate, sixth soil formation factor. Very few soils have been unaffected by human activities. Effects may be as subtle as the deposition of air pollutants distant from any human habitation to as massive as earthmoving during road construction. The latter resets the time clock for this new soil material to zero, and the earth moved by the machinery is the parent material for this new soil. Chapter 19 describes traits of urban soils, those most modified by humans.

The Soil Profile

Soils change over time in response to their environment, represented by the soil-forming factors. Soil scientists have classified the causes of those changes into four soil-forming processes:

ADDITIONS: materials may be added to the soil; some examples are fallen leaves, wind-blown dust, alluvium, and man-made materials like air pollutants and compost.

LOSSES: materials may be lost from the soil, as a result of deep leaching, erosion from the surface, or as gases filtering out of the soil.

TRANSLOCATIONS: materials may be moved within the soil, by leaching deeper into (but not out of) the soil, being carried upward with evaporating water, or by being moved by animals like ants or earthworms.

TRANSFORMATIONS: materials may be altered in the soil; for example, organic matter decay, weathering of minerals to smaller particles, or chemical reactions.

Each of these processes occurs differently at different depths. For instance, organic matter tends to be added at

FIGURE 2-17 The top 30 inches of a soil profile. *(Courtesy of USDA NRCS)*

or near the surface, not deep in the soil. Some material moves from high in the soil to be deposited lower. As a consequence, different changes occur at different depths, and horizontal layers develop as a soil ages (figure 2-17).

These layers are known as **soil horizons,** visible wherever the earth is dug deep enough to expose them. The **soil profile** is a vertical section through the soil extending into the unweathered parent material and exposing all the horizons. Each horizon in the profile differs in some physical or chemical way from the other horizons.

In a very young soil, weathering and plant growth produce a thin layer of mineral particles and organic matter atop parent material. The thin layer of soil is labeled the A horizon, a surface mineral horizon enriched with organic matter. The parent material below the A horizon of this young soil is termed the C horizon. It is defined as a subsurface mineral layer only slightly affected by soil-forming processes. Thus, this young soil has an AC soil profile.

As the young soil ages, the soil increases in depth. In addition, clay-sized particles and certain chemicals leach out of the A horizon (figure 2-18), moving downward in the profile to create a new layer, the B horizon.

Master Horizons. The A, B, and C horizons are known as **master horizons.** They are part of a system for naming soil horizons in which each layer is identified by

FIGURE 2-18 The dark staining in the light horizon was leached (eluviated) from the A horizon. *(Courtesy of Howard Hobbs, Minnesota Geological Survey)*

a code: O, A, E, B, C, and R. These horizons are shown in figure 2-19, and are described as follows.

O The O horizon is an organic layer made of wholly or partially decayed plant and animal debris. The O horizon generally occurs in undisturbed soil, because plowing mixes the organic material into the soil. In a forest, fallen leaves, branches, and other debris make up the O horizon.

A The A horizon, called **topsoil** by most growers, is the surface mineral layer where organic matter accumulates. It is darker than the horizons below. Over time, this layer loses clay, iron, and other materials to leaching. This loss is called **eluviation.** Materials resistant to weathering, such as sand, tend to remain in the A horizon as other materials leach out. The A horizon provides the best environment for the growth of plant roots, microorganisms, and other life.

E The E horizon, the zone of greatest eluviation, is very leached of clay, chemicals, and organic matter. Because the chemicals that color soil have been leached out, the E layer is very light. Many soils have no E horizon; it is mostly likely to occur under forest vegetation in sandy soils in high rainfall areas.

B The B horizon, or **subsoil,** is often called the "zone of accumulation" where chemicals leached out of the A and E horizon accumulate. The word for this accumulation is **illuviation** (figure 2-18). The B horizon has a lower organic matter content than the topsoil and often has more clay. The A, E, and B horizons together are known as the **solum,** the portion of the soil profile most affected by soil-forming processes and that usually contains most plant roots.

C The C horizon lacks the properties of the A and B horizons. It is the soil layer little touched by soil-forming processes and is usually the parent material of the soil. It may also include very soft, weathered bedrock that roots can penetrate.

R The R horizon is underlying hard bedrock, such as limestone, sandstone, or granite. It may be cracked and fractured, allowing some root penetration. The R is identified only if near enough the surface to intrude into soil.

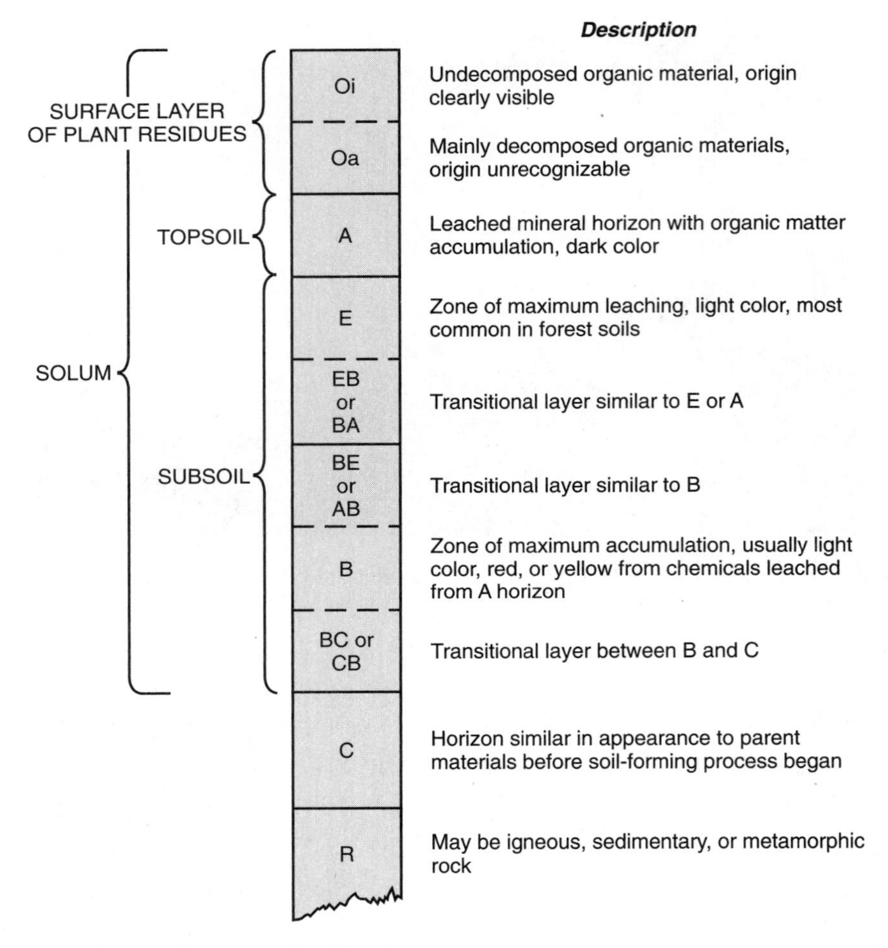

FIGURE 2-19 Main horizons of the soil profile. Bold lines divide O, A, E, B, C, and R horizons; broken lines show the subhorizons.

Subdivisions of the Master Horizons. As soils age, they may develop more horizons than the basic master horizons. Some of these layers are between the master horizons both in position and properties. These transitional layers are identified by the two master letters, with the dominant one written first. Thus, an AB layer lies between the A and B horizons and resembles both, but is more like the A than the B. Figure 2-19 shows these layers.

A soil layer can be further identified by a lowercase letter suffix that tells some trait of the layer. Appendix 4 lists these suffixes but two will serve as examples here—the Ap and Bt. An Ap layer is a surface layer disturbed by humankind, so that the old layers were mixed up. For instance, plowing would mix up an O, A, and AB horizon if they were all in the top eight inches. The Ap horizon is the same as the **plow layer,** the top seven or eight inches of soil in a plowed field. A Bt horizon is a B horizon in which clay has accumulated, usually by illuviation.

Further subdivisions are noted by a number following the letters. Thus, one could have a soil with both a Bt1 and a Bt2 horizon. This means that the Bt horizon of the soil has two distinct layers in it.

Now for an example. Figure 2-17 has the profile Ap-E-Bt-C. The top seven inches are an old plow layer or Ap. A strong, light-colored E horizon extends from the seven- to fourteen-inch depth, showing a leaching of clays and chemicals. Those clays then settled in the Bt horizon, lying between fourteen and twenty-two inches deep. Below this is the C horizon of sand and gravel. Notice the rodent hole in the E horizon.

SUMMARY

Soils form from minerals broken up by the action of weathering and plant roots and from the addition of decaying plant parts. Young soils continue to age—growing deeper, being leached by rainfall, developing layers, and changing over time. This soil-forming process involves the addition, loss, translocation, or transformation of soil materials, and is governed by the five factors of parent material, climate, life, topography, and time. Some authorities would add humans as a sixth factor in soil formation. Over time soil deepens and develops recognizable horizons, and may finally become severely weathered and highly leached.

Residual soils develop directly from bedrock (igneous, sedimentary, or metamorphic). Most mineral soils come from parent materials moved from one area to another by ice, water, wind, or gravity. Organic soils are composed of decaying plants. Each type of parent material is responsible for a different soil.

Parent materials are acted on by climate and living organisms. Soils develop quickly in warm areas with high rainfall, then age into heavily weathered soils low in organic matter. In cooler regions, organic matter accumulates and weathering is less extreme. In arid climates, sparse plant growth inhibits the formation of organic matter. Grassland soils tend to be high in organic matter, forest soils lower, and dryland soils lowest of all.

Topography affects soil formation by changing water movement and soil temperature. Low areas often have deep, rich soils that drain slowly. Erosion causes thin soils on slopes.

Time is a factor because soil development is a continuing process. Young soils tend to be thin with little horizon development. Mature soils are deeper and productive with several recognizable horizons. Old soils are severely weathered, highly leached, and less productive.

Soil profiles, which develop over time, are divided into master horizons. These, in turn, may also contain layers. Each layer is named by a code system that identifies its position in the profile and provides some information about it.

REVIEW

1. Name the five soil factors and give an example of the effect of each on soil formation.
2. Draw a soil profile containing seven distinct horizons in the correct order, and label them. Indicate topsoil, subsoil, and solum. Hint: You will use more than just the regular master horizons, and there are many possible configurations.
3. Describe the major parent materials and vegetation as well as climate that contributed to the soils of your state. Describe how they influenced your soils.
4. What do alluvial fans, floodplains, deltas, and terraces have in common? How are they different?
5. Would you be likely to read a soil description that includes an At or As horizon? Explain your answer.
6. Discuss the four soil-forming processes and give examples of each.
7. Consider the five soil-forming factors. Which are most likely to account for soil variation on a local scale? Which are more likely to operate on a larger scale? Explain your answer.

8. Running water removes soil from the surface. Which of the four soil-forming processes does this exemplify, and how might the soil-forming factors affect it?
9. Organic matter tends to increase from west to east in the United States because of increasing rainfall. Yet, some of the highest organic matter soils are in the plains states, which are relatively dry. Explain why.
10. A case study: Eighteen thousand years ago, during the peak of the Wisconsin Glaciation, northern Kentucky was probably free of ice. However, the climate was cold, and the likely vegetation was taiga, which is relatively open conifer forest. How do you think today's soils of northern Kentucky would compare with those of 18,000 years ago? Use information found in this chapter.

ENRICHMENT ACTIVITIES

Study the history of the soils in your state or vicinity.

1. Dig a soil pit and study the soil profile. See if you can name the layers.
2. Obtain samples of common soil-forming rocks and minerals. Find more information about each from a simple field guide to rocks and minerals. What plant nutrients does each contain (see chapter 10 for a list)? Using one of the several available laboratory exercises, experiment with the various weathering processes. For instance, try to scratch feldspar with quartz, and vice versa. Which is harder?
3. To observe the effects of freezing on physical weathering, pat a handful of clay soil into a ball. Inject water into the ball with a syringe, then freeze overnight. Observe the results.
4. This Web site from Alberta discusses soil-forming processes: <http://www.rr.ualberta.ca/Courses_Teaching/SOA/process.html>. It has one additional factor not included here. Which one is it and how would you fit it into this text's five factors?
5. This chapter skimmed over the complex deposits left behind by glaciers that became parent materials for many states. Type the search phrase "glacial landforms" into your favorite Internet search engine and find out more.

CHAPTER 3

Soil Classification and Survey

OBJECTIVES

After completing this chapter, you should be able to:

- describe the current USDA soil classification system
- explain how soil surveys are prepared and used
- list soil capability classes

TERMS TO KNOW

diagnostic horizons
families
great groups
land capability classes
land capability subclasses
mapping units
phases
soil association
soil classification
soil order
soil series
soil survey
soil taxonomy
subgroups
suborders

At the end of the 1800s, public leaders began to realize that land in the United States was being damaged by poor land policies. This realization led to public efforts to conserve soils—efforts that continue today. A start was made in the early 1900s when the government began to survey and classify the soils of the United States.

Soil Classification

Soil survey depends on a system of grouping soils of like properties. **Soil classification** helps us to understand, remember, and communicate knowledge about soils.

The Russian soil scientist V. V. Dokuchaev first suggested a way to classify soils around 1880. He proposed that soils were natural bodies created by soil-forming factors. This proposal formed the basis of a classification system that soil scientists began using to survey United States' soils.

Over the years, the United States has used several, constantly evolving soil classification systems. Early in the 1900s soils were grouped based on the soil-forming factors that created them, using terms like "brown forest soil" or "black prairie soil." Further systems were developed in 1938 and 1949, and familiar names like podzols and chernozems came into use. The USDA introduced the current classification system in 1960 with the publication of *Soil Classification, a Comprehensive System.* Continued modifications lead to its republication in 1975 as *Soil Taxonomy: A Basic System of Soil Classification for Making and Interpreting Soil Surveys,* the basis for modern soil classification in the United States. This system too continues to evolve, with changes like the addition of a twelfth soil order, the Gelisols, in 1998.

Earlier systems generally were based on the presumed history of a soil, that is, its process of formation under the five soil-formation factors. The system of the **Soil taxonomy** is based, rather, on the properties of the soil as it can be observed in the field or laboratory.

It resembles the way plants and animals are grouped according to a system known as taxonomy—a grouping of objects at several levels to show how they relate. Figure 3-1 compares the taxonomy of plants with the taxonomy of soils. However, unlike taxonomy for plants and animals, soil classification is not universal. While the USDA's soil taxonomy can be applied outside the United States, other nations employ their own systems to serve their own purposes.

Soil Classes	Plant Classes
Order (12)	Kingdom
Suborder (66)	Division
Great Group (>320)	Class
Subgroup (>1,400)	Order
Family (>8,000)	Family
Series (>19,000)	Genus & species
(Phases)	(Variety)

FIGURE 3-1 The USDA soil classification system lists six levels of soil classes *(left)*. The approximate number of each is provided but continues to increase. Phases are also listed but are not an official level of the soil taxonomy. The classes of plants taxonomy are listed for comparison.

As shown in figure 3-1, the new system has six levels of classification. The highest level, the **soil order,** is the broadest group. The system recognizes twelve soil orders, described in figure 3-2. These orders are based mainly on the presence or absence of certain key horizons in the soil profile, called **diagnostic horizons,** and on average temperatures and rainfall. An Alfisol, for instance, has a subsurface horizon with a clay accumulation, a medium to high base supply, and moisture at least ninety days of the growing season. An Entisol shows few signs of soil development, with little horizon development. Note that the names of all soil orders end in the suffix "ol." Appendix 3 presents a map of the soil orders of the United States.

Each order is divided into several **suborders,** the next highest level of the soil taxonomy. Suborder members of the same order differ most often in soil moisture or temperatures but may differ by other factors. A Psamment, for instance, is a suborder of Entisols that is highly sandy. The name of a suborder includes a Latin or Greek root that provides information about the suborder and ends in several letters that identify the order to which it belongs. A Psamm*ent,* for instance, is an *Ent*isol. The letters "psamm" come from the Greek word for sand.

Suborders, in turn, are divided into **great groups,** which are often based on the presence of certain key horizons but may differ by other traits like soil moisture and temperature. A great group is named by adding a prefix to the suborder name. A Udipsamment

Soil Order	Description	Use
Alfisol	Usually deciduous forest soils of temperate moist climates, light colored, slightly to moderately acid with illuvial layer high in silicate clays. Medium to high base saturation (>35%). Fertile soil. Especially north central states. *Typical profile:* O-A-E-Bt-C	Cropland, forest, range
Andisol	Geologically recent volcanic materials. Dark, fertile, high CEC and OM, low density, often on volcanic slopes and high altitude. Pacific Northwest, Hawaii, Alaska.	Cropland, forest
Ar**id**isol	Arid soils of cool to hot deserts and dry shrublands, often alkaline with salted horizons, thin or no O or A. High base. Western states. *Typical profile:* A-Bt-Ck or Ckm, Cy, Cz	Range, irrigated cropland
Entisol	Soils lacking well-developed horizons, often young, or under conditions that inhibit horizon development like being sandy, wet, alluvial, or steeply sloped. Least developed soil order. Often difficult to use. Widely scattered in US. *Typical profile:* A-C	Range, cropland, forest, wetlands
Gelisol	Very cold soils of the tundra, cold deserts, or high peaks with subsoil permafrost. Often with muck or peat surface soil. Alaska mostly. Very fragile. *Typical profile:* O-A-Cf	None safely except wildlife
Hi**st**osol	Organic soils, usually of wetlands. Organic matter >20-30%. Very low density. Must be drained for use, then prone to subsidence and fire. Northern Midwest and Atlantic/Gulf coastal areas. *Typical profile:* O1-O2-O3-C	Wetlands, forest, horticulture fuel
Inc**ept**isol	Soils with minimal horizon development, but more than Entisols. Often young. May have weak B horizon visible by color or structure; no illuviation. Extremely variable, and widely scattered in US. *Typical profile:* A-Bw-C	Cropland, forest, range
M**oll**isol	Mostly grassland soils. Dark, thick, high organic matter and base A horizon. Low to moderate rainfall. May have illuvial or calcareous subsoil. Highly fertile and productive. Great Plains and Northwest states. *Typical profile:* A1-A2-A3-Bw-C	Cropland, range
Oxisol	Highly weathered tropical soils, often under rainforests. Subsurface horizon low in weatherable minerals but high in aluminum or sesquioxide clays. Low native fertility, but can be fertilized. Hawaii and Puerto Rico. *Typical profile:* A- Bo (or Bv) - C	Cropland, forest, shifting agriculture
Sp**od**osol	Light colored, acid coarse soils, typically under coniferous forest. Usually of cool humid regions, but not always. Illuviation of iron or aluminum-humus complexes in B horizon. Low base saturation, infertile. Upper Midwest to Northeastern states. *Typical profile:* A-E-Bs (or Bhs) - C	Forest, pasture, cropland
Ultisol	Highly weathered soils of humid warm climates, often under forest. Low base saturation (<35%), acid, leached. Subsoil layer with illuviated silicate clays. Surface layer light colored, subsoil often red clay. Can be productive if properly fertilized and limed. Southeast states mostly. *Typical profile:* A-E-Bt-C	Forest, cropland
Ertisol	High in swelling clays in climates with a dry season. When dry, large, deep cracks form that surface soil falls into, mixing the soil. Unstable for engineering uses. Most common in Southcentral states, especially Texas, some in upper plains states. *Typical profile:* A-AC-C	Range and pasture, cropland

FIGURE 3-2 This simplified listing of the soil orders provides the order names, description, and main uses in order of importance. The boldfaced letters in the order names are used in suborder names. Some terms are explained in later chapters.

is a Psamment (sandy Entisol) that is usually moderately moist, expressed by the prefix "udi."

Great groups are further divided into **subgroups,** based on how close a soil is to the "central concept" of its great group. That is, there is a core image of what that great group should be, but there are gradations within it. A subgroup that matches the central concept is called Typic, while other words express the variations. We affix the subgroup name as a separate word in front of the great group name, for instance, a Typic Udipsamment.

Subgroups are themselves divided into **families,** which are units of a subgroup with similar properties important to the growth of plants and soil use, such as subsoil particle sizes or the minerals found in the soil. Family names are composed of a string of descriptive words placed in front of the subgroup name, for example, a frigid, mixed Typic Udipsamment. This is now the full, taxonomic name of this soil, and you will find such names in soil surveys. The naming system for families is quite complex and will not be further discussed here.

All the words and syllables used to create these names are listed in *Soil Taxonomy,* and the reader should refer to that publication for further details about these levels of the system. Those who use soils at the local level, like growers, builders, or county extension agents, are more concerned about the lowest soil grouping, called the **soil series.**

Soil Series. Soil scientists divide soil families into smaller units called soil series. The soil series is the taxonomic unit with the narrowest range of features, and all pedons within a series have very similar soil profiles. Each of these units is distinct from other units and is the same as the polypedon described in chapter 2.

In the United States, each series is given the name of the town, county, or other location near where the series was first identified. The Mahtomedi soil is named after a town in east central Minnesota and is an example of the soil family just classified above. Other examples of soil series include the Saybrook (a Mollisol), found near the central Illinois town of Saybrook, or the Ontario (an Alfisol), named after a town in New York. A series name may be followed by the surface texture of the soil, as in the Saybrook silt loam.

The series is the lowest official category in the soil taxonomy. However, in practice, a series may be subdivided further into **phases.** A phase is a variation of a series based on some factor that affects soil management, such as slope, degree of erosion, or stoniness. One might have, for example, an Ontario loam, 3 percent–6 percent slope phase. Soil series with their phases become mapping units for the most detailed soil surveys—the next topic in this chapter.

Soil Survey

The USDA developed the soil classification system for use in soil surveys. Soil surveys classify, locate on a base map, and describe soils as they appear in the field. These soil surveys are performed under the auspices of the National Cooperative Soil Survey Program, a joint effort of the USDA Natural Resources Conservation Service, or NRCS (formerly Soil Conservation Service), and state Agricultural Experiment Stations (see chapter 20). Most of the actual surveying is done by soil scientists of the NRCS.

Field Mapping. For detailed mapping, a soil scientist walks the land to survey it (figure 3-3). Frequently he or she stops to probe the soil. By study-

FIGURE 3-3 This soil scientist conducting a soil survey probes the soil to find diagnostic horizons and other soil factors. Then he marks the boundaries of the mapping unit on an aerial photograph.

FIGURE 3-4 A base map being recorded on an aerial photograph during a soil survey.

ing the soil profile, that spot can be placed in the correct series. The surveyor also notes slope, evidence of erosion, and other interesting features. With this information, the surveyor draws the soil series or phase on a base map.

The NRCS uses aerial photographs as a base map (figure 3-4). Aerial photographs make good base maps because they show landscape features, including ponds, woods, and sand pits. Figure 3-5 shows a map of a small farm as it was hand-drawn by an NRCS surveyor. Note that the aerial photograph base map shows much of the farm as wooded.

When the survey is complete, the resulting map is copied neatly. Maps show the boundaries of the mapping units, with each unit identified by codes that vary from state to state. Note the codes shown in figure 3-5. These codes may have one, two, or three parts. The main group of digits refers to the soil series. The mapping unit labeled 169, for instance, is a member of the Braham series. In addition, a unit may be labeled with codes to indicate the slope and erosion. If the latter two codes are absent, one assumes a nearly level relief with no erosion. Figure 3-6 gives common codes and other symbols that indicate different features in the field. The following codes are used on the map in figure 3-5:

158 Zimmerman series, no slope or erosion

158C Zimmerman series, 6 percent–12 percent slope, no erosion

Modern technology is becoming an increasingly useful tool for soil surveys. Even satellites are employed for remote sensing of earth forms. Such technologies aid, but do not replace, the activities of a soil surveyor doing field mapping.

Mapping Units. Different **mapping units** are used in soil surveys, depending on how large an area the map or survey covers. For small areas, the mapping units are detailed phases of soil series. For larger-scale maps, the units may be higher levels of the soil taxonomy, like families or great groups. Soil orders are the mapping units on the national soil map in appendix 3. For most county maps, phases of soil series are the basic mapping unit.

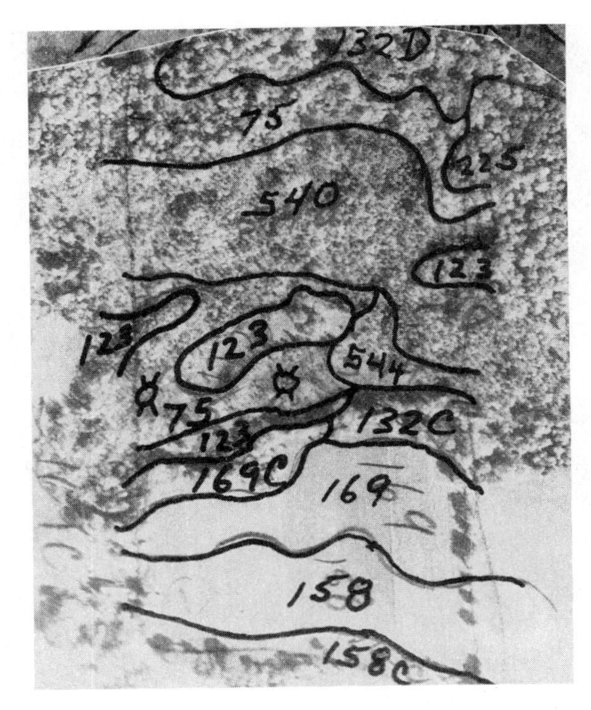

FIGURE 3-5 A soil map of a farm in Minnesota prepared by an NRCS surveyor. The numbers are codes for the following soil series:

75	Blufton	169	Braham
123	Dundas	225	Nessel
132	Hayden	540	Seelyville
158	Zimmerman	544	Cathro

(Courtesy of USDA, NRCS)

For some land, different soils are so mixed that many cannot be separated on the scale of the map. For instance, tiny pockets of one soil may be mixed into a larger soil unit. Therefore, many mapping units contain more than one series, family, or whatever level is being used. One such mapping unit is the **soil association.** An association consists of one or more major soils and one or more minor soils. For instance, Zimmerman soils usually appear beside two other series called the Isanti and Lino. In this area, glaciers carved out a landscape of fairly level outwash soils (Zimmerman), with scattered poorly drained low spots (Isanti), and other soils. Because they are so mixed with each other, they appear as one mapping unit on some maps as the Zimmerman-Isanti-Lino association.

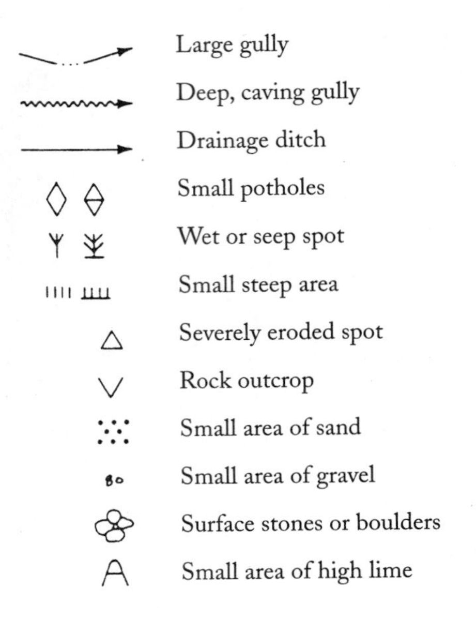

Slope		
Legend	**Percentage of Slope**	**Description**
A	0–2	Nearly level
B	2–6	Gently sloping
C	6–12	Sloping
D	12–18	Strongly sloping
E	18–30	Very strongly sloping
F	30–60	Steep
	More than 60	Very steep

Erosion	
Legend	**Description**
0	No erosion
1 or P	Slight, 0 to 1/3 topsoil gone
2 or R	Moderate, 1/3 to 2/3 topsoil gone
3 or S	Severe, 2/3 or more topsoil to 1/3 subsoil gone
4	Heavy subsoil erosion, deposition of eroded soil

FIGURE 3-6 A sample of soil-mapping symbols that give information about slope, erosion, and landscape features.

Soil survey maps are drawn to scale appropriate to the area and detail needed. The larger the scale, the larger the area covered but the less detail offered. Most county surveys are drawn to scales of 1:12,000 to 1:31,680. At the smaller scale, there are about 0.2 miles per map inch (5.28 inches per mile), and the smallest size area that can be practically noted is about one and a half acres. Therefore, small pockets of soil may well differ from what is mapped, a warning to be considered when interpreting a map. Where finer-textured detail is needed, a more exhaustive survey and a map drawn to finer scale may be required.

Soil Survey Reports. A completed soil map becomes part of a soil survey report. A soil survey report has four major parts: a set of soil maps, map legends that explain the map symbols, descriptions of the soils, and use and management reports for each soil. All these parts provide much useful information about the soils, including:

- Taxonomy of the soil—telling the order, suborder, and other classes.
- A brief description of the soil. For instance, for mapping unit 544 in figure 3-5, the Cathro series, the description reads: "The Cathro series consists of very poorly drained soils formed in deposits of herbaceous organic material over loamy sediments in depressions. This soil is black muck 23 inches thick. The substratum is a grayish brown sandy loam. Slopes are less than 2 percent. Most areas are used for woodland."
- Soil properties of each horizon, including texture, bulk density, permeability, available water, pH, salinity, and other features. Engineering properties are also listed.
- Rating of suitability for engineering projects like landfills, buildings, and roads. Problems are mentioned. For instance, the Cathro is listed as poor for most projects because of ponding.
- Suitability for water-management projects like reservoirs, drainage, and irrigation. Problems are mentioned. The Cathro, for instance, is poor for digging aquifer-fed ponds because of a low refill rate.
- Suitability for recreational development like playgrounds and campgrounds. Problems are mentioned. The Cathro is listed as poor because of ponding.
- Potential for cropping, including capability class and projected yields for common crops grown under high management. The Cathro, for instance, cannot be cultivated unless drained. If drained, one can expect 50 bushels of corn per acre, or 55 bushels of oats per acre.
- Woodland suitability, including problems and suggested trees to plant. The Cathro is rated as poor for woodlands, but certain trees that are tolerant of wet soil may be planted.
- Information about good plants for windbreaks.
- Potential as a habitat for wildlife. The Cathro is rated as good for wetland plants and animals, but poor for others.

Survey Report Uses. Soil maps are the heart of good land-use planning. Soil maps give the information needed to make good land-use decisions—whether the decision maker is a national planner or a farmer, home builder, or landscape designer. At the national level, for instance, the USDA has inventoried soil resources of the United States and kept track of them from soil maps.

Engineers also need soil maps. Civil engineers planning a new road will study maps to find routes with good soils for roadbeds. Planning commissions searching for new landfill sites will begin with soil maps.

New growers or growers planning to expand find soil maps useful for choosing new land. Instead of driving all over a region searching for the right land, one can target certain prime areas on soil maps.

Growers can use soil maps in many other ways. The information in soil surveys helps in planning irrigation or other engineering projects. For instance, the grower who owns the farm in figure 3-5 dug a pond in the wet Cathro soil in hopes of irrigating out of the pond. Had he read a soil report first, he would have known the pond would refill too slowly to be used for this purpose. Surveys also give farmers guidance as to what yield they should be getting and other useful information.

For a grower, an important use of soil maps is to prepare the field map, a most useful planning tool.

Field Maps. An NRCS map makes a good base for a grower's own field map. He or she traces the NRCS map on a sheet of clean paper, redrawing it at a larger scale if needed. With this map, the farmer divides the farm into fields and labels each part. By basing the fields at least partially on soil mapping units (or capability classes, to be covered next), fields will be uniform for cropping and for soil sampling. The grower can now use these blank maps as record-keeping and planning tools for numerous uses, including:

- noting crop rotations
- tracking manuring or other practices that affect fertilization
- recording pesticide applications
- making notes of problem spots in the field, like wet areas or large rocks
- mapping locations of irrigation and drainage systems

Land Capability Classes

Soil maps provide the basis for placing soils into **land capability classes.** This system indicates the best long-term use for land to protect it from erosion or other problems. The uses include cropping, pasture, rangeland, woodland (for lumber), recreation, and wildlife. The classes are not designed for all horticultural crops or crops that need very special management.

For example, flat land with deep rich soil can sustain long-term heavy cropping without erosion. It has few limitations and can be used for any of the listed uses. Sloping land, on the other hand, must be managed carefully to avoid destructive erosion and should not be "overfarmed." Sloping land has more serious limitations.

Capability Classes. The United States NRCS recognizes eight land capability classes. These are numbered by Roman numerals I to VIII. Class I soils have the fewest limitations and Class VIII soils are so limited as to be totally unsuitable for agriculture. Erosion hazard due to slope is the main criterion, but other criteria are used as well. Figure 3-7 shows sample uses for each class. Note that there are fewer safe uses for each succeeding class.

Class I soils have few limitations. They can be heavily cropped, pastured, or managed for woodlands or wildlife. Crop cultivation is the most profitable use of Class I soils (figure 3-8). These soils are well-drained and nearly level (0 percent to 2 percent slope). They have good water-holding capacity and are fertile. Ordinary cropping practices such as liming, fertilizing, and crop rotation keep these soils productive.

Class II soils are also suitable for all uses, but they have mild limitations that need moderate soil conservation or other measures when cropped (figure 3-9). Problems include:

1. gentle slopes (2 percent to 6 percent slope);
2. moderate erosion hazards;
3. shallow soil;
4. less than ideal tilth;

	Class							
Use	I	II	III	IV	V	VI	VII	VIII
Row crops	X	X	/					
Hay, small grains	X	X	X	/				
Pasture	X	X	X	X	X	X		
Range	X	X	X	X	X	X	/	
Woodland	X	X	X	X	X	X	/	
Recreation, wildlife	X	X	X	X	X	X	X	X

FIGURE 3-7 Suitable uses for soil capability classes. The higher the class number, the more limited is the number of safe uses. A single slash indicates that very careful management is needed or that the soil cannot be safely used for this purpose every year.

5. slight alkali or saline conditions; or
6. slightly poor drainage.

Class III soils can grow the same crops as Class I and II soils. However, serious problems need to be addressed, such as:

1. moderately steep slopes (6 percent to 12 percent slopes);
2. high erosion hazards;
3. poor drainage;
4. very shallow soil;
5. droughtiness;
6. low fertility;
7. moderate alkali or saline conditions; or
8. unstable structure.

Special conservation methods are needed. Growers should limit the number of row crops grown and favor close-growing crops. This is the lowest soil class that can be used safely for all crops, but only if used carefully.

FIGURE 3-8 Class I land consists of level, deep, rich soil that is excellent for cultivated crops. *(Courtesy of USDA NRCS)*

FIGURE 3-9 Class II land is often gently rolling. Here it is being used as wooded pasture. *(Courtesy of USDA NRCS)*

FIGURE 3-10 A Class IV soil in Idaho. This field is too steep for safe culture of row crops. The large circle on the hill is a slump, caused when a section of saturated soil slides downhill. *(Courtesy of University of Idaho Cooperative Extension Service)*

Class IV soils are marginal for cultivated crops (figure 3-10). Limitations are those listed for Class III but are more severe. Slopes may be 12 percent to 18 percent. Row crops cannot be grown safely but close-growing crops may be. Crops that cover the soil completely, like hay crops, are best. Careful erosion control measures must be practiced.

Class V soils are not suited to cultivated crops but may be used for range, pasture, woodlands, and recreation. These soils are level, have little erosion hazard, but are limited by factors such as (1) flooding; (2) short growing season; (3) rockiness; or (4) wet areas that cannot be drained.

Class VI soils are unsuitable for cultivated crops but may be used for pasture, range, wildlife, and woodland. Problems may include (1) steep slopes (18 percent to 30 percent slope); (2) severe erosion hazard; (3) established severe erosion; (4) stoniness; (5) shallowness; or (6) drought.

Class VII soils have the same problems as Class VI but are more severe. It is difficult to maintain high-quality pasture, but the land may be used for range, woodlot or forest, recreation, or wildlife if it is carefully managed. Slopes may be greater than 30 percent.

Class VIII soils cannot support any commercial plant production, even timber. They may only be preserved for recreation, wildlife, or for beauty (figure 3-11). Sandy beaches, rock outcroppings, and heavily flooded river bottoms are examples of Class VIII land.

A soil class may be upgraded if the problem is removed. For instance, a Cathro soil (figure 3-5, mapping unit 544) is so wet as to be placed in Class V. An artificially drained Cathro, however, may be moved to Class IV. Permanent irrigation, land leveling, and other practices may also upgrade a soil class.

The eight classes can be simplified to soils that can be used for cultivated crops (Classes I–III), marginal land for cropping (Class IV), and lands not suitable for cropping (Classes V–VIII).

Land Capability Subclasses. All classes except Class I have one or more limitations. **Land capability subclasses** indicate factors that limit soil use by means of a single letter code added to the class number. A Class IIe soil, for instance, is slightly limited by erosion hazards; a Class VIe soil is very limited by erosion hazards. The letter codes are as follows:

- e—Runoff and erosion. Land with slopes greater than 2 percent are those that need some form of water control.
- w—Wetness. These soils may be poorly drained or occasionally flooded (figure 3-12). Some such soils may be drained; others are classed as wetlands and are best left as is.
- s—Root zone or tillage problems. These soils are shallow, stony, droughty, infertile, or saline. Wind and water erosion may be problems.
- c—Climatic hazard. Areas of rainfall or temperature extremes make farming difficult. Examples include deserts or the Far North.

Soil Use Maps. Land capability classes rate soils for their use in agriculture. One could modify a soil map to replace soil series identifications with use classes and

FIGURE 3-11 The North Shore of Lake Superior in Minnesota is typical of Class VIII land. It cannot be farmed or lumbered safely but is a major recreational area.

FIGURE 3-12 This Class IIIw land could be drained for farming, but a better use may be to conserve it as wetlands. *(Courtesy of USDA NRCS)*

have a map that would show suitability of the area soils for agriculture. This would be a type of soil-use, or interpretive, map.

Similarly, a variety of other soil-use maps could be derived from soil surveys. For instance, one could draw a map of suitability for home drain fields, or woodlot production. The soil scientist, using the survey, rates each soil for the amount of hazard involved in that soil use. The map is then drawn and coded, for example, as low, medium, or high hazard for the intended use.

Appendix 5 shows how one might judge soils for a number of uses. From this appendix, students could judge soils as a lab exercise or even try to prepare their own land-use maps.

Computer Applications. Soil-mapping information can be assembled in a computer for faster, more knowledgeable natural resource and land management decisions. Information on mapping units in a state or county, for instance, can be computerized for easy access, or even put on the Internet.

Soil researchers are beginning to use computer software called Geographic Information Systems (GIS) in studying soils and making management decisions. GIS, as used in soil work, pulls together information about soils, topography, watersheds, and geology from such sources as soil surveys, the United States Geologic Survey, and even satellite images, to create a database. The GIS integrates all this information to generate a variety of customized interpretive maps. These maps are invaluable for land-use planning.

Lands of the United States. The United States is fortunate to have a great deal of good farmland. Figure 3-13 summarizes the capability of U.S. soils, excluding Alaska. Approximately 43 percent of our soil is rated in Classes I to III. This is soil on which nearly any crop can be grown. Most of the rest of U.S. land is suitable for

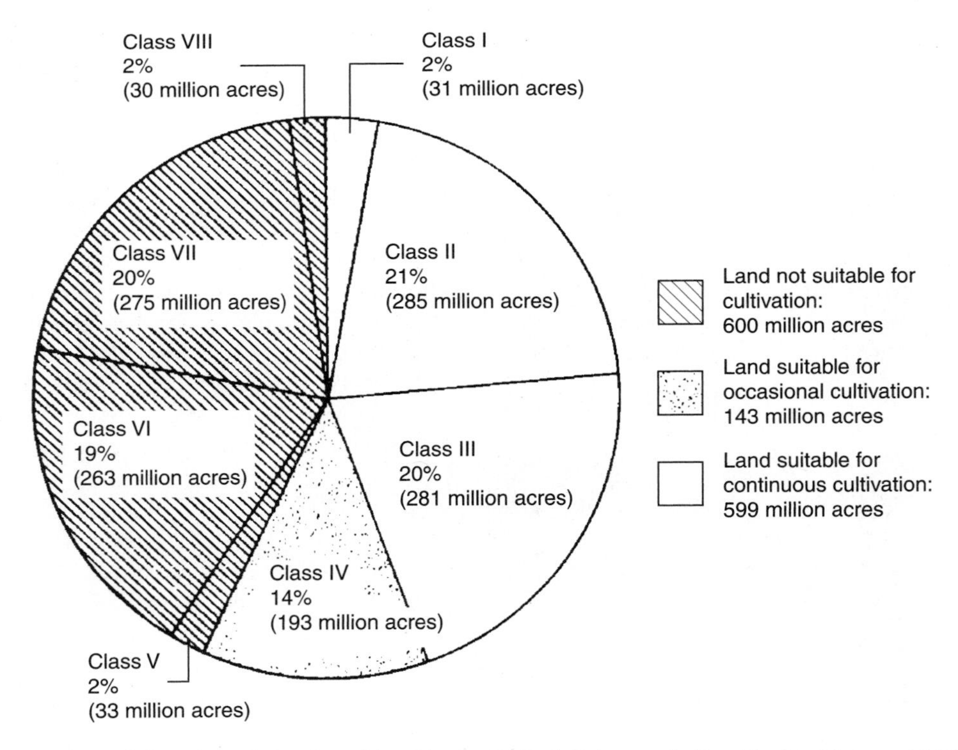

FIGURE 3-13 Land capability of nonfederal rural land in the United States, excluding Alaska. *(Source: USDA 1997 National Resource Inventory)*

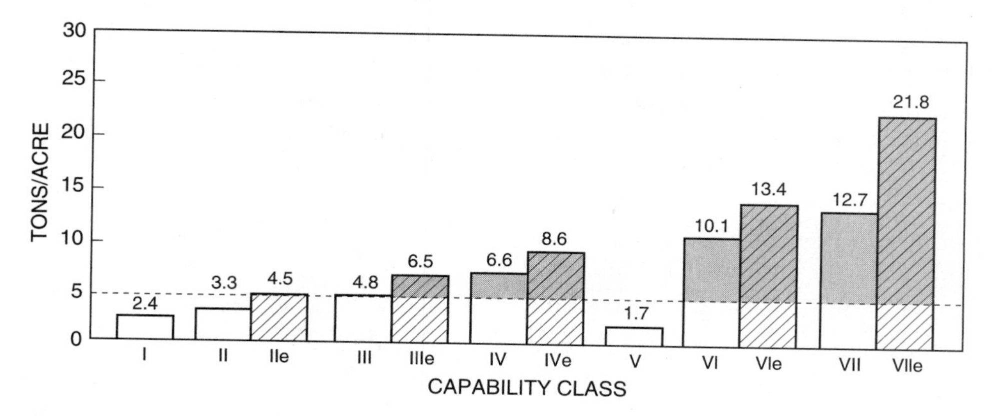

FIGURE 3-14 Average rate of sheet erosion by capability class in 1982 for cultivated cropland. A loss of about 5 tons per acre is generally the highest acceptable level. Most erosion occurs when farmers cultivate marginal land. Note that even Class IIe land nearly exceeds the acceptable rate. *(Source: USDA Preliminary Report 1982, National Resource Inventory 1984)*

some form of commercial production like grazing or woodlands.

Good farmland is not evenly distributed over the United States. Corn Belt states have the highest percentage of good farmland, followed by Northern Plains states and Delta states. Much of the land of the West is too mountainous to be useful for cultivated crops.

Generally, land use in the United States follows capability. Land in the top three classes is used primarily for cropland. The remaining land is used for pasture, range, or forest. However, the most severe erosion in the nation occurs on marginal land that is cropped by farmers. As figure 3-14 shows, soil loss increases when farmers use less suitable land.

SUMMARY

Soil survey efforts in the United States began around 1900. At first, a simple soil class system based on soil formation factors was used. This system was refined over time until the current system came into use.

Soil scientists presently classify soils according to their soil properties and profiles. This soil taxonomy has six levels. The top level consists of twelve soil orders. Each order is further divided into sub-orders, great groups, subgroups, families, and series. The important level to an individual grower is the soil series and its subdivision, the soil phase.

Soil scientists survey land and prepare a soil map based on this classification system. The surveyor studies the soil profile and notes slope, erosion, and other features. A soil survey report includes the map plus printed information about the soils on the map and their suitable uses. These reports are then used by regional planners, engineers, growers, and others.

The information in a soil survey places the land into one of eight capability classes. Classes I, II, and III are suitable for cultivated crops. Class IV is marginally useful for cultivation. Classes V to VIII are restricted to noncultivated uses. A number of factors are used to classify the soil, principally erosion hazard. Other factors include drainage, droughtiness, and extreme climates.

REVIEW

1. How do the five soil-forming factors interact to produce an Alfisol?
2. Explain what a soil interpretive map is. How could you draw one from a soil survey map?

3. The Saybrook silt loam, mentioned in the text, is a "fine-silty, mixed, superactive, mesic Oxyaquic Argiudoll." Try to identify which syllables or words apply to family, subgroup, great group, suborder, and order. What soil order is it?
4. At the time colonists arrived in the New World, most of the northern New England states were covered by forests. They were cut down and replaced by farms. Later, forests grew back when the center of agriculture shifted to the Midwest. Explain differences in soil that could have contributed to this shift. Hint: look at the soil map of the United States.
5. As soils develop over time, they may move from one soil order to another. What might be examples of young, "middle-aged," and older soil orders? Explain.
6. What are soil survey maps drawn on? What are the advantageous of this as a base map?
7. Describe the suitability of soils of the United States for crop growing. Give numbers.
8. What are the major soil orders of your state? How did the soil-forming factors interact to put them there?
9. Go back to Review Question 10 in chapter 2. Speculate on what soil order that soil would have been. Is there more information you need to be certain? What soil order is there now?

ENRICHMENT ACTIVITIES

1. The official inventory of soil series of the United States is maintained at the National Soil Survey Center at Iowa State University (NSSC), and is accessible on the Internet at <http://www.statlab.iastate.edu/soils/nsdaf/>. This site carries a wealth of data, including the classification and description of all the soil series. Search for the Ontario series under "Official Soil Series Descriptions." Examine the classification and information. What would you say about this series? Much of this information will be more clear after chapter 4.
2. The NSSC's Soil Science Education Web site has links to a listing of the order, suborder and great group "formative elements"—the combinations of letters that make up names. It is under "Soil Formation and Classification." It also has a link to a glossary of soil terms, which you may need to interpret the lists. So for deeper study of the naming system, go to: <http://www.statlab.iastate.edu/soils/nssc/educ/Edpage.html>.
3. For photos and descriptions of soils of the twelve orders, visit <http://soils.ag.uidaho.edu/soilorders> or <http://www.geobop.com/paleozoo/Soils/index.htm>. Compare and contrast the appearance of a Spodosol and a Mollisol. Where are they mostly located?
4. Examine the soil survey report for your county. What are the main chapters and what information do they contain? If you don't have a copy of your county report, the National Soil Survey Center has several on-line (note: many are PDF files). Pick one on the list found at <http://www.statlab.iastate.edu/soils/soildiv/surveys>.
5. For a detailed description of soil survey, study this on-line booklet from the USDA (a lengthy PDF file) at <http://www.statlab.iastate.edu/soils/nssc/nsscprod/surdown.pdf>.

CHAPTER 4

Physical Properties of Soil

OBJECTIVES

After completing this chapter, you should be able to:

- describe the concept of soil texture and its importance
- identify the texture of a sample of soil
- describe soil permeability and related properties
- describe structure and its formation and importance
- explain other physical properties
- discuss soil compaction and tilth

TERMS TO KNOW

aeration pores
biopore
blocky structure
bulk density
caliche
clay
claypan
clods
compaction
cone penetrometer
duripan
fragipan
friable
gleying
granular structure
hydraulic conductivity
infiltration
loam
macropores
massive soil
mechanical analysis
micropores
mottling
Munsell system
oven-dry soil
pan
particle density
peds
percolation
permeability
physical properties
platy structure
plinthite
prismatic structure
puddling
redoxymorphic feature

continued

sand
silt
single-grain soil
soil aggregates
soil consistence
soil separates
soil strength
soil texture
soil triangle
subsoiling
tilth
total pore space

Physical properties are soil characteristics a grower can see or feel. Physical properties greatly affect how soils are used to grow plants or for other activities. Is the soil loose so roots can grow easily through it or water seep in easily? Or is the soil tight, preventing root growth and water absorption? How well does the soil supply air, water, and nutrients? A knowledge of physical properties helps to answer these questions.

Soil Texture

The most fundamental soil property, one that most influences other soil traits, is texture. **Soil texture** describes the proportion of three sizes of soil particles—sand (large), silt (medium), and clay (small). The size of soil particles, in turn, affects such soil traits as water-holding capacity and aeration. Let us first describe why particle size affects these properties.

Effect of Particle Size. Soil particle size affects two important soil features: internal surface area and the numbers and sizes of pore spaces. The internal surface area of a soil is the total surface area of all the particles in the soil. Figure 4-1 uses children's alphabet blocks to demonstrate that the smaller the particles in a soil, the larger the internal surface area. Since soil contains many small particles, a handful of soil may hold many thousand square feet of internal surface area.

Internal Surface Area. Soil surface area is important because reactions occur on the surface of soil particles. Picture pouring water over a pile of marbles. Most of the water runs quickly away. Droplets clinging to the surface of the marbles are the only water retained in the pile, since water cannot soak into the marbles. Following the rule about particle size, a pile of small beads holds more water than a pile of marbles because it has more surface area for water to cling to. Because soils with the smallest particles, like silt

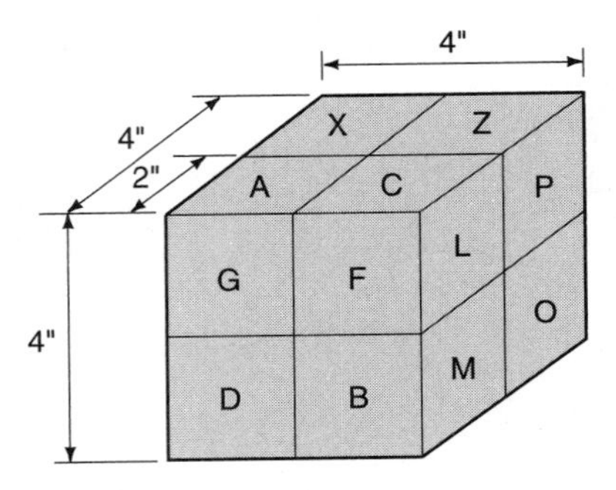

(*A*) Eight blocks have been put together to make a single large block. Each side measures 4 × 4 inches. The total surface area of this large block is 96 square inches:

Total area = area of each side × number of sides
96 sq. in. = 4 × 4 × 6

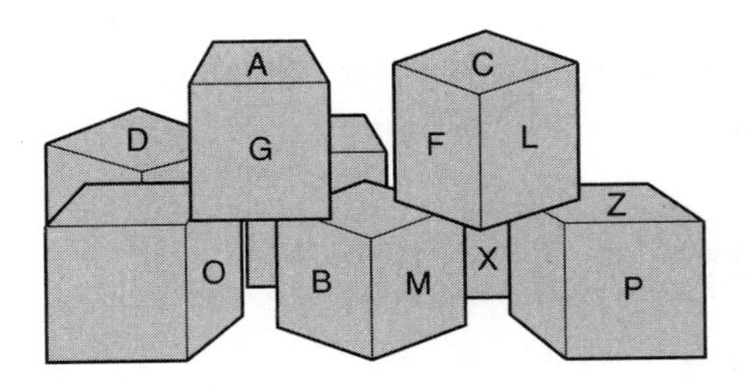

(*B*) The large block has been cut into eight equal blocks. The total surface area of all the blocks is now 192 square inches:

Total area = area of each block × number of blocks
192 sq. in. = 2 × 2 × 6 × 8

By halving the size of the blocks, the total surface area is doubled. In the soil, small particles create a large surface area for water and nutrients to hold on to.

FIGURE 4-1 The smaller the soil particles, the greater the internal surface area.

and clay, have the largest surface area, they hold the most moisture.

Reactions that hold plant nutrients in the soil also occur on particle surfaces. Therefore, we can make the rule that the smaller the particles in a soil, the more water and nutrients the soil can retain.

Soil Pores. Pore size and number depend on particle size. Figure 4-2 suggests that more pores are found between small particles than between large ones. However, the figure also shows that pores are larger between larger particles. Thus, soils high in clay have many small pores, while soils high in sand have fewer but larger pores. Water drains rapidly through large pores, called **macropores** or **aeration pores.** As water drains, it pulls air in behind it, filling up the spaces. Small pores, or **micropores,** tend to retain water. Both pore sizes are important, because soil needs micropores to hold water and macropores for air.

Soil Separates. Soil scientists divide mineral particles into size groups called **soil separates** and define three broad classes: sand, silt, and clay. The largest size separate, sand, is further divided into four subcategories. Figure 4-3 names the separates and gives their sizes according to the system adopted by the United States Department of Agriculture. Figure 4-4 gives some idea of the relative sizes of the separates.

Sand, the largest soil separate, is composed mainly of weathered grains of quartz. Individual sand grains, except for very fine ones, are visible to the eye. All are gritty to the touch. Sand grains do not stick to one another, so they act as individual grains in the soil. Enough sand in a soil creates large pores, so sand improves water infiltration (rate at which water enters the soil) and aeration. On the other hand, large amounts of sand lower the ability of the soil to retain water and nutrients.

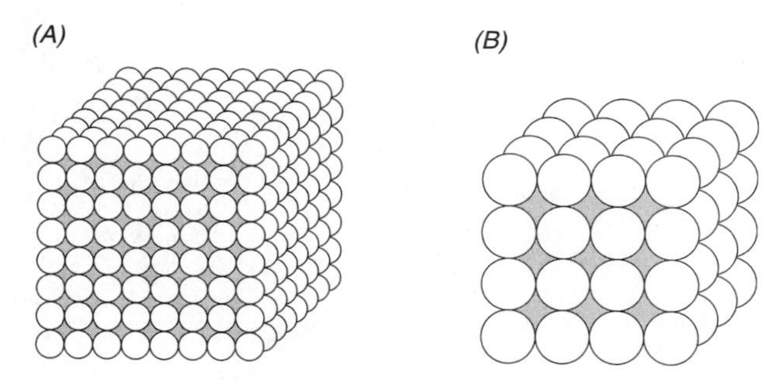

FIGURE 4-2 Soil particle size affects pore size. *(A)* Small particles create many small pores. *(B)* Pores are larger but fewer in number between large particles. Micropores usually hold water, macropores hold air.

Separate	Diameter (mm)	Comparison	Feel
Very coarse sand	2.00–1.00	36″	Grains easily seen, sharp, gritty
Coarse sand	1.00–0.50	18″	
Medium sand	0.50–0.25	9″	
Fine sand	0.25–0.10	4 1/2″	Gritty, each grain barely visible
Very fine sand	0.10–0.05	1 3/4″	
Silt	0.05–0.002	7/16″	Grains invisible to eye, silky to touch
Clay	<0.002	1/32″	Sticky when wet, dry pellets hard, harsh

FIGURE 4-3 The USDA system of soil separates. The comparison shows the differences by setting a very coarse sand equal to three feet in diameter.

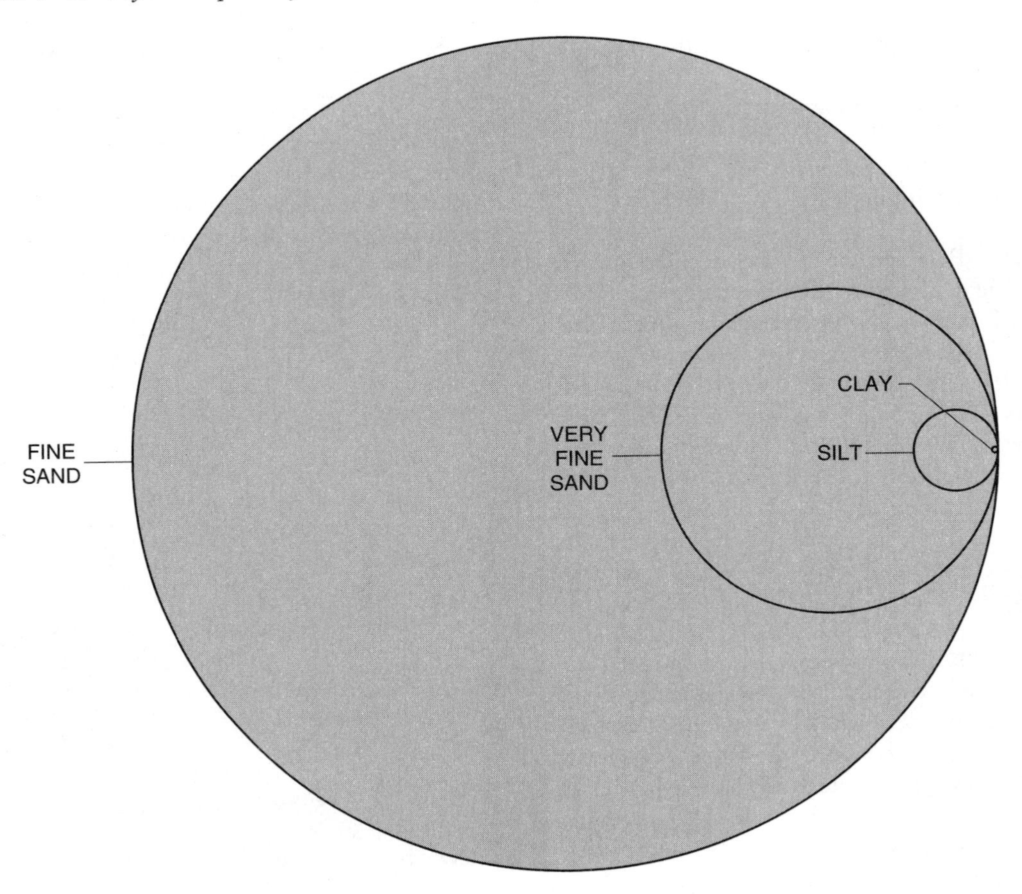

FIGURE 4-4 Comparing the relative sizes of soil separates. On this scale, very coarse sand would be three feet across.

Silt is the medium-sized soil separate. Silt particles are silky or powdery to the touch, like talc. Like sand, silt grains do not stick to one another. Of all the soil separates, silt has the best ability to hold large amounts of water in a form plants can use.

Clay is the smallest soil separate and consists of mostly tiny sheet-like crystals. While sand and silt result simply from rock crumbling into small particles, clay results from chemical reactions between weathered minerals to form tiny particles of new minerals. These new minerals are able to bond nutrients chemically to their surfaces, holding plant nutrients in the soil.

Clay particles stick to one another and so do not behave as individual grains in the soil. Wet clay is usually sticky (figure 4-5) and can be molded. Some types of clay swell when wet and shrink as they dry.

As mentioned earlier, internal surface areas influence a number of soil properties. A handful of sand may have a surface area the size of a ping-pong table, while a handful of clay could reach the area of a football field. It is not surprising that soils high in clay better retain water and nutrients. Conversely, clays are less well aerated, and water seeps into them more slowly.

Gravel and other pieces of stone larger than two millimeters are not considered to be part of soil texture. They often are, however, part of the soil and affect its use, as anyone who has picked rocks out of a field can

FIGURE 4-5 Treading on this wet soil near Winnipeg, Canada, proves how sticky clay can be.

Class	Diameter Range (mm)	Diameter Range (in.)
Gravel	2–75	1/12–3
Cobbles	75–250	3–10
Stones	250–600	10–24
Boulders	>600	>24

FIGURE 4-6 USDA size classification for stones in the soil.

testify. Figure 4-6 lists the USDA size classifications of rock fragments in the soil.

Textural Classification. Soils usually consist of more than one soil separate; all three are found in most soils. The exact proportion, or percentage, of the three separates is called soil texture. Obviously, any number of combinations of the three are possible, so soil scientists simplify texture by dividing soils into textural classes. Soils in the same textural class are similar.

The twelve textural classes are shown in the **soil triangle** in figure 4-7. Each side of the triangle represents the percentage of one soil separate. A person can measure the amount of sand, silt, and clay in a soil sample and simply read the class off the triangle. An example included in figure 4-7 shows how to read the triangle.

Examine the soil classes carefully. Each corner of the triangle is a class dominated by one soil separate: sand, silt, or clay. The largest class is clay soil, because clay has the most powerful effect on soil properties. With as little as 40 percent clay a soil is classified as a clay soil. Another important textural name is **loam,** a soil in which sand, silt, and clay contribute equally to the soil's properties. The remaining classes have properties between those of the four major classes and their names suggest the difference. For example, a loamy sand is a sandy soil containing enough clay or silt to make it more loamy.

Growers can usually manage soils without knowing the exact soil texture, so a broader classification is often adequate. One can simply classify soils as sandy, loamy, or clayey, as described previously. Figure 4-8 shows another approach. The twelve classes are divided into three broad categories—coarse, medium, or fine—based on the size of the soil separates.

Determining Soil Texture. The amount of sand, silt, and clay in a soil can be measured by **mechanical analysis.** Mechanical analysis is based on the fact that the larger a soil particle, the faster it sinks in water. For instance, it takes only forty-five seconds for very fine sand to settle through four inches of water, while it takes about eight hours for large clay particles. In mechanical analysis, one stirs soil into water and notes how fast the soil particles settle out. Appendix 2 provides instructions for a simple type of mechanical analysis called a sedimentation test.

An even simpler test, which can be done on-site, is the ribbon or feel test. The test is based on the feel of damp soil and how easily it can be molded. All those who work with soil should be able to do ribbon testing. The procedure for the test is as follows:

Step 1 Obtain a large enough sample of soil to form a half-inch ball. The sample should contain no gravel or bits of debris. If needed, run the sample through a sieve to remove such material.

Step 2 Moisten the sample to a medium moisture level, like workable putty. Work the soil between the fingers until it is uniformly moist and dry lumps are wetted. Note any grittiness

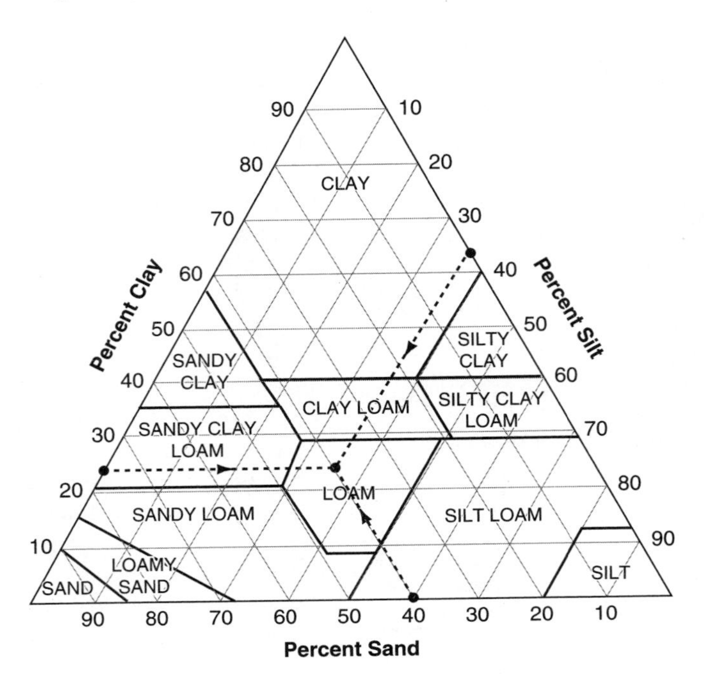

Example: Identify a soil that is 40% sand, 22% clay, and 38% silt

1. Find 40 on the side for sand.
2. Draw a line in the direction of the arrow.
3. Do the same for clay (22%) and silt (38%).
4. The spot where the three lines come together is the soil texture. In this case, the soil is a loam.

A textural name may include a prefix naming the dominant sand size, as in "coarse sandy loam."

FIGURE 4-7 The soil triangle. Each side of the triangle is a soil separate. The numbers are the percentage of soil particles of that type.

that indicates sand or the stickiness of clay. Clay also stains the fingers.

Step 3 Mold the sample into a half-inch ball and try to lightly squeeze it. If the ball breaks at the slightest pressure, the soil is a sand or coarse sandy loam. If the ball stays together but changes shape easily, it is a sandy loam, loam, or silt loam. Finer-texture soils resist molding.

Step 4 Squeeze out a ribbon between the thumb and forefinger, noting how long a ribbon can be formed before it breaks (figure 4-9). Use this guide to narrow down the choice of textures:
no ribboning: loamy sand
ribbon shorter than one inch: loam, silt, silt loam, sandy loam
ribbons one to two inches long: sandy clay loam, silty clay loam, clay loam
ribbons two to three inches long: sandy clay, silty clay, clay

Step 5 Put all the observations together to decide the textural class. Sand feels gritty, silt feels

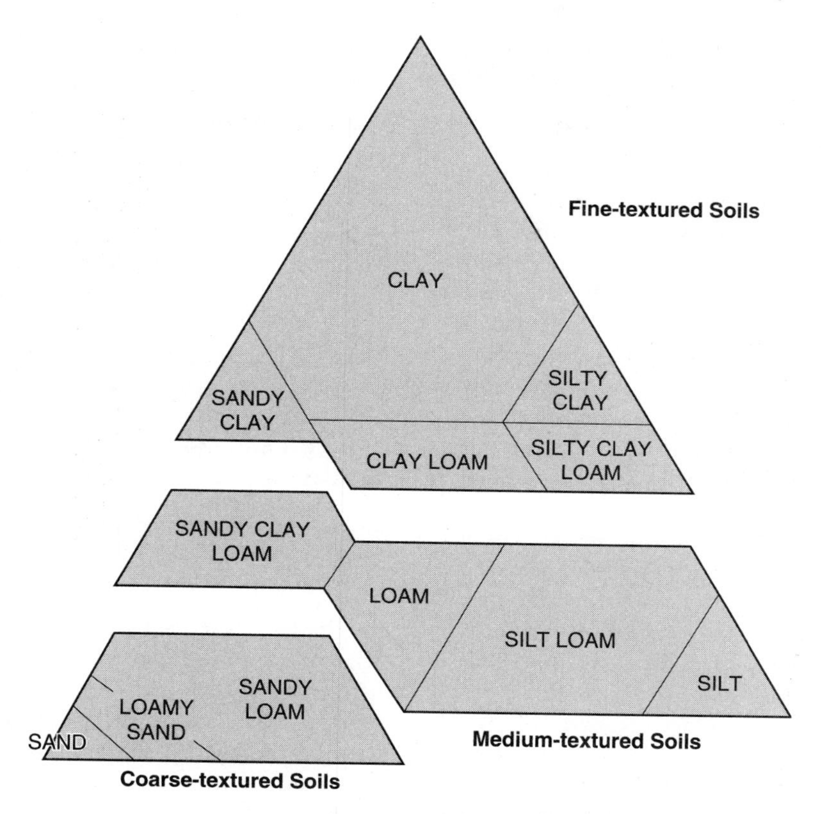

FIGURE 4-8 The soil triangle is redrawn to show fine-, medium-, and coarse-textured soils. An exception is very fine sandy loam, which is considered medium textured.

smooth, clay feels sticky. So, for instance, sandy clay forms a long ribbon yet feels slightly gritty. A short ribbon that feels smooth is a silt loam.

The ribbon test is only useful if one has practiced it enough to "get a feel" for it. Try it out a few times.

Characteristics of Textural Classes. Soil scientists place soils in textural classes because each class has properties important to its management. One can only generalize about textural effects because other properties also affect the soil. However, here are a few useful guidelines.

Texture governs the way water behaves in the soil. For instance, water enters (**infiltration**) and drains through (**percolation**) coarse soil most rapidly because of the large pore spaces. Thus, a coarse soil dries out most quickly after a heavy rain or in the spring, allowing a

FIGURE 4-9 Ribbon test to determine soil texture.

grower to get into the field more quickly. Similarly, coarse soils are more likely to need frequent irrigation. Growers with fine soils are likely to worry about the opposite problem—dealing with excess water.

Fine soils retain plant nutrients better than coarse soils. This is true partly because the rapid percolation of water through coarse soil leaches out nutrients. Also, clay particles have the best ability to retain nutrient chemicals.

Soil texture influences how easily a soil can be worked. Because clay particles stick together, it takes more horsepower and fuel to pull tools like plows through a fine soil. Landscapers also find that it is harder and slower to dig holes in fine soils for tree planting—so texture can even affect what should be charged for landscaping services.

The stickiness of clay also affects the physical condition of the soil. For instance, fine soils often form clods when they are tilled. A crust may also form on the surface and interfere with seedling emergence. A fine soil tends to be "tight," meaning it has mostly small pores that are difficult for air and roots to penetrate. In contrast, coarse soils are "loose" and well aerated.

For most purposes, growers consider medium soils to be ideal. They hold water, but they don't stay wet too long. They are neither sticky nor hard to work. In general, medium soils have the good traits of both coarse and fine soils without their bad traits.

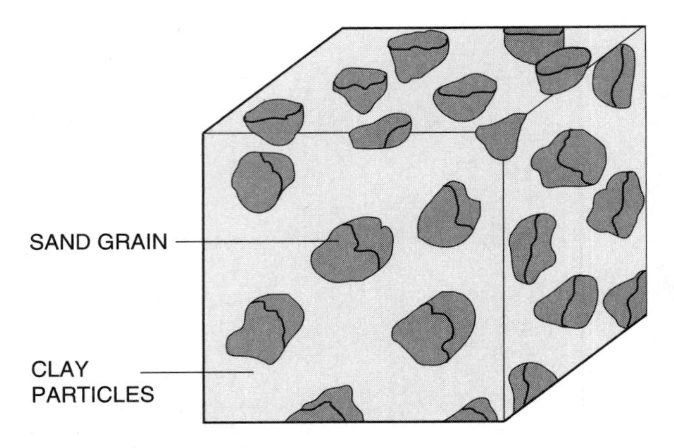

FIGURE 4-10 When sand is mixed into clay, clay particles surround the sand grains and large pores are not formed. Very large quantities of sand are needed to loosen clay soils–enough that the sand grains touch and there isn't enough clay to fill all the gaps.

Modifying Soil Texture. Growers and engineers use soils for many purposes. For each purpose, a different soil texture may be best. For example, corn tends to be most productive on a loam, potatoes on a sandy loam, and black walnuts on fine-textured soils. Loamy soils are easiest to use for landscapes and gardens.

Can a grower change soil texture to improve it for the crop being grown? Except in very small areas, like golf greens or potting soils, changing texture is impractical. The amounts of clay or sand to be added are too large. Figure 4-10 shows the effect of adding sand to a clay soil to loosen it—clay particles surround the sand grains and fill in any pores that may be created. As a result, the soil continues to behave much like clay. To greatly modify clay, enough sand must be added to make the sand grains touch each other and form bridges that exclude clay from the pores between the sand grains.

There are three ways for growers to take texture into account. First, select a crop to fit the soil, or purchase land that suits the crop. For example, an apple grower may purchase land with the fine soil on which apples grow best. Second, manage the soil in a manner that fits the texture. For example, with proper fertilization and irrigation, coarse soils can be very productive. Third, organic matter can improve texture extremes by making sandy soils less droughty and by loosening clay soils.

Soil Density and Permeability

As stated earlier, important physical properties relate to the spaces between soil particles. For a better understanding of soil pore space, this discussion will work through a series of related properties, beginning with soil density. We begin here because the density of soil—its mass per volume—is related to the amount of empty space in the soil.

Particle Density. One could ask how much soil would weigh if there were no pore space. This is **particle density,** the density of solid particles only. As an example, the particle density of a soil made wholly of quartz sand would be the same as the density of a solid block of quartz or 2.65 grams per cubic centimeter (166 pounds per cubic foot).

Mineral	Density (grams/cm^3)	Density (lbs/ft^3)
Water	1.0	62.5
Quartz	2.65	166
Feldspars	2.5–2.7	156–169
Micas	2.7–3.0	169–188
Clay minerals	2.0–3.0	125–188

FIGURE 4-11 Densities of several soil-forming minerals.

Particle density varies according to the type of minerals in the parent material and the amount of organic matter in the soil. Figure 4-11 lists the density of several soil-forming minerals. Note that the densities are very similar. In fact, there is surprisingly little variation in the particle densities of most mineral soils. Most soils average about 2.65 grams per cubic centimeter, a value used as a standard density in soil calculations. High amounts of organic matter reduce the value because organic matter is much lighter than mineral matter.

Bulk Density. Because soil does contain pore spaces, the actual density of a soil is less than the particle density. This measurement is **bulk density,** or the mass of a volume of undisturbed oven-dry soil.

To measure bulk density, a core of soil of known volume is carefully removed from the field. The soil core is then dried in an oven at 105°C until it reaches a constant weight. This is called **oven-dry soil.** The core is then weighed, and the bulk density is calculated. The example that follows is for a core of 500 cubic centimeters (cm^3) that weighs 650 grams (g):

$$BD = \frac{\text{weight dry soil}}{\text{volume dry soil}} = \frac{\text{g}}{\text{cm}^3}$$

$$BD = \frac{650\text{ g}}{500\text{ cm}^3} = 1.3\text{ g/cm}^3$$

The bulk densities of mineral soils depend mostly on the amount of pore space in the soil, since particle weight is fairly constant. Bulk densities of mineral soils usually range from 1.0 grams per cubic centimeter (62.5 lbs/ft^3) for "fluffed-up" clay soils to 1.8 grams per cubic centimeter (113 lbs/ft^3) for some sandy soils. Organic soils are much lighter, with values of 0.1 to 0.6 grams per cubic centimeters (6–38 lbs/ft^3) being common.

Soil Porosity. **Total pore space** is a measure of the soil volume that holds air and water. This value is usually expressed as a percentage and is known as porosity. Thus, a soil with a 50 percent porosity is half solid particles and half pore space.

Porosity can be measured by placing an oven-dry soil core in a pan of water until all of the empty pore space is filled with water. The water volume, which fills the pore space, divided by the total core volume, is porosity. The water volume, of course, would be difficult to measure directly. However, the metric system defines one cubic centimeter of water as weighing one gram. Thus, if one measures porosity metrically, water volume and weight are the same. Therefore, the difference in weight between the dry and the wet cores is the total pore space. This number is converted to a percentage to get porosity. The soil core used as an example before had a volume of 500 cubic centimeters and weighed 650 grams when dry. When wet, the same core weighs 900 grams. Porosity is calculated as follows:

$$\text{Porosity} = \frac{\text{wet weight (g)} - \text{dry weight (g)}}{\text{soil volume (cm}^3\text{)}} \times 100$$

$$= \frac{900 - 650}{500} \times 100 = 50\%$$

Porosity can also be calculated from bulk density and particle density. If there were no pore space, then bulk density (BD) would be the same as particle density (PD). The ratio BD/PD would be equal to one. The more pore space, the smaller the bulk density and the smaller the ratio BD/PD. In fact, the ratio BD/PD is simply the percentage of the soil that is solid matter. If one subtracts that percentage from 100 percent, the difference is the percentage of pore space. To make the calculation, one can usually assume that PD is 2.65 grams per cubic centimeter. The following equation can be used to calculate porosity:

$$\text{Porosity} = 100\% - \left(\frac{\text{BD}}{\text{PD}} \times 100\right)$$

If we substitute the values for the bulk density just calculated:

$$\text{Porosity} = 100\% - (\frac{1.3}{2.65} \times 100) = 50\%$$

The porosity of sand (about 30 percent) is lower than the porosity of clay (about 50 percent). Figure 4-12 shows that porosity increases at finer textures. Yet common sense tells us that water seeps into sand very rapidly, but it seeps only slowly into clay even though it has higher porosity. The next section explains why.

Permeability. **Permeability** is the ease with which air, water, and roots move through soil. In highly permeable soil, water infiltrates soil rapidly, and aeration keeps roots well supplied with oxygen. Roots grow through permeable soil with ease. We can think of permeable soil as being "loose" and impermeable soil as being "tight."

Permeability depends partially on the number of soil pores, but it depends more on size and continuity of the pores. The movement of air, water, and roots can be likened to walking a maze. If paths are too narrow, progress is difficult. Progress is even more difficult when paths come to a dead end. Like a maze with dead ends, soils lacking large, continuous pores limit the flow of air and water.

Large, continuous pore spaces in the soil, or macropores, occur between large particles. Therefore, the number of macropores depends on texture, as shown in figure 4-12, and permeability must also depend partially on texture. Permeability is not a soil property that can be measured directly. However, the movement of water, which reflects permeability, can be measured. **Hydraulic conductivity** is a measure of the rate of water movement through a soil. Coarse-textured soils might have conductivities of one and one-half inches per hour or more, while fine-textured soils might measure a hundredth of that, or even less. Fortunately for those who crop such soils, another physical property, structure, influences permeability.

Soil Structure

Heavy soils would make very poor root environments, except that structure can alter the effects of texture. Struc-

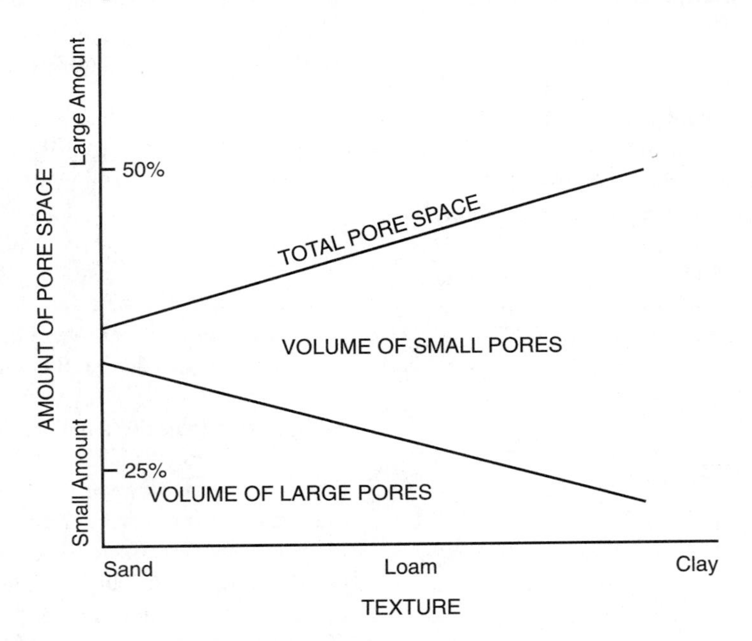

FIGURE 4-12 Texture affects soil pores. The top line shows the total pore space in the soil. Clay has the greatest total pore space. The lower line shows how much space is in large pores. Sand has the most large pore space. The amount of small pore space lies between the two lines. Clay has the most small pore space. Loam has a balance of large and small pores.

FIGURE 4-13 Soil aggregates. The peds create large spaces in the soil to improve infiltration, aeration, and root growth. This is an example of granular structure.

ture refers to the way soil particles clump together into large units (figure 4-13). These large units are called **soil aggregates.** Aggregates that occur naturally in the soil are **peds,** while clumps of soil caused by tillage are called **clods.**

Peds are relatively large, ranging from the size of a large sand grain to several inches. Spaces between clay particles may be tiny, but the spaces between peds may be large. Inside the peds are small, water-holding micropores; between the peds are large air-filled macropores. Well-aggregated soils contain the large, continuous pores that promote good air and water movement and that provide easy paths for root growth. Good water-holding capacity is maintained within the peds.

There are many different kinds of structure, and some are better at improving permeability than others. Soil scientists classify structure according to three groups of traits.

- Type refers to the shape of the soil aggregates (figure 4-14). These shapes are described in detail following.
- Class is the size of the peds, which can be very fine, fine, medium, coarse, or very coarse.
- Grade refers to how distinct and strong the peds are. One grade, structureless, applies to soils that have no peds. Weak grades are barely visible in a moist soil, whereas strong peds are quite visible and can be easily handled without breaking. Moderate grade is in between.

Structureless Soil. Sand behaves as individual grains, so sandy soils seldom have much structure. These soils are called **single-grain.** Sandy soils are naturally permeable, so single-grain soils have good infiltration rates and aeration.

Finer soils lacking structure are a solid mass stuck together like molding clay. These **massive soils,** as they are called, lack permeability. Massive soil is typical of some C horizons. Tillage of wet soil may result in massive soil in the A horizon.

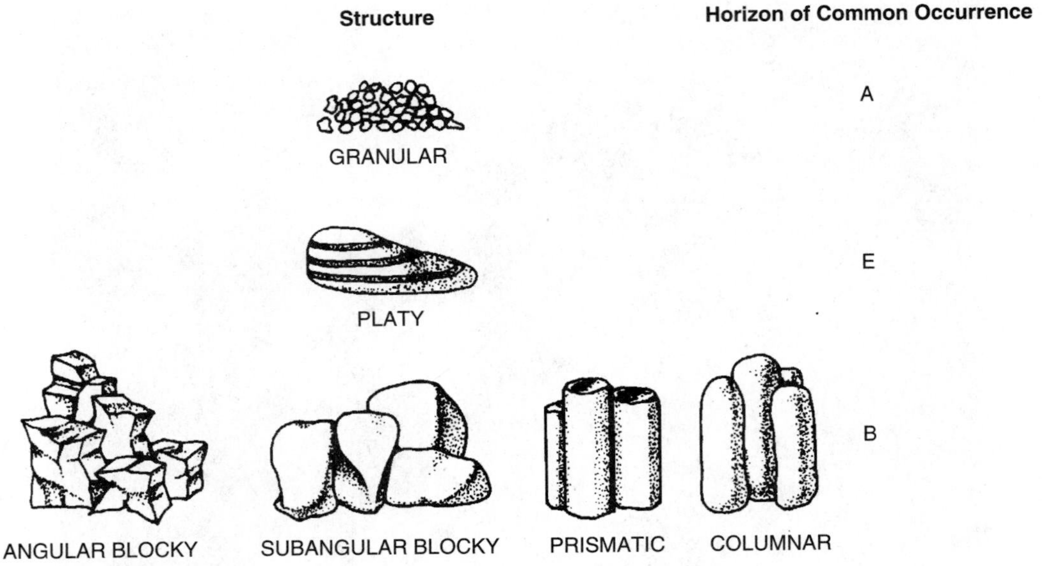

Structure | Horizon of Common Occurrence

GRANULAR — A

PLATY — E

ANGULAR BLOCKY, SUBANGULAR BLOCKY, PRISMATIC, COLUMNAR — B

FIGURE 4-14 Types of soil structures.

Types of Soil Structure. **Granular structure** is commonly found in A horizons (figure 4-13). The peds are small, usually between one to ten millimeters (1/25 to 2/5 inch), rounded in form, and considered the most desirable of structures. Such structure increases total pore space and lowers bulk density compared with a soil lacking structure.

Platy structure is usually found in E horizons. Peds are large but thin, platelike, and arranged in overlapping horizontal layers. The arrangement makes discontinuous pores that reduce penetration of air, water, and roots. Soil compaction can create platy structure in the A horizon when granules of topsoil are crushed into thin layers.

Blocky structure is typical of many B horizons. The peds are large, five to more than fifty millimeters (1/5 to 2 inches) and blocklike in shape. If the ped is very angular, it is termed "angular blocky." More rounded peds are described as "subangular blocky." Blocky structure has medium permeability.

Prismatic structures also occupy the B horizon of some soils. Peds are large, usually from ten to more than 100 millimeters (1/5 to 4 inches), forming angular columns that stand upright in the soil. If the top of the ped is pointed or flat, the structure is called "prismatic." If the top of the ped is rounded, it is termed "columnar." Prismatic structure is moderately permeable; columnar structures are slowly permeable.

Formation of Soil Structure. Soil structure forms through a two-step process; the first creates a loose ped, the second cements it. First, a clump of soil particles sticks loosely together to form a loose aggregate that is weak and easily crushed. This is usually by a process of localized compression. That is, a force compresses a small mass of soil together, separating it slightly from the surrounding soil mass. Compression may repeat itself several times to the same soil mass, increasing the coherence of the aggregate. Roots, especially of grasses, can surround a soil mass, and as the roots expand, create the compression. Fungal bodies can do the same. This mechanism tends to dominate in A horizons. Alternate freezing and thawing, and wetting and drying, also fractures and compresses the soil into separate soil masses. These mechanisms tends to dominate in B horizons.

Second, weak aggregates are cemented to make them distinct and strong. Clay, iron oxides, and organic matter may each act as cements. In most soils, microorganisms provide the best cement. When soil microbes break down plant residues, they produce gums that glue the peds together. Biological cementation tends to glue the granules of the A horizon, while chemicals cement the blocks and prisms of the B. Therefore, the best way to enhance stable soil structure in the topsoil is to frequently add organic matter to the soil.

Large amounts of sodium in the soil reverse the process, causing soil aggregates to disperse. Sodium occurs naturally in some soils or may build up because of use of high-sodium irrigation water. This will be discussed in more detail in chapter 11.

Soil Consistence

Soil consistence refers to the behavior of soil when pressure is applied. It relates to the degree that soil particles stick to one another and mostly results from certain types of clay.

The effect of consistence can be best explained by some examples. Loose sand, for instance, shifts easily under pressure, so that vehicles may get stuck in sand along a beach. Preparing a seedbed for planting is another example. A grower wishes to break apart large chunks of soil to get a fine surface to plant seeds in. It is the consistence of the soil that determines how easily those chunks can be broken down.

Consistence depends on how moist a soil is, so it can be measured at three different moisture levels. Each level has its own descriptive terms (figure 4-15).

- **WET SOIL.** Wet soil is checked for stickiness and plasticity. Plasticity is how easily soil can be molded between the fingers. To determine stickiness, some soil is pressed between thumb and forefinger, and the amount that sticks to the fingers is noted.
- **MOIST SOIL.** The terms friable and firm apply to soils in the moist state. **Friable** means that soil materials can be crushed easily under pressure. Technically, a soil is termed friable if a one-inch block of moist soil can be easily crushed between thumb and forefinger.
- **DRY SOIL.** Determined by trying to crush an air-dried mass of soil in the hand. Very hard, dry soil, for instance, can be crushed between two hands.

By rating a soil for consistence, one can infer such information as suitability for plowing, likelihood of erosion, or texture. The feel test for texture works because of the consistence of different soil textures. Loose soil is coarse textured, friable soil is medium textured or well aggregated, and firm soil is tight or fine textured. A firm soil may lack good structure or be compacted.

Soil Tilth

Tilth is a general term for the physical condition of a tilled soil. It suggests how easy the soil is to till, how good a seedbed can be made, how easily seedlings can come up, and the ease of root growth. Tilth is actually a combination of other physical properties, including texture, structure, permeability, and consistence.

Tillage improves soil tilth for a time, improving soil-air-water relations for new seedlings. It does so by loosening the soil and stirring air into it. Fine-textured soils are most improved by tillage because coarse soils are already well-aerated.

Tillage tends to cause a year-by-year decline in soil structure, however. Compare the topsoil of a cultivated field with that of a nearby fencerow that has not been

Wet		Moist	Dry
Stickiness	Plasticity		
Nonsticky	Nonplastic	Loose	Loose
Slightly sticky	Slightly plastic	Very friable	Soft
Sticky	Plastic	Friable	Slightly hard
Very sticky	Very plastic	Firm	Moderately hard
		Very firm	Hard
		Extremely firm	Very hard

FIGURE 4-15 Consistence terms for soil at three different moisture levels.

cultivated. Peds from the fencerow will be more numerous, stronger, and of a better type. The weakening structure of the tilled soil, in turn, lowers water infiltration, aeration, and ease of root growth.

Compaction. **Compaction** results when pressure is applied to the soil surface. Light compaction in aggregated soils squeezes soil aggregates together, reducing the size of interped pores. Further compaction begins to crush the aggregates. In single-grain or massive soil, pressure forces individual particles closer together. Compaction primarily alters soil traits related to pores and soil strength.

Compaction can profoundly alter soil traits like porosity and permeability. Not only can peds collapse and particles be squeezed together, but small soil particles can also be forced into spaces between larger soil particles. Total pore space declines, while the number of micropores increases and macropores are lost. Some micropores shrink enough to become so tiny as to make the water in them unavailable to plants. Pores become disconnected, and the pathways for gas and water movement more tortuous.

Compaction increases **soil strength,** the degree to which a soil can resist movement due to pressure. Recall from chapter 1 that roots must push aside soil to grow. Put another way, roots grow well only if root pressure exceeds soil strength. Compaction therefore increases the resistance that roots must counter as they grow through the soil.

As a result, severe compaction has the following unfortunate results:

- reduced porosity and permeability
- reduced air exchange, and a longer residence time for air in the soil. Oxygen content falls and carbon dioxide content increases.
- decreased infiltration rate of water
- increased erosion associated with greater runoff
- reduced water percolation, which can leave soils excessively wet
- reduced availability of the water that is in the soil
- restricted root growth

Interestingly, roots try to compensate by changing their growth habit. In compacted soils, roots tend to be less branched, shorter, and thicker. This shape better protects roots from the higher soil pressure and increases the root pressure that can be exerted to deform the soil enough to allow root growth. While these modifications aid growth in compacted soil, they also reduce root effectiveness in gathering water and nutrients. Roots also become very efficient at finding and following channels in the soil like fractures and areas left by worms or decaying roots.

FIGURE 4-16 Tillage compaction. This cross-section of a corn field displays compaction from both between-row cultivation and a deeper plow pan. *(Courtesy of American Society of Agronomy)*

A number of agricultural activities induce compaction. For instance, cultivation and other equipment operations during the growing season compact the soil between crop rows (figure 4-16). If severe enough, it can restrict root growth between the rows.

Annual plowing breaks up this compaction. However, just below the plow layer, a compacted zone develops. This plow pan, or tillage pan (figure 4-16), restricts root growth and drainage of water deeper into the soil.

The worst compaction occurs when heavy equipment compresses the subsoil, leading to subsoil compaction. Harvest and transport equipment can exert loads up to forty tons per axle (figure 4-17). Such a heavy load, if

FIGURE 4-17 Harvest equipment can cause subsoil compaction. Grain carts such as this can carry loads of up to 40 tons per axle. *(Courtesy of American Society of Agronomy)*

applied to wet soil, can cause deep compaction that seriously reduces crop yields. Tillage equipment, which generally exerts only about five tons per axle, does not create subsoil compaction.

The degree of compaction relates to the natural compressibility of a soil, its moisture content, axle weight and the inflation pressure of tires, and how often equipment is driven on the soil. Fine-textured soils and wet soils are most compressible. Large rounded grains of sand contact each other at several points, so they do not compress easily. Silt is more compactable. Under wet conditions, flat clay particles may realign horizontally, rather than be randomly positioned, packing tightly. Further, a high water content weakens the binding of soil particles within a ped.

Compaction has varying effects on crop yield. Research indicates these effects:

- Any degree of compaction decreases yields when conditions are wet.
- Slight compaction may improve yields under normal conditions, especially on sandy soils that could use a few more micropores.
- Moderate compaction can improve yields in dry soils by increasing micropore space to hold water.
- Severe compaction always inhibits production.
- The finer the soil texture, the more damaging is compaction.

Compaction is a problem for all soil users. Virtually all landscape sites are badly compacted by construction equipment. Parks and recreational areas suffer from the pressure of countless footsteps, mowers, and off-road vehicles. Logging operations compact forest soils. Compaction is most severe in these nonfarming areas, like football fields, because the land cannot be plowed.

Other forces as well as surface-applied pressure creates compaction. Vibration increases soil settling and sifts small particles into larger pores; this can be a problem near large construction sites or roads. Repeated heavy rainfall and flooding does the same, and may be one of the most damaging, long-term results of a flood.

The natural processes that create structure can slowly repair a compacted soil. Repeated freezing/thawing cycles, for instance, may fracture compaction—but not where the soil never freezes, nor where it freezes solid for the whole winter. Rather than wait for natural action, it is best to avoid compaction or to break it up with machinery.

Compaction can be measured two ways. The most accurate is to compare the bulk density (*BD*) of the compressed soil to that of nearly unaffected soil. An average uncompacted tilled loam might have a *BD* of about 1.3 gm/cc. Compaction by farm equipment can raise the *BD* to 1.8. This amounts to a reduction of pore space from 50 percent to almost 30 percent.

Simpler, but less precise, is the use of a **cone penetrometer** (figure 4-18). A rod with a cone-shaped tip is pushed into the soil, and a dial reads the pressure it took. The result is an index that can be compared to a nearby unaffected soil. Unfortunately, soil type, moisture level, and stoniness also affect the result. Use cone penetrometer indices best by comparing nearby plots of the same soil type and moisture level. They are also handy for detecting subsoil compaction, because the reading will jump when a compacted layer is hit.

Aggregate Destruction. Plowing tends to create large aggregates as it flips the soil over. However, other tillage operations, such as cultivating, tend to crush soil aggregates. High-speed rototillers are especially destructive because they batter the aggregates apart. Many gardeners favor the fluffy, loose bed created by rototillage, but over the long term, structure is hurt.

FIGURE 4-18 A cone penetrometer in use to measure soil compaction.

There are two reasons why tillage destroys aggregates. First, by stirring oxygen into the soil, tillage speeds up organic matter oxidation. This loss, in turn, reduces the amount of "organic glue" that holds the peds together. Second, the tillage tools smash the now-weakened peds.

Puddling and Clods. Working wet soil greatly harms tilth, especially in soils high in clay. When pressure is applied to very wet soil aggregates, they fall apart. This results in a condition known as **puddling**—the conversion of aggregated soil into massive soil. The puddled soil is very dense and tight. In fact, it is done on purpose in some rice paddies, canals, and reservoirs to keep water from leaking away.

Working soil that is either too wet or too dry can also break up soil into large, seemingly indestructible clods. Soils with a hard consistence are most likely to form such clods.

Surface Crusts. Most forms of tillage bare the soil until the crop grows large enough to cover the soil. When raindrops hit this bare surface, their impact breaks apart peds on the surface. The free soil particles then splash around, washing into spaces between large particles and sealing the surface. When the soil dries, a crust forms that sheds water and inhibits aeration and the emergence of seedlings. The more clay in the soil, the harder the crust.

Improving Tilth. As noted earlier, tilth relates to the properties of texture, structure, permeability, and consistence. However, texture and consistence cannot, in most cases, be changed. Therefore, improving tilth is largely a matter of improving structure and avoiding compaction. The following practices can help protect or improve tilth:

- Never work wet or quite dry soils.
- Avoid unnecessary traffic over the soil. If possible, set aside paths through fields and nurseries to limit driving on the soil. If possible, drive only on dry or frozen soils.
- Employ controlled traffic in the field by setting the wheel base on all equipment to the same width. Then always drive in the same rows. While the wheel tracks will compress more severely, the remaining rows will be compaction free.

- Use equipment with the lowest practical axle weight. Large or dual tires do not seem to greatly reduce compaction, but spread it over a wider area. Flotation tires with low inflation reduce the pressure applied to the soil surface.
- Reduce the number of tillage operations. One can reduce trips across the field by combining operations or by simply not repeating them so many times. Some modern tillage practices, called minimum or reduced tillage, use less tillage. This is covered in chapter 16.
- Deep plowing, or **subsoiling** (figure 4-19), can break up tillage or natural hardpans, resulting in deeper penetration of water and plant roots and improved yields. The benefits may be short-lived, however, because further tillage reforms the compacted layer.
- Wherever possible, keep the soil covered by vegetation or mulch. Crops that fill in between rows quickly and crops that form complete cover, like alfalfa, protect soil from raindrop impact. Tillage that leaves a lot of crop residue on the soil surface also helps by creating a mulch.
- Incorporating grasses into a crop rotation improves structure, and as little as three years of continuous grass cover can improve structure on a degraded soil.
- Avoid, if possible, bare-fallow treatments, which destroy soil structure.
- Frequently add organic matter to the soil. Growers can leave crop residues in the field and spread manure. Gardeners and some organic farmers use compost. Lawn clippings, leaves, and any other source of organic matter can be useful.
- Plow under "green manure" crops of grasses or legumes. The plants' roots help create the loose aggregates, while decaying organic matter glues them together. Taprooted plants like alfalfa also help break up hardpans.
- Add lime where needed. Many soil organisms that decay organic matter require adequate levels of lime.

FIGURE 4-19 A subsoiling chisel plow. A tool like this can break up hardpans in the soil to temporarily improve penetration of water and roots. *(Courtesy of Year-A-Round Cab Corp)*

- Treat high-sodium soils with gypsum to remove excess sodium (see chapter 11) and manage high-sodium irrigation water correctly (chapter 9).

Soil Channels. To micropores and macropores we could add a third type of soil pore: large, continuous channels that begin at or near the surface and lead deeper into the soil. These may be created by cracks in the soil, but more often are biological in origin. Earthworms create most of the channels as they drag food from the surface into their burrows, and decayed roots leave soil channels behind. These channels caused by soil life have been termed **biopores,** and they are very important in undisturbed soil.

Biopores greatly enhance soil aeration and water infiltration. Water on the soil surface, as well as air, can quickly move deeply into the soil down these channels. We consider this generally beneficial, but such paths can also speed downward flow of pollutants like pesticides and nutrients, possibly endangering groundwater. Roots also tend to follow these channels, particularly in hard soil.

Tillage erases the tops of these channels, reducing their effectiveness in improving soil conditions. No-till soil management (chapter 16) preserves the channels. They are also promoted by leaving organic matter on the soil surface for earthworms, and by other means of enhancing earthworm populations (chapter 5).

Soil Pans

The preceding section mentioned plowpans. Any layer of hardened soil is called **pan.** Pans restrict deep rooting of crops and deep percolation of water, and can be a serious hindrance to cropping.

Growers create plowpans, but other types are natural. Here are some examples:

- **Claypans** occur where extreme illuviation has caused a very high clay content in a subsoil layer. The layer is quite dense.
- **Fragipans,** like claypans, result from clay accumulation. Here the clay binds soil particles into a hard, brittle layer.
- **Plinthite** layers are cemented by a special type of clay common to the tropics. When plinthite dries, it hardens to a bricklike substance; the process cannot be reversed by later wetting. Plinthite commonly renders tropical soils poor for agriculture.
- **Caliche** and **duripans** are layers of soil in which chemicals cement soil particles together. Lime cements caliche, typically a white, hardened layer found in arid regions. Many soils of the American Southwest contain caliche.

Soil Temperature

As indicated in chapter 1, soils in the growing regions of the world keep a temperature balance over the year that is satisfactory for plant growth. Short-term temperature changes, between seasons or night and day, can be dramatic. For instance, in the summer, soil temperatures can easily rise 30°F in the top inch of soil during the day. Below twelve inches, temperature varies little day to day.

Soil temperature is critical to the grower. Seed germination, for instance, is affected by soil temperature. Each crop has a temperature range that is best for seed growth. Corn germinates between 50°F and 105°F with an optimum at 95°F, while peas germinate between 40°F and 85°F, with an optimum of 75°F. Crops are planted outside in the spring when soil has warmed to the proper temperature for rapid germination.

In the greenhouse, most bedding plant seeds germinate well at 70–75°F, but pansies prefer a cooler 65°F, while fibrous begonia seeds germinate best at a warmer 80°F. Proper temperatures are maintained by warming the medium.

Root growth also depends on soil temperature. Roots have a minimum, optimum, and maximum for good growth which is species specific. Warm season crops such as corn or tomatoes grow best at relatively high soil temperatures, while peas and other cool season vegetables grow best at cooler temperatures. Root injury due to very high soil temperatures often occurs in plants grown in containers when sunlight strikes the pot on hot summer days.

Soil temperature depends not only on weather conditions but on several soil factors. Sunlight striking the earth is partially absorbed by the soil and partially reflected into the atmosphere. Dark soils absorb more sunlight, so they tend to be warmer than pale-colored soils.

Sunlight absorbed by the soil raises the temperature of both mineral particles and soil water. It takes five times as much energy to warm water than to warm an equal volume of mineral particles. As a result, it takes far less energy to warm dry soils. Sandy soils, which hold the least amount of water, tend to warm up most quickly in the spring and remain warmer during the season.

Managing Soil Temperature. The most obvious effect of soil temperature is in determining planting dates, which, in turn, affect harvest timing. The value of perishable crops, like vegetables, is at its highest when supplies are low, usually very early in the season. Vegetable growers, therefore, often favor coarse soils that allow early planting. Most annual flowers, or bedding plants, are native to warm climates and their roots will not thrive in cold soils. They should not be transplanted into the garden until the soil warms sufficiently in the spring.

Management of landscape turfgrass is strongly related to soil temperatures. Crabgrass, a common turf weed, germinates when soil reaches 55°F—so herbicides are applied just before then. Because cool season grass roots cease to grow above 85°F soil temperature, such turf will be severely stressed in hot summers and require special care.

Gardeners and growers change soil temperature by using mulches. Light-colored organic mulches like straw insulate the soil and reflect sunlight. This lowers average soil temperature, improving the growth of many crops. Crops that grow best in warm soil, like melons, are often mulched with black plastic. Plastic absorbs sunlight and raises the average soil temperature, improving production and speeding growth so the melons come to market early.

Recent trends in tillage have made soil temperatures a greater concern. Conservation tillage leaves crop residues on the soil surface that reflect sunlight and reduce soil drying. Thus, soil temperature can be several degrees cooler under conservation tillage, affecting the production and harvest dates of several crops. Conservation tillage will be discussed in chapter 16.

Frost damage to crops can sometimes be avoided by taking advantage of heat stored in the soil. At night, when damaging frosts occur, heat absorbed during the day warms the air immediately above the soil. Growers can heighten the effect by keeping the soil bare of debris that cools and insulates it. Irrigation is also helpful, because moist soil stores far more heat than dry soil.

Soil Color

While soil color is easily noted, it does not itself greatly affect the soil. However, color is an indicator of soil conditions, so growers can learn about a soil by its color. The main soil colorants are iron oxides and organic matter.

Color as a Guide to Soil Use. Soil color can be a useful guide to the suitability of the soil for various uses. White or light-colored soils usually have low fertility, either because they are leached or high in salts. Proper irrigation, fertilization, or treatment to remove salts may render these soils usable. Very dark topsoils that are high in organic matter may be quite fertile. However, one should check the subsoil for **gleying**—the high organic matter content may result from lack of oxygen needed to decay organic matter.

Subsoil **mottling** and gleying indicates poor soil drainage. Figure 4-20 shows a simplified classification system for soil drainage based on soil color.

DARK BROWN TO BLACK. Dark soil colors result from organic matter or dark parent materials, usually the former. The organic matter accumulation that creates dark colors arises from three situations, which can often be distinguished by smell:

- Organic matter can reach very high levels in soils that are usually waterlogged. Such soils often have a sour, oily smell.
- Organic matter can also reach high levels in adequately aerated soils, especially prairie soils. These soils have the earthy smell of good soil.
- Dark parent materials will affect the color of young soils. A faint chalky odor often describes these soils.

Drainage Class	Description
Very poorly drained	In level or low spots, black topsoil with gray color under the A or AB horizon, water table very near surface much of the year
Poorly drained	High water table part of the year or impermeable subsurface layer, gray or black surface, gray B horizon with brownish mottles at 6–20 inches
Somewhat poorly drained	Gray or brown A horizon with brownish upper B horizon, gray and rust mottles between 6- to 20-inch depth
Moderately well drained	Fairly bright colors in upper B horizon, few mottles between 20- to 40-inch depth
Well drained	Free of mottles above 40-inch depth, may be a few mottles below 40 inches
Excessively well drained	Sandy soils with rapid permeability, shallow soils on steep slopes, soil free of mottles

FIGURE 4-20 Guide for determining soil drainage class using soil color. *(Source: USDA)*

WHITE TO LIGHT GRAY. This color may indicate that the chemicals that color soil have leached out. It may be seen in heavily leached sandy soils and E horizons. White color may also be due to accumulations of lime, gypsum, or other salts.

LIGHT BROWN, YELLOW TO RED. These are the colors of oxidized iron minerals, chemically similar to rust. Red color indicates good drainage because there is enough oxygen in the soil to form the oxides.

BLUISH-GRAY. This is the color of reduced iron and indicates a lack of oxygen in the soil. The lack of oxygen results from waterlogging, so bluish-gray color in the subsoil indicates poor soil drainage. The occurrence of this color is called gleying.

MOTTLED COLORS. The soil shows patches of different colors, often spots of rust, yellow, and gray. Mottling in the subsoil suggests that the soil is waterlogged for part but not all of the year. Some wetland plants transport oxygen to their roots, leaving rust-colored zones in an otherwise gleyed soil around their roots. Because mottles result from oxidation-reduction reactions involving iron, the preferred technical term for mottling is now **redoxymorphic feature.**

Describing Soil Color. A simple description of a soil as "dark" would not be adequate for a soil survey. Soil surveys rely on a system that provides a precise description of soil color, the **Munsell system** of color notation.

FIGURE 4-21 The Munsell color chart is used to identify soil color. Each page is a hue, and the rows of color chips are values while columns are chromas.

Using the Munsell method, the surveyor matches the soil to standard color chips (figure 4-21). The Munsell system identifies each chip with three variables:

- Hue is the color, such as red or yellow.
- Value is the lightness or darkness of the hue. Value is denoted by the numbers 0 to 10, where 0 is black, each subsequent number represents a lighter color, and 10 is white.
- Chroma is the purity of the dominant color, and is also denoted by a number. Low chroma suggests muddy colors.

Using the Munsell system, a soil might be labeled 10YR 3/6. This soil has the hue 10YR, a yellow-red; the value 3 (dark), and the chroma 6. This soil is described as a dark yellowish brown.

SUMMARY

The most basic physical property of soil is texture—the proportion of sand, silt, and clay. All the possible variations of the three are divided into twelve textural classes. These classes range from the coarsest, which is sand, to the finest, which is clay. Medium-textured soils are called loams.

Texture strongly affects how growers use soil. Coarse soils are easy to work and dry out quickly. They warm up early in the spring, but do not hold nutrients well. Therefore, coarse soils perform best with irrigation and proper fertilization. Fine soils hold water and nutrients well, but they are more poorly aerated unless of good structure. Fine soils tend to stay wet later in the spring and are more difficult to work.

The ease with which air, water, and roots move through soil is called permeability. Permeability largely relates to the number of large, continuous pores. Pore spaces lie between the textural units of sand, silt, and clay, or between the structural units of grains and blocks, and in soil channels like biopores. Coarse soils, because of their large sand content, are naturally permeable. Finer soils depend on structure, the aggregation of soil particles into peds, to create large pore spaces. Structure forms when root masses, freezing/thawing cycles, or other forces create loose aggregates, which are then cemented by chemicals or gums exuded by soil organisms. Biopores are larger channels like earthworm burrows and decayed root channels.

Consistence measures traits such as stickiness, plasticity, and friability. Soil tilth, the physical condition of the soil for growing plants, results from the interaction of consistence, texture, structure, and permeability. Tillage, done correctly, improves the tilth of a seedbed. In the long term, tillage can cause compaction, crusting, puddling, or deterioration of structure. Minimizing tillage, adding organic matter, and working soil at the proper moisture level preserve tilth. Some soils contain hardened soil pans such as plowpans or caliche.

Seed germination and plant growth are strongly affected by soil temperature. Some crops, like corn, germinate best in warm soil while other crops, like peas, accept cooler soils. Growers can change soil temperatures by using mulches. Organic mulches lower the average soil temperature while plastic mulches raise it.

Soil color is an indicator of soil conditions. For instance, dark color in the topsoil suggests a high amount of organic matter. Gray or mottled colors suggest slow drainage. Soil scientists employ the Munsell system to identify soil color.

REVIEW

1. Which are likelier to have a higher oxygen content—pores within soil peds or pores between soil peds? Explain your reasoning.
2. What should professional gardeners do to maintain or improve structure in the gardens they care for?
3. Does puddling change particle density, bulk density, or texture? Explain your answers.
4. What effect would you expect the addition of large amounts of organic matter to have on bulk density? Explain.
5. What would be the effect on roots if bulk density of a soil increased from 1.3 to 1.8 g/cc? What could cause this increase?
6. Looking at figure 4-12, what soil textures give the best balance between micro and macropores? What would be the effects?
7. In choosing a site to build a home with a basement, would you prefer one with a gray subsoil or a light brown subsoil? Explain your answer.

8. Draw three soil matrices, one each for a soil layer with granular, blocky, and platy structure. Draw in the path that water would have to follow to percolate through these soils. What does this say about effect of structure on percolation rate?
9. The ideal soil is about 50 percent porosity. Using one of the formulas presented here in the discussion of porosity, calculate the porosity of a soil that has been compacted to a bulk density of 1.9 gms/cc. How does this compare to the ideal?
10. A case study: In 2002 there was a lot of discussion in the local media of the author's home city about damage to state parks caused by all-terrain vehicles (ATV) riding off designated paths, particularly the practice of "mudding"—purposely driving in muddy areas. Analyze possible effects of heavy ATV use on the soils of forested parks.

ENRICHMENT ACTIVITIES

1. Obtain soil samples of different known textures and practice the ribbon test. Then try to identify some unknown samples. Practice the Munsell color system on the same samples.
2. Try measuring soil texture using the sedimentation test in appendix 2.
3. Obtain soil samples from an old fencerow and the neighboring field. Observe the difference in structure.
4. Try this method of measuring bulk density. Dig a small hole in the ground, saving the soil for later measurement. Line the hole with a plastic bag, and measure the volume of water it takes to fill the hole. Now oven-dry the soil, weigh it, and calculate bulk density.
5. For more information about the Munsell Color System and color pictures, check the Web site at <www.munsell.com> and select "color order system." This site is for the whole Munsell system, not just the soil application of it.
6. The Soil Science Education Homepage (SSEH), designed for high school students, has good drawings and photographs of soil structure at <http://ltpwww.gsfc.nasa.gov/globe/basics.htm>. Go to chapter on soil structure.
7. Check out the soil profile of the month through the SSEH page in activity 6. Using the information you have learned so far, analyze the soil. Describe the soil. How hard would it be to plow? What about drainage? How fast would it warm up in the spring? How droughty would it be? Come up with other questions of your own.
8. The University of Minnesota has a good publication on compaction on the web at <http://www.extension.umn.edu/distribution/cropsystems/DC7400.html>.

CHAPTER 5

Life in the Soil

OBJECTIVES

After completing this chapter, you should be able to:

- define the carbon cycle and explain its importance
- briefly describe soil organisms
- list ways that soil organisms are important
- describe how to promote populations of beneficial soil organisms

TERMS TO KNOW

actinomycetes
aerobic
algae
anaerobic
antagonism
arthropods
autotrophs
bioremediation
carbon cycle
decomposers
denitrification
detritus
fungi
heterotrophs
humus
hyphae
immobilization
inoculation
macrofauna
mesofauna
methane
microfauna
microflora
microorganisms
mineralization
mycelium
mycorrhizae
nematodes
nitrogen fixation
nonsymbiotic nitrogen fixation
organic matter
parasites
predators
primary consumers
primary producers
protozoa
rhizosphere
saprophytes
secondary consumers
solarization
symbiosis
symbiotic nitrogen fixation

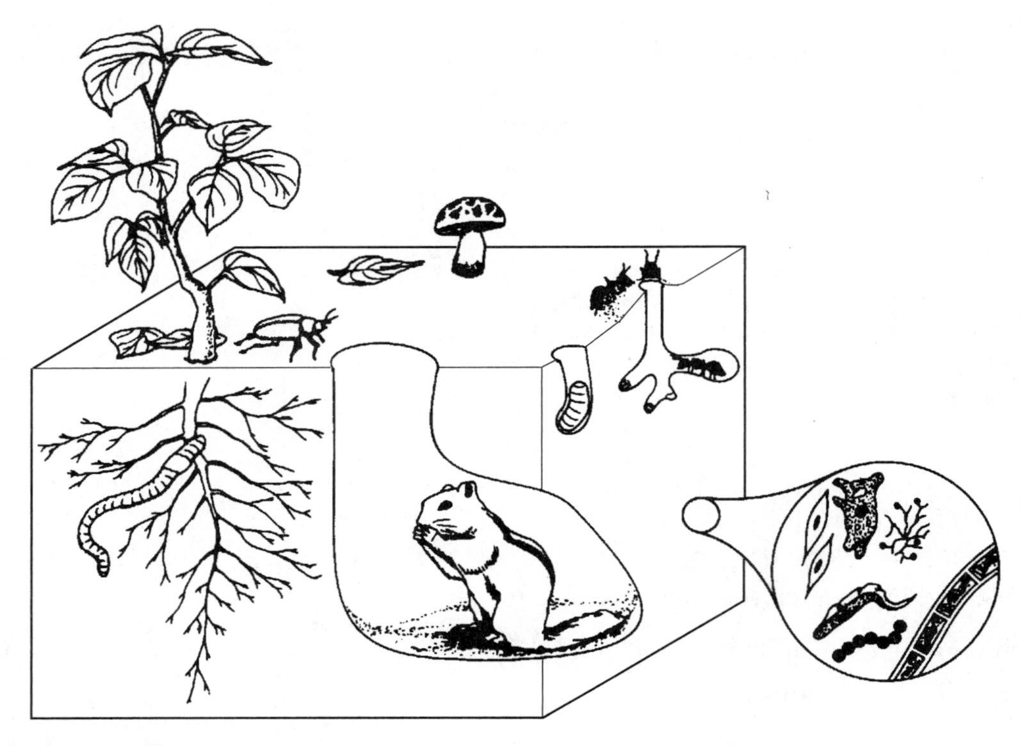

FIGURE 5-1 The soil teems with life. Life helps make soil the way it is.

We live in a world teeming with life, yet few people know the creatures inhabiting the soil beneath our feet. Every acre of soil is home to two or more tons of living things (figure 5-1). Soil results from physical, chemical, and biological forces acting on the earth's surface; the activities of biological organisms makes soil be a soil, rather than, say, sand in a sandbox. What are these organisms, and why are they so important to agriculture and ecosystems? It has been suggested that a teaspoon of fertile soil hosts thousands of separate species of such organisms. That teaspoon could contain:

- 100 nematodes
- 250,000 algae
- 300,000 amoeba
- 450,000 fungi
- 11,700,000 actinomycetes
- 100,000,000 bacteria

This chapter will describe these organisms and their important functions. We will begin with the soil food chain and the carbon cycle.

The Soil Food Chain and Carbon Cycle

A food chain is a model of how food—which is to say, carbon and energy—moves from one organism to the next. All organisms are part of some food chain. Food chains operate above the ground and in the soil, and they interconnect. The main base of food chains is photosynthesis by plants. During photosynthesis two events occur that make life possible. First, carbon dioxide in the air is changed to organic carbon, the building block of living tissue. Second, photosynthesis converts solar energy to chemical energy stored in sugars and other compounds. Plants, then, make the food that is the base of the food chain, and so are called **primary producers.** Some other creatures eat the plants, and so are **primary consumers.** Predators are **secondary consumers,** and so on up the chain. Carbon and energy move up the food chain from plants to the highest level.

If the food chain contained only producers and consumers, the chain would collapse and all the world's carbon would be fixed in the bodies of dead life. This does not happen because another type of food chain breaks down and recycles dead organisms, called the detrital food chain. **Detritus** is simply dead organisms or their products, such as crop residues, fallen leaves, or animal wastes, and when it enters the soil, we call it soil organic matter. While the regular food chain is based on plants, the detrital food chain is based on detritus. The soil is the main location of the detrital food chain.

Decay organisms, or **decomposers,** consume **organic matter** as a food source, returning most of the carbon to the atmosphere as carbon dioxide by their own respiration, leaving behind a residue called **humus.** In the process, plant nutrients are released that were tied up in the bodies of plants and animals. This is the most grand function of soil life: the recycling of carbon and nutrients. If we combine both food chains, we have a grand cycle of carbon on earth, called the **carbon cycle,** shown in simplified form in figure 5-2.

In the complex ecology of the soil, there are five roles of interest to those who study soil:

- Producers include mostly plants and a few microorganisms. They produce their own food from inorganic carbon (like carbon dioxide) by photosynthesis or by certain reactions with soil chemicals. The technical term for producers is **autotrophs,** from a Greek term meaning to supply one's own food. All other organisms are **heterotrophs,** meaning they get their food from others.
- **Parasites** feed on plant roots and are often responsible for plant diseases.
- **Predators** prey on other soil life. They help keep parasite populations in check and perform other functions.
- **Saprophytes,** or decomposers, feed on dead organic matter.
- Symbionts are organisms that live with another organism in a partnership that is helpful to both. Ecologists would call this relationship "mutualism," but people who work with soils tend to use the broader term **symbiosis.**

The above list classifies organisms as to ecological function. Soil organisms can also be classified in other ways. They can be classified taxonomically, that is, as animals, plants, fungi, bacteria, and so on. They can also be classified as to their preferred environment. For instance, we call soil organisms that need oxygen **aerobic.** A few need no oxygen. These **anaerobic** organisms exist in low oxygen sites, like water-filled pores, in all soils but are most plentiful in poorly drained soil.

We often broadly classify soil life by a system that involves both size and taxonomy. This classification is based on the older idea that all life is either flora (plant) or fauna (animal). **Microflora** are mostly microscopic organisms that were once classified as primitive plants, including bacteria, fungi, and algae. While they have been reclassified into their own kingdoms, the category is still useful when discussing soil life. **Microfauna** are microscopic creatures that were once considered single-celled animals, also now placed in their own kingdom. The larger, but still very tiny **mesofauna** are multicelled animals, like the tiniest insects. **Macrofauna** are animals large enough for us to see, like earthworms and woodchucks. Each of these categories plays a different role in the soil food chains.

Now let's examine organisms that live in the soil.

Microorganisms

Microflora are **microorganisms,** organisms too small to be seen with the naked eye. They are observed by soil scientists under an optical or an electron microscope. While there are other microorganisms, here we will consider bacteria, fungi, actinomycetes, algae, and microfauna.

Bacteria. Bacteria are simple, one-celled organisms that lack a nucleus and are the most abundant inhabitants of the soil. A more recently identified, similar kingdom called the Archaeae will be included here as bacteria. Common soil bacteria are rod-shaped, though

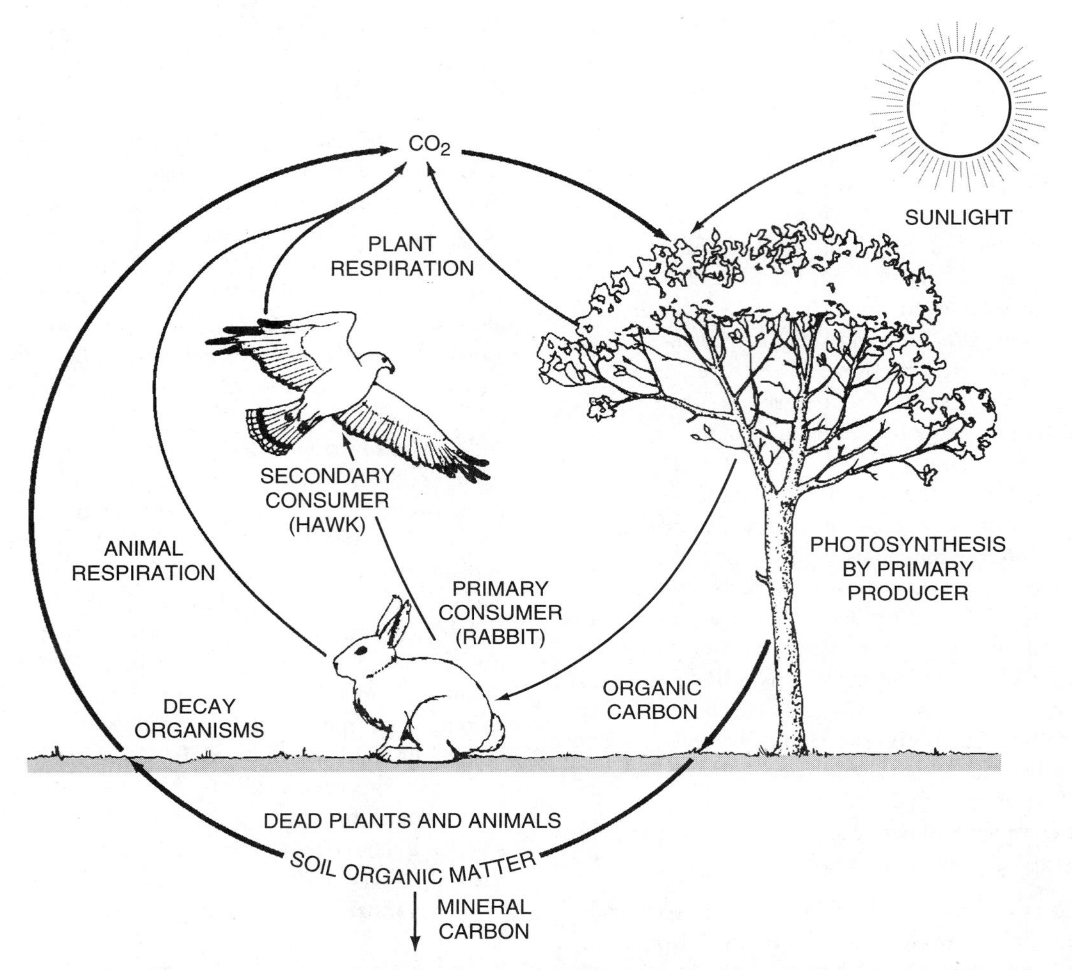

FIGURE 5-2 The carbon cycle begins when plants change inorganic carbon (carbon dioxide) into organic carbon in plant tissue. The cycle closes when microorganisms decompose organic matter to carbon dioxide.

many assume other shapes (figure 5-3). They are about 1/25,000 inch wide and slightly longer. While they are single-celled, many may cling together to form chains. Bacteria usually grow as small colonies on the surface of soil particles and in smaller pores. They are most dominant in nonacid and grassland soils.

Bacteria are the most variable of soil organisms. While most are aerobic, many thrive in anaerobic soil. Most are heterotrophic, yet many obtain energy autotrophically from chemical reactions with certain soil substances.

Most soil bacteria are saprophytic. They comprise one of the groups most responsible for breaking down organic matter in the soil, especially easily decayed materials. A few species are parasites, causing plant diseases such as crown gall (*Agrobacterium tumefasciens*), which causes a tumorlike growth on roots of many plants.

Fungi. Most **fungi** resemble a mass of tangled threads (**hyphae**) called a **mycelium** (figure 5-3). Fruiting bodies grow from the mycelium. These bodies release spores that may be considered the "seeds" of fungi. Some fungi grow to become quite large—the common mushroom is a fungus. The mushroom is the fruiting body of a fungus whose hyphae feed on decaying material in soil (figure 5-4). However, much of the fungi in the soil must be examined under a microscope.

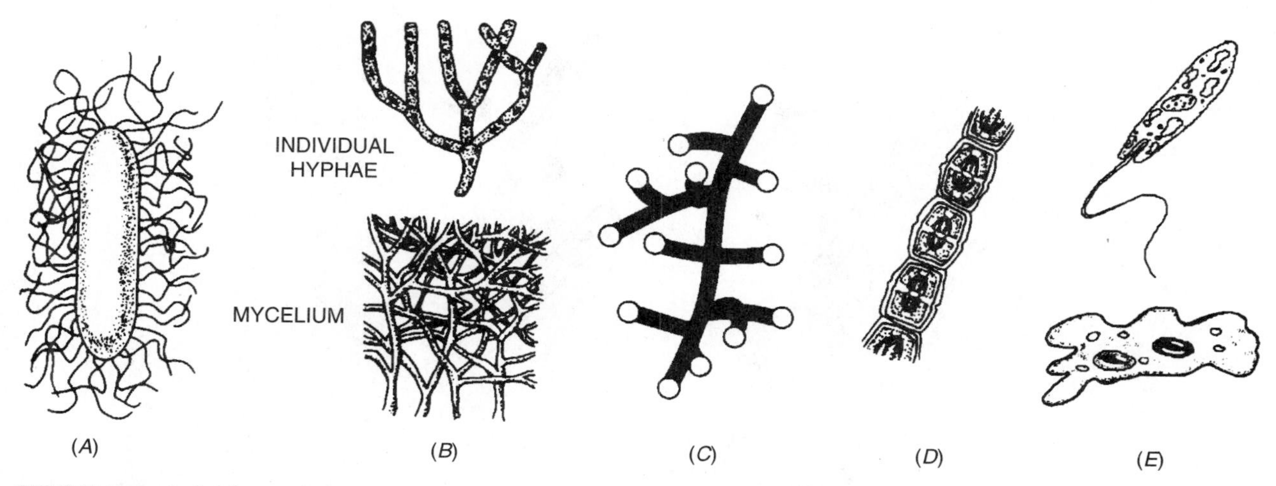

FIGURE 5-3 Soil microorganisms show a great diversity. *(A)* Rod-shaped bacteria with "hairs" that can move the bacteria. Not all bacteria have hairs. *(B)* Fungal mycelium, composed of individual hyphae. *(C)* An actinomycete. *(D)* Algae. *(E)* Protozoans, including *Euglena* (top) and *Amoeba* (bottom).

FIGURE 5-4 *(A)* These mushrooms are the fruiting bodies of a fungus whose *(B)* hyphae are underground, rotting the wood poles.

While fungi are less numerous than bacteria, because of their larger size, they generally make up the largest microbial mass in the soil. Fungi are entirely heterotrophic and aerobic, and occupy larger pore spaces. Fungi tend to dominate in acid and forest soils.

Along with bacteria, fungi act as the main soil decomposers. Fungi can attack matter that resists breakdown by bacteria, partly because hyphae can grow into the material. Many fungi are plant parasites, such as the wilt fungus (*Verticillium spp*) that attacks potatoes, several important landscape plants, and other plant species. A group of soil fungi called "damping-off fungi," like *Rhizoctonia,* attack seeds and seedlings and cause root rots, particular problems for greenhouse and container nursery growers.

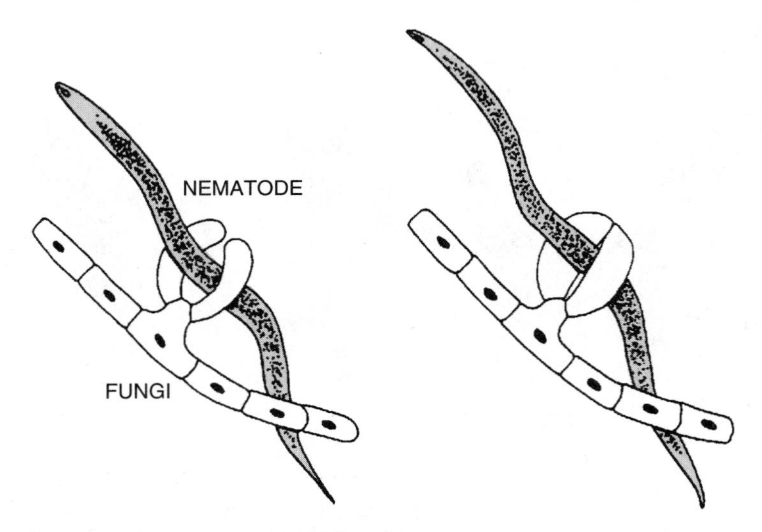

FIGURE 5-5 A fungus is shown trapping a nematode. Once the nematode enters the ring, the ring constricts, trapping the animal. The nematode is then digested by the fungus.

A few odd fungi are predators. For instance, certain fungi capture and consume nematodes (a microscopic worm). These fungi trap nematodes either by growing rings that can tighten around the body of a nematode (figure 5-5) or by growing knobs covered with a sticky substance. After the nematode is trapped, hyphae grow into its body until it is consumed.

Actinomycetes. **Actinomycetes,** also called mold bacteria, while resembling fungi in appearance, are classified as bacteria. They look like fungi because they grow a threaded network. Like fungi, actinomycetes can work on resistant organic matter. Actinomycetes are particularly tolerant of dry soil and may dominate soils after prolonged moisture stress. They also thrive in more alkaline soils or in high-temperature conditions.

Many actinomycete species produce chemicals that stop the growth of other microorganisms, a phenomenon called **antagonism.** Many useful antibiotics used in modern medicine are derived from actinomycetes. In fact, the characteristic odor of damp, well-aerated soil comes from the most important genus of antibiotic-forming actinomycetes, the *Streptomyces.* In the soil, these natural antibiotics sometimes protect plant roots from attack by disease organisms. All but a few species of actinomycetes are saprophytes, though a few produce such plant diseases as potato scab (*Streptomyces scabies*).

Algae. While algae growing in water may be quite large, most soil **algae** are single-celled. Some algae are simple chlorophyll-containing plants that live in high-water environments. Others, the "blue-green algae," are actually autotrophic bacteria called *cyanobacteria.* In the soil, most live in water films. Like higher plants, algae can photosynthesize and are considered primary producers.

As producers, algae add slightly to the organic matter of soil. Certain algae combine with fungi to form lichens. Lichens growing on rocks release mild acids that dissolve minerals, adding to soil formation (chapter 2).

Microfauna. The microfauna of the soil are **protozoa,** the kingdom Protista, and while they are not truly animals, we tend to think of them as simple, single-celled animals. *Amoebae* and the interesting *Euglena,* a chlorophyll-containing protozoan capable of photosynthesis, are examples (figure 5-3). These microorganisms live in water films in the soil and are able to move about cap-

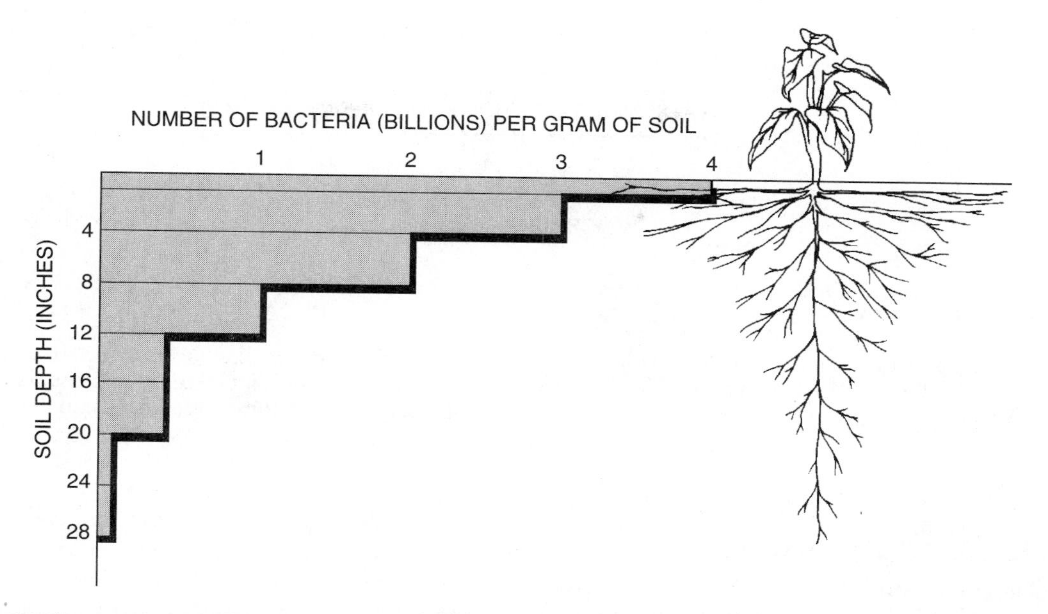

FIGURE 5-6 The population of bacteria decreases from a maximum near the soil surface to a minimum near the greatest soil depth for a fertile loam.

turing prey. Protozoa are mostly predators, and they capture their prey by engulfing them and absorbing them into their bodies. While protozoa require water to be active, they form protective bodies or cysts when conditions are unfavorable, and can survive for long periods of time in this inactive state.

Most protozoa graze on soil bacteria so play an important role in regulating bacterial populations. They also perform a function in the release of nitrogen into the soil during organic matter decay. The bodies of bacteria are very high in nitrogen, while bacteria-grazers need less nitrogen. Protozoans and other bacteria-feeders excrete excess nitrogen as ammonium ions, a form of nitrogen immediately useful to plants.

Distribution and Functions of Microorganisms

Organic matter decay is an important task of soil organisms. However, many organisms perform other tasks that also are important to agriculture. Before examining these tasks in detail, let's look at where organisms live in the soil, because their location affects their function.

Distribution in the Soil. Most microorganisms need air, water, and food to thrive. These materials are best supplied in the top two feet of soil, especially the A horizon. Here, organisms find the most oxygen, the most organic matter for food, good soil structure, and water storage. Thus, most soil organisms live near the soil surface (figure 5-6), as do most plant roots.

Plant roots exude, or "leak," a variety of chemicals into the surrounding soil, including sugars and other organic compounds. In addition, roots slough root caps, bits of bark, and old root hairs. All this material acts as food for microorganisms, which, in response, multiply in great numbers. This area of high biological activity surrounding plant roots, called the **rhizosphere,** extends up to about 5 millimeters or a quarter-inch from plant roots.

The effect of the rhizosphere and the preference of microbes for the top layer of soil mean that microbe populations concentrate near plant roots. As a result, the

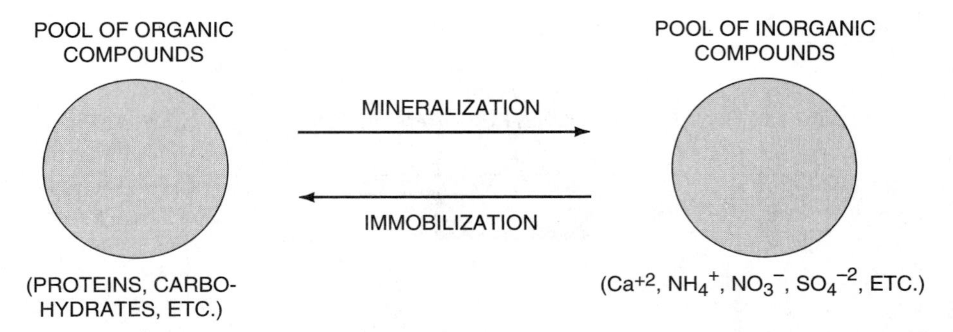

FIGURE 5-7 Nutrients can be viewed as occupying two pools of compounds and can pass back and forth between them. When organisms die, their parts create a pool of organic compounds. Decay microbes mineralize these compounds to inorganic forms. These, in turn, are taken up by plants or other organisms, and so are immobilized back into the organic pool.

desirable activities of microbes reach their peak near plant roots—to the benefit of plants. In fact, most root-soil-microbe interactions described in this text take place in the rhizosphere, which is essentially an interface zone between root and soil.

Nutrient Cycling. Nutrients taken from soil by plants cannot be used by other plants, nor can chemicals in the bodies of living microorganisms, animals, or fresh organic matter. The nutrients in living bodies or fresh organic matter are said to be **immobilized.** These nutrients are bound in complex organic forms.

Unlike animals, plants need nutrients in simple, inorganic and ionic forms. Thus, plants cannot use immobilized nutrients until they have been changed to simple, inorganic forms by microbial decomposers. This process is called **mineralization,** and the microbes that do this are abundant in the rhizosphere. Immobilization and mineralization are opposite processes (figure 5-7). The sulfur cycle is an example of these processes (figure 5-8).

Most soil sulfur comes from the weathering of sulfur-containing minerals. Some of it comes from industrial pollution as sulfur dioxide in acid rain. Sulfur is changed by microorganisms in the soil to sulfate ions. Plants absorb sulfate to make protein and other compounds, thus immobilizing the sulfur. When leaves fall, they decay and soil flora mineralize the sulfur in the leaves to sulfate ions. Some sulfate is taken up again by plants, some is again immobilized in the bodies of microorganisms, and some leaches away.

The sulfur cycle shows how immobilization and mineralization lead to elements being recycled by plants and microbes. An essential element of this recycling process is the storage of nutrients for plant use. Many mineralized nutrients easily leach from soil. This loss is reduced by soil flora, which capture nutrients for their own use. When they die in the rhizosphere, the nutrients are available again to plants. Therefore, soil organic matter and life can be seen as a means of nutrient storage.

We could say that the soil food web preserves nutrients, and that unhealthy soils with an impoverished microbial population leak nutrients. This includes farm fields. In natural ecosystems, it is the recycling of nutrients through decay of organic matter that primarily defines nutrient availability and productivity of the system. Conifer needles, for instance, decay more slowly then regular leaves, so conifer forests are less productive than hardwood forests.

Microorganisms are involved in another important cycle, the nitrogen cycle. The nitrogen cycle is covered in detail in chapter 12, but we can look here at the role microbes play in the cycle.

NITROGEN FIXATION. Nitrogen comes from nitrogen gas (N2) in the atmosphere—approximately 34,500 tons of it over every acre of the earth's surface. Higher plants cannot use even one molecule of this nitrogen gas. However, certain bacteria, blue-green algae, and actinomycetes can use it. They absorb the gas and convert it to

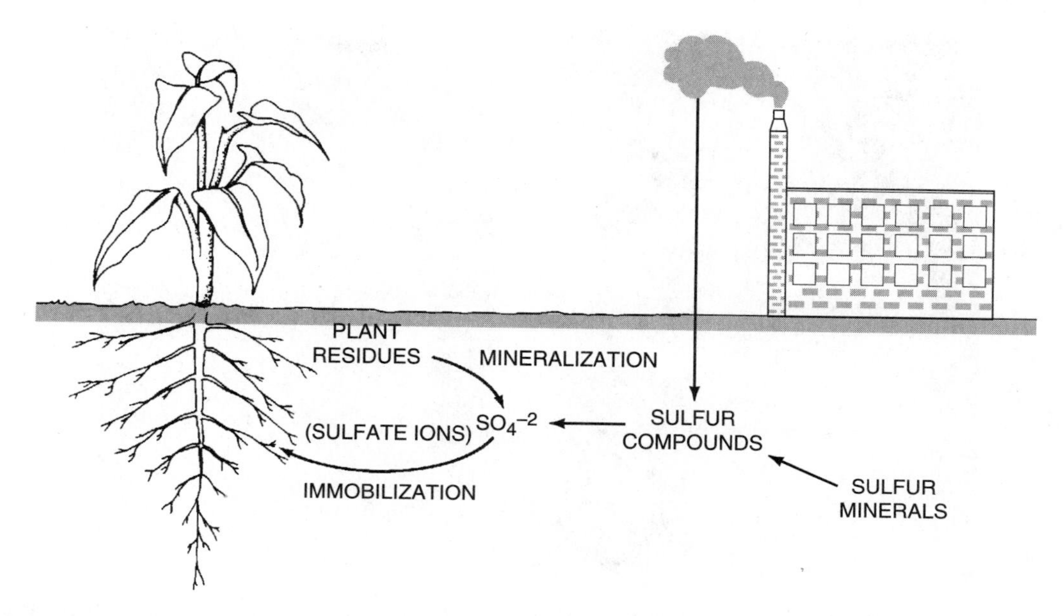

FIGURE 5-8 The sulfur cycle is an example of the way microorganisms recycle plant nutrients.

ammonia that plants can use. This process is called nitrogen fixation.

Legume plants (family *Fabaceae*), like alfalfa, host an important group of nitrogen-fixing bacteria in the genera *Rhizobia* and *Bradyrhizobia*. The bacteria occupy the soil as free-living organisms feeding off rhizosphere carbon, but if they contact legume roots, an infection occurs. These bacteria invade the root hairs, which respond by surrounding the bacteria with plant cells to form a lump, or nodule, on the roots (figures 5-9 and 5-10). The bacteria get minerals and food from the plant roots while the nitrogen it has fixed can be used by the plant. Thus, this association is useful for both the bacteria and the plant. The way the bacteria gather nitrogen is thus called **symbiotic nitrogen fixation.** Up to three hundred pounds of nitrogen per acre can be added to the soil yearly by the legume-*Rhizobium* association.

A number of noncrop plants also host nitrogen-fixing bacteria. Some trees, such as locusts, are also legumes. Many nonlegume plants, such as alders, host nitrogen-fixing actinomycetes of the genus *Frankia*, and many cycads, often used as indoor foliage plants, host nitrogen-fixing cyanobacteria. Alders can add between seventy and one hundred fifty pounds of nitrogen per acre each year. Such trees add to the nitrogen status of woodlands and can be used to good effect in efforts to replant forests. Some have been useful in reclamation of surface mines, dumps, and other heavily disturbed areas.

A few genera of free-living bacteria (*Clostridium*, *Azotobacter*, and others) also fix nitrogen. These nonsymbiotic bacteria do not live on plant roots. **Nonsymbiotic nitrogen fixation,** for the most part, is not considered important to agriculture. Under the best conditions, it adds about forty pounds of nitrogen yearly to the acre.

Another interesting group of free-living nitrogen fixers is blue-green algae, an order of bacteria called *cyanobacteria*. Blue-green algae grow in aquatic environments; thus, they thrive in water films in the soil. Their small numbers in well-drained soil limit the amount of nitrogen they can add. Blue-green algae can achieve high populations in rice paddies and have been an important source of nitrogen for rice production.

NITRIFICATION. The nitrogen fixed by soil microbes is, of course, immobilized in the bodies of microbes or host plants. When these die, they decay to form a pool

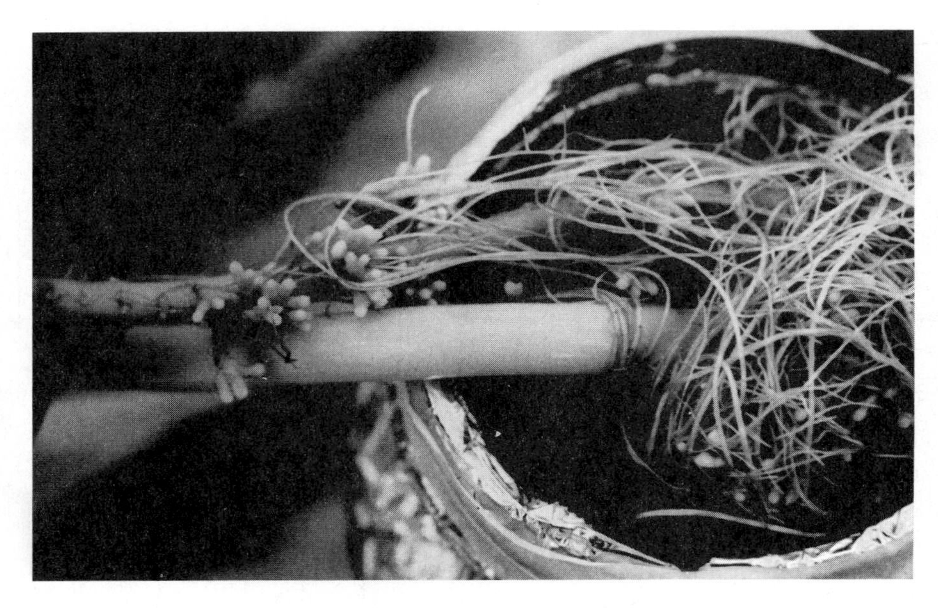

FIGURE 5-9 Nitrogen-fixing nodules on the roots of an alfalfa plant. *(Courtesy of Dr. J. Burton, The Nitragin Company)*

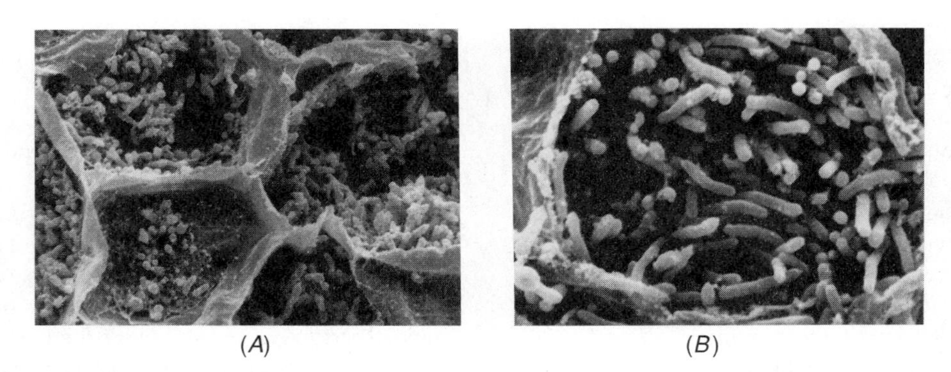

FIGURE 5-10 *(A)* Root nodules broken open to reveal the bacteria (magnification 640X). *(B)* Detail of *Rizobium* bacteria (magnification 5,000X). The photographs were taken with a scanning electron microscope. *(Courtesy of Dr. Carroll Vance, University of Minnesota)*

of organic nitrogen. The nitrogen is mineralized to ammonium ions ($NH4^+$), which may be absorbed by plants for growth. Protozoans and other bacteria-feeders also excrete ammonium ions as they graze on bacteria. While plants use ammonium, most is oxidized by a group of bacteria (*Nitrosomonas*) to another form of nitrogen—nitrite ions (NO_2^-). Nitrite ions are then quickly oxidized by other bacteria (*Nitrobacter*) to nitrate ions (NO_3^-), the favored form of nitrogen for many plants. Nitrites are toxic to plants but reside in the soil for a very short time. The net reaction of these processes looks like this:

$$2NH_4^+ + 3O_2 \rightarrow 2NO_3^- + 8H^+$$

Nitrification strips hydrogen off the nitrogen atom, producing lots of hydrogen ions. Since hydrogen ions make soil acid, nitrification is an acidifying process.

Conditions that promote the activity of the nitrifying bacteria naturally increase the rate of nitrification.

These include a warm, moist, nonacid and well-aerated soil. Nitrification will be inhibited by the opposite conditions and can also be artificially slowed by the application of chemicals called nitrification inhibitors.

Nitrate ions are taken up by plants or other microbes. This action completes a cycle in the soil: from living matter to organic matter to ammonium to nitrites to nitrates and back to living matter. Some nitrates, however, are changed by other bacteria to nitrogen gas again. It then escapes back to the atmosphere. This process is called **denitrification.**

Denitrification. Denitrification completes the nitrogen cycle by converting nitrate ions to nitrogen gas, which filters out of the soil. Certain anaerobic, heterotrophic bacteria use nitrate instead of oxygen to oxidize organic matter during respiration. This reduces nitrates through a series of steps:

$$NO_3^- \rightarrow NO_2^- \rightarrow NO \text{ (gas)} \rightarrow N_2O \text{ (gas)} \rightarrow N_2 \text{ (gas)}$$

While nitrogen gas is the usual result, intermediate nitrogen oxides also make it into the atmosphere. Nitrogen oxides are greenhouse gases much more potent than carbon dioxide in trapping heat in the atmosphere so likely contribute to global warming. This is another example of important interactions between the soil and the atmosphere.

Because denitrifying bacteria are anaerobic, the process occurs most rapidly in wet soils, wetlands, or wetland agriculture like rice paddies. Flooded soils, like fields under water in the spring and overly irrigated turf, lose nitrogen by denitrification to the detriment of plant growth. However, even well-drained soils have anaerobic and other oxygen-depleted sites, like ped interiors. Thus some denitrification occurs in all soils: the wetter the soil, the faster the rate.

All these nitrogen transformations make up the biological portion of the nitrogen cycle (figure 5-11). Note that the nitrogen cycle is really two nested cycles. In the outer cycle, nitrogen enters the soil by fixation and leaves by denitrification, recycling nitrogen between the earth and air. The inner cycle recycles nitrogen inside the soil.

Soil Structure. Soil microbes are important agents of soil aggregation. Fungi and actinomycetes are the most effective of these organisms. First, their thread-like hyphae twine between soil particles, pulling them together to form loose aggregates. Both organisms also produce gummy substances that glue the aggregates together. These substances resist wetting, and so peds do not fall apart when they get wet. The improved strength and wetting resistance of these soil aggregates keep soil structure sound during tillage, rainfall, and irrigation.

Mycorrhizae. **Mycorrhizae** are fungi that form a symbiotic relationship with plant roots, colonizing plant roots to obtain food and nutrients. In fact, infected plants shunt about 15 percent of sugars produced by leaves to their mycorrhizal symbiont. In return, the host plant gains a number of benefits:

- Roots are better able to absorb phosphorus—probably the most certain and important benefit. Other gains for the plant may derive from better phosphorus nutrition.
- Roots are better able to absorb water, making plants more drought resistant.
- Roots are better able to absorb zinc, copper, and other nutrients.
- Infected rootlets live longer than uninfected ones.
- Some mycorrhizae protect roots from disease and probably from low levels of the toxins aluminum and heavy metals.

Mycorrhizae growing on many forest trees create a thick growth, or "mantle," of fungal hyphae on the outside of the root (figure 5-12). These fungi, called *ectomycorrhizae,* penetrate between the outer few cells of the plant root. Ectomycorrhizae, most common on woody plants, are presently used in the production of tree seedlings by artificial infection in the greenhouse. When planted in the field, these infected seedlings have better survival rates, become established more quickly, and grow faster than uninfected seedlings. This proves especially useful in planting heavily disturbed soils like mine tailings.

Most plants, including common crops, host mycorrhizae that grow inside root cells These fungi are called *endomycorrhizae.* While there are several varieties of endomycorrhizae, crop plants most commonly host VAM, or vesicular-arbuscular mycorrhizae. This name simply means that certain fungal bodies—vesicles and arbuscles—develop inside, rather than between, host root cells, and hyphae extend from them out into the soil.

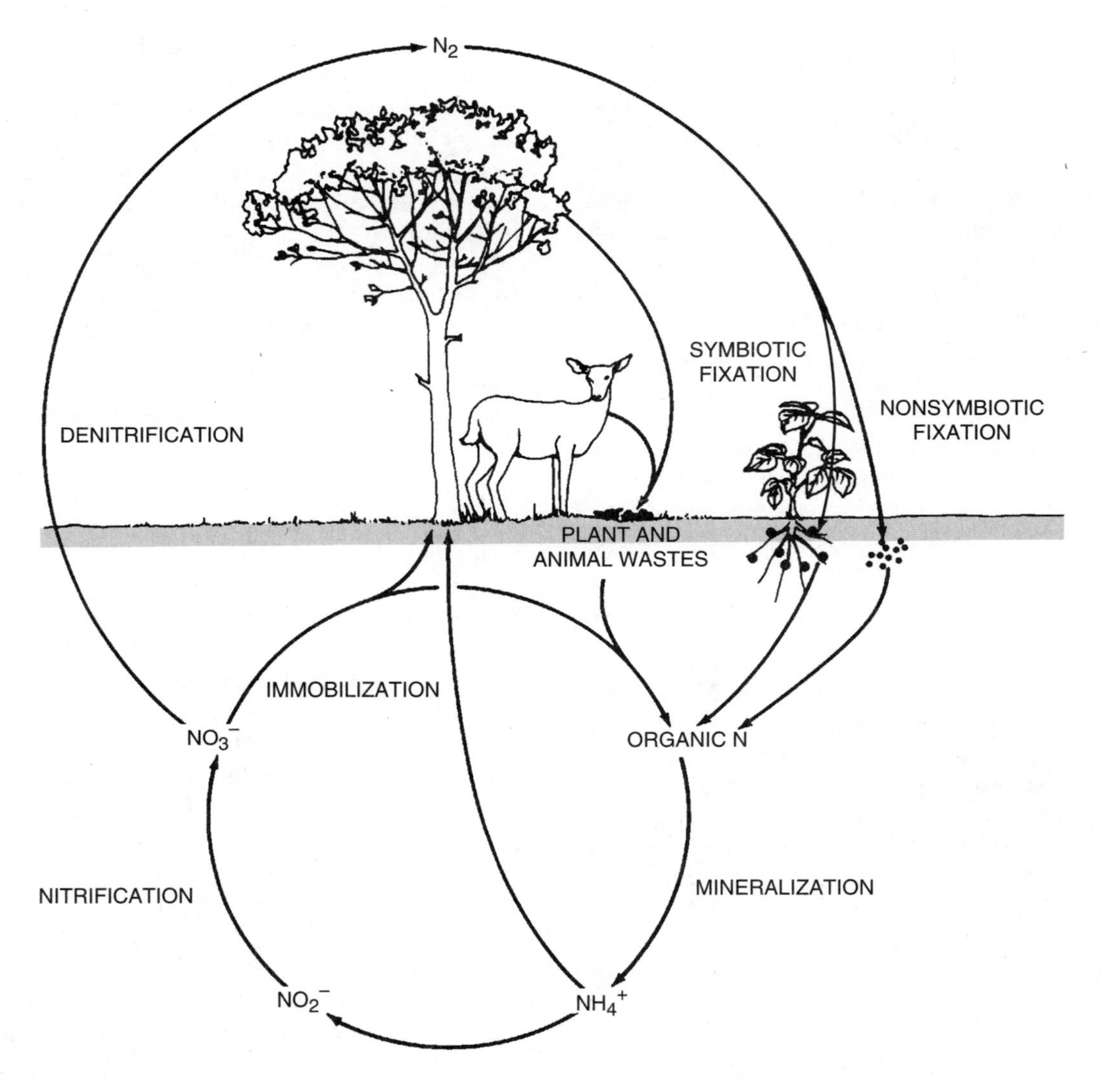

FIGURE 5-11 This is the core of the nitrogen cycle—the natural biological transformations of nitrogen. See figure 12-2 for the complete cycle.

The association improves crop productivity. For instance, the fungus improves the growth and yield of soybeans. Beans are larger on infected plants, and plant tissues contain larger amounts of plant nutrients.

Both types of mycorrhizae act as extensions of the root system of the plant. Indeed, mycorrhiza means "fungus root." Hyphal strands extend from plant roots into the surrounding soil, greatly increasing the absorbing area of the root system (figure 5-13). Because the strands are much finer than plant roots, they can grow into tinier pores than can roots. There they can forage for water, phosphorus, and other nutrients that plants cannot reach. Some mycorrhizae are also able to actively release phosphorus from insoluble sources like organic matter and minerals. As a result, mycorrhizae greatly enhance plant growth in low-phosphorus soils.

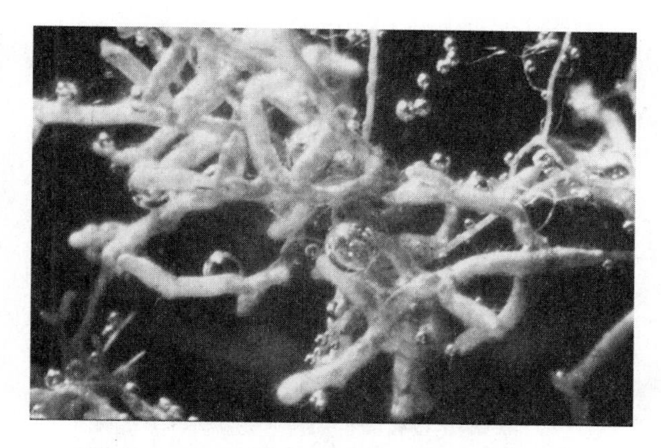

FIGURE 5-12 A mantle of ectomycorrhizae surrounding roots of its host plant. *(Courtesy of Jo Anna Hebberger)*

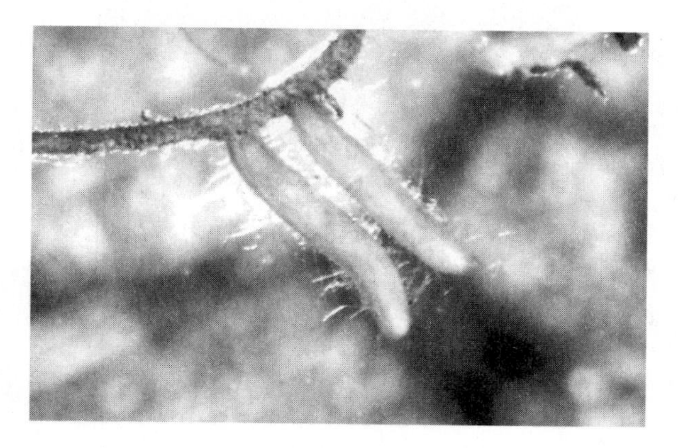

FIGURE 5-13 Individual mycorrhizal hyphae grow out of the mantle into the soil, acting as extensions of the host's root system. *(Courtesy of Jo Anna Hebberger)*

Research is being done to find ways to use mycorrhizae more often in plant production. While most plant species host mycorrhizae—estimates range from 65 percent to 95 percent—agricultural systems often work against their use. Ectomycorrhizae can be grown artificially in the laboratory. However, endomycorrhizae cannot be grown artificially, making it difficult to produce large quantities of agriculturally important organisms. High phosphorus levels in soil, a common condition in farm soils because of phosphorus fertilization, suppress the growth of mycorrhizae, as does severe soil disturbance.

Interestingly, mycorrhizae can infect and connect neighboring plants, allowing nutrient and maybe other transfers between the two. Even plants of different species may be connected through hyphae bridges between roots. Some species of plants, like the forest wildflower Indian pipe (*Monotropa uniflora*), obtain food not by photosynthesis but from mycorrhizae attached to a nearby tree—the tree feeds the fungus, and the fungus feeds the Indian pipe.

Breaking Down Chemicals. Fortunately for modern society, organisms inhabit the soil that can break down chemical products and refuse deposited in the soil. Chapter 1 mentioned the importance of soil in waste disposal. The cleanup of increasingly frequent oil leaks and spills is also aided by organisms that digest oil. Research is identifying strains of microorganisms that are quite effective in digesting chemical wastes in soil, making possible **bioremediation,** the biological cleanup of contaminated soils.

Farms avoid a buildup of agricultural chemicals in the soil mainly because of microorganisms. While some chemicals leave the soil by leaching or by evaporation, biological decomposition is the most important means of removing chemicals. The ability of a soil to degrade a chemical depends upon the substance. Some pesticides disappear quickly, while others, such as DDT, persist in soil for years.

Interestingly, some microbes have adapted to soil chemicals so well that an herbicide or soil insecticide fails altogether because organisms break them down so fast. Until more is known, researchers suggest that where the problem has occurred, growers should use soil pesticides as little as possible, rotate crops, and rotate chemicals.

Methane bacteria. Certain anaerobic organisms produce methane (CH_4) while breaking down organic matter in oxygen-free sites. **Methane,** also known as swamp-gas, is the major component of natural gas fuel. Other aerobic bacteria in the soil oxidize the methane to carbon dioxide (CO_2). Therefore, in most soils, methane may be produced in oxygen-free sites in the soil (like ped interiors), but as the gas filters into aerobic pores, it is consumed before reaching the atmosphere. However, large areas that are entirely oxygen-free, like wetlands, rice paddies, or landfills, can generate large amounts of methane. Methane acts as a greenhouse gas much more powerful than carbon dioxide and thus contributes to global warming.

Other Functions. Microbes are active in other useful ways. Some of these activities are well-known and understood. The exact impact of others is uncertain, but they are interesting activities that are probably helpful to plant growth.

- Organic acids produced during decay help dissolve soil minerals to make plant nutrients more available to plants. Certain autotrophs live by "feeding" on soil minerals, releasing their elements in forms useful to plants.
- Some rhizosphere microbes, besides those already mentioned, actively improve plant growth, include a group called "plant growth promoting rhizobacteria." These may act by producing plant hormones and vitamins, by improving nutrient uptake, or by suppressing root disease. Some of these are now available as products that can be applied to roots and soil. These microbes need organic matter as a food source.

Promoting Soil Organisms

Some growers talk about having a "healthy" soil—a soil with a large population of beneficial microorganisms. How is this condition achieved? Let us look at some suggestions for populating the soil with healthy numbers of organisms.

Inoculation. **Inoculation** is purposely infecting soil with useful organisms. As an example, there have been many attempts to speed up nonsymbiotic nitrogen fixation by soil inoculation. The market is now supplying preparations of mycorrhizae as an inoculant for landscape plants to improve transplant success. Researchers continue to explore ways to infect soil or plants with "friendly" microorganisms, but microbes introduced into a soil often do not survive in competition with or under attack from native flora.

FIGURE 5-14 Good nodule formation on legumes, such as this alfalfa, is obtained when plants are inoculated with the right bacteria. *(Courtesy of Dr. J. Burton, The Nitragin Company)*

Inoculation of legume seed with *Rhizobium,* however, has long been an important farming practice (figure 5-14). Inoculants can be purchased and applied to seeds to ensure good nodule growth on the roots of host crops. They are also available for native prairie legumes, important in the increasingly common practice of planting or restoring native prairies.

Soil Conditions. "Healthy" soils provide a good place for the growth of microorganisms. Here are good soil conditions:

- A constant supply of fresh organic matter is needed as a food source for most organisms. Other factors being adequate, this is the single most important factor for successful microbe populations.

- Good aeration supplies soil flora with the oxygen they need. While some organisms are anaerobic, most of the agriculturally important organisms need oxygen.
- Adequate moisture is important. Many organisms survive periods of drought by going dormant. Most, however, need moist soil to multiply and function actively.
- Soil temperature affects soil flora. Few organisms grow actively below 41°F, and most grow best between 77°F and 99°F. In cold soils, many beneficial activities slow down, including mineralization of phosphorus and nitrogen and nitrification.
- Most organisms grow best at neutral pH. Acid soils suppress growth, so liming of acid soils is a helpful practice. Potato growers, in contrast, prefer acid soil because it slows growth of scab-forming actinomycete. Of the soil organisms, fungi are most acid-tolerant.
- Proper nutrients are as important to microorganisms as they are to plants. For example, legume root nodules grow best in high-phosphorus soil. The number of mycorrhizae, on the other hand, is kept down by a high phosphorus content. Microbes have a high enough need for nitrogen that they can rob plants of that needed nutrient (see chapter 6).

Controlling Harmful Organisms. Part of making a soil healthy for growing crops is to control harmful organisms like nematodes and parasitic fungi and bacteria. In cases of severe infestation, soil sterilant chemicals can be used. Some soil sterilants, such as methyl bromide (slated to be removed from the market by 2005 as an ozone-depleting chemical), are injected into the soil. Some chemicals are solids, washed into the soil by rain or irrigation.

Solarization is a nonchemical method to sterilize soil. A clear plastic sheet covering the soil for several weeks during warm, sunny weather heats the soil enough to destroy undesirable organisms. Solarization works best in hot, sunny climates.

Sterilizing an entire field is very expensive; therefore, the practice is often reserved for high-value crops like strawberries. In addition to the cost, sterilizing soil kills the good along with the bad. The key to controlling harmful organisms is to prevent their occurrence and to use other control methods where possible. A few suggestions for alternative methods of control follow:

- Practice sanitation. Start with disease-free seeds and plants. Don't drag infected soil into a field on tillage equipment.
- Take advantage of certification programs where they exist, such as the certified seed potato program. The programs certify growers to grow disease-free plants and seeds for other growers.
- Obey quarantines. Quarantines are intended to prevent the transportation of parasitic nematodes or other diseases on plants into uninfected areas.
- Control soil pH. For instance, potato scab is not a serious problem in acid soils. Some wilts are more of a problem in heavily limed soil.
- Crop rotation can suppress diseases that are fairly limited in their hosts. For example, soybean cyst nematode feeds on soybean roots. During years in which corn is grown in rotation with soybeans, the nematode population declines because of a lack of a host. However, some harmful organisms are able to survive for years living as saprophytes.
- Incorporating large amounts of organic matter may help control parasites by promoting the growth of decay organisms. These organisms compete with the parasites and many, like some actinomycetes, "fight off" harmful flora.
- Rhizosphere microflora of one crop may antagonize the parasites of another crop following in rotation, as has been observed in peas following oats. This further suggests the usefulness of crop rotation.
- Some plants suppress soil pathogens. For instance, both marigold roots and sesame stalks contain chemicals that inhibit nematodes.
- Efforts are being made to develop "living pesticides," organisms that are antagonistic to soil pathogens. A few are being marketed.
- Select crop varieties resistant to soil pathogens.
- Simple diagnostic test kits are available for detecting soil pathogens in greenhouses, turf, and other situations.
- Greenhouse and container nurseries may use special growing media that suppress soil pathogens.

Soil sterilization is a must in the greenhouse, unless media are used that contain no soil. This practice is discussed in detail in chapter 17. Even in a greenhouse, sanitation remains one of the best defenses against soil-borne diseases.

Soil Animals

Many animals, from tiny **nematodes** to larger animals like badgers, make their home in soil. Animals affect cultivated soil less than microorganisms. Undisturbed soil, which provides a better habitat, can be heavily changed by soil animals. Let us look at some soil animals, starting with nematodes.

Nematodes. Nematodes, the dominant mesofauna, are microscopic eel-like worms. Many inhabit the soil or plant roots. They can swim short distances through water films on soil particles. Those that infest plants puncture the roots with their needle-like mouthparts (figures 5-15 and 5-16).

Farmers are most concerned with parasitic nematodes. These animals infest plant roots, sapping plants of their strength and reducing yields (figure 5-17). The tiny puncture wounds also provide entry for other fungal or bacterial diseases. For that reason, nematode feeding is often related to infections by other soil-borne disease.

Most nematodes actually graze on soil bacteria and fungi. Like protozoans, they affect microbe populations and also release plant-usable nitrogen into the soil. A few prey on other soil fauna, including other nematodes. In fact, an exciting advance in insect control is the harnessing of insect-attacking nematodes, which are now commercially available as "living insecticides."

Arthropods. Mites, millipedes, centipedes, billbugs, and insects are the most common soil **arthropods.** The smallest arthropods, like the smallest mites, are classified as mesofauna, while the readily visible ones we classify as macrofauna. Arthropods are easily recognized because they have jointed legs and a hard outside skeleton. Many arthropods, like some mites, millipedes, and insects, feed on decaying organic matter and the bacteria and fungi growing on it. Such mesofauna play an important role in the decay of organic matter by shredding raw organic materials. Others, like centipedes and some mites and insects, feed on other soil fauna. A number of arthropods feed on plant roots. June beetle and Japanese beetle larvae, or grubs, often injure crop and turf areas by feeding on plant roots.

Ants and some termite species alter soil more than other insects by their tunneling behavior (figure 5-18). Tunneling mixes the soil, a subject to be considered later. Ant and termite burrows also aid soil aeration.

Earthworms. Earthworms, the dominant macrofauna of most soils, feed on organic matter and its load of fungi and bacteria, and excrete it as digested matter

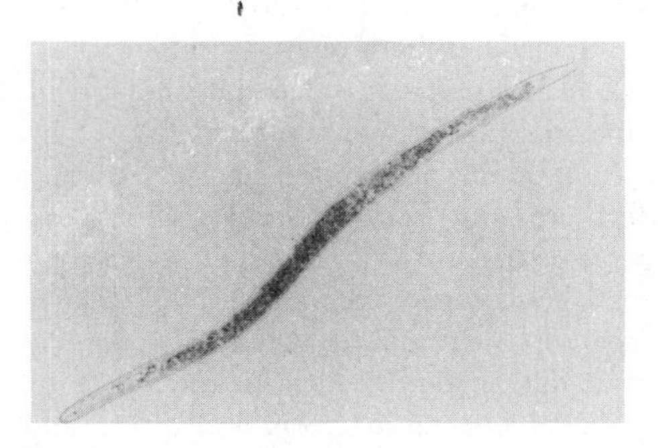

FIGURE 5-15 This parasitic nematode is the root lesion nematode (*Pratylenchus penetrans*). This adult female is eel-like and about 0.7 mm long. *(Courtesy of Dr. D. H. MacDonald, University of Minnesota)*

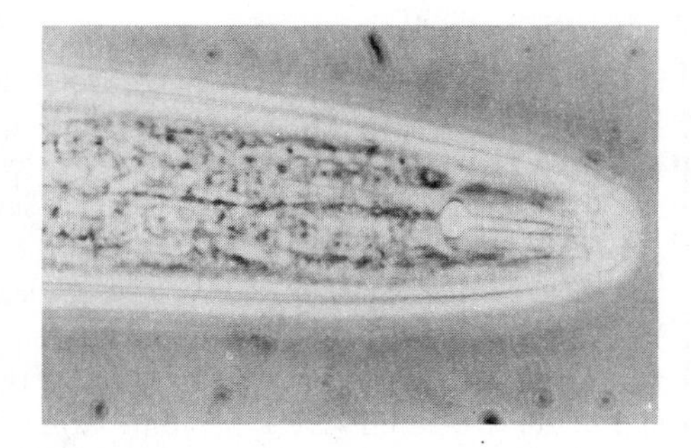

FIGURE 5-16 In the anterior ("head") portion, one can see the bulb-like muscle that controls the stylet, or needle, that punctures plant cells for feeding. *(Courtesy of Dr. D. H. MacDonald, University of Minnesota)*

FIGURE 5-17 Field damage by the Soybean Cyst nematode. Notice the stunted plants. *(Courtesy of Crops and Soils Magazine, American Society of Agronomy)*

FIGURE 5-18 Ants have almost covered up this low-growing evergreen in a nursery with soil dragged up from below.

called "casts." In doing so, decay and nutrient cycling is greatly accelerated and nutrients become more available to plants. Earthworm burrows aerate the soil and both burrows and casts improve soil structure. Species of worms that feed on surface debris dramatically improve moisture infiltration by extending tunnels to the surface. Earthworms cannot, however, by themselves, improve a problem soil.

Earthworms develop best in moist fine loams with a good supply of fresh organic matter and neutral pH. Between two hundred to a thousand pounds of earthworms may occupy an acre of soil. Earthworm populations are a visible sign of soil health, because conditions in which earthworms thrive are also good for other soil flora and fauna and plant roots.

Farming practices that promote soil organic matter levels and structure are good for earthworms. These practices include crop rotation, minimum tillage, and additions of organic matter. Liming acid soils also helps. Organic mulches benefit earthworms by providing a continuous food source, preventing rapid deep-freezing of soil in the winter, and preventing high soil temperatures in the summer.

While earthworms are beneficial in agricultural soils, they may be problematic in other settings. In lawns, earthworms prevent thatch buildup (a layer of dead and living stems and roots between green vegetation and the soil surface) by feeding on it. However, homeowners often object to casts that mar the surface of the turf, and on golf courses, casts hinder playability. Earthworms may also attract moles, whose burrows can severely damage a lawn.

Recent evidence suggests that earthworms may harm northern forests. The last glacial advance eradicated earthworms in their path, so earthworms in glaciated areas are foreign, mostly from Europe and Asia. Forests of these areas developed in the absence of the worms. When one of these forests is invaded, detritus recycling accelerates rapidly, removing their deep layer of litter. This sets off a cascade of effects, including a serious decline in reproduction of many forest plant species and accelerated erosion.

Mammals. Mammals affect soil by burrowing. The greatest number of burrowing mammals are rodents, such a gophers, woodchucks, and prairie dogs. Their population is highest in undisturbed soils such as pastures, forests, and prairie.

FIGURE 5-19 Prairie dogs mixed the soil of the western prairies, renewing the soil.

Rodent digging, like that of ants, alters soil by mixing the layers. The mixing counteracts, to some degree, the natural soil aging process in which clay particles and nutrients leach into the B horizon. Mixing has the effect of rejuvenating soil.

The native prairie of the western United States, before its use as agricultural land, was continuously churned by the soil-mixing activities of burrowing animals. An important part of the animal community was the prairie dog town. Prairie dog burrows extend some five feet into the ground. In digging the burrows, prairie dogs carried a lot of subsoil to the surface and piled it on the topsoil (figure 5-19). As a result, the subsoil was mixed with the topsoil. The soils of these areas are surprisingly high in clay.

SUMMARY

More than two tons of living creatures inhabit an acre of soil. An important role of microorganisms is the decay of organic matter. In this process, nutrients are returned to the soil, and the level of humus is improved. An even more essential result of decomposition is the return of carbon from plants to the atmosphere.

The most numerous soil microflora are bacteria. Most bacteria decay organic matter, though a few cause plant diseases. Important bacteria include the symbiotic nitrogen fixers *Rhizobia,* nonsymbiotic nitrogen fixers, and other bacteria involved in the nitrogen cycle. Both fungi and actinomycetes are excellent decay organisms and help preserve soil structure. Some actinomycetes fix nitrogen and some produce antibiotics. Mycorrhizae, which help plants absorb water and nutrients, are fungi. Algae add organic matter to the soil by being primary producers. Protozoans, the soil's microfauna, regulate soil bacterial populations and release plant-usable nitrogen.

Microorganisms need proper soil conditions to grow and multiply. The most basic requirements are a constant supply of fresh organic matter, good soil aeration, and enough moisture. The growth of organisms is further influenced by pH, nutrient levels, and warmth.

Common soil mesofauna and macrofauna include nematodes, arthropods, and earthworms. Some nematodes are helpful, but growers are concerned primarily about the types that cause plant disease. Earthworms feed on fresh organic matter, making the nutrients more available to plants. They also improve soil permeability and structure, but may damage northern forests. The burrowing of worms, ants, and larger soil mammals mixes the soil layers. This slows the "aging" of a mature soil and helps keep it fertile.

REVIEW

1. How do protista and many nematodes fit into the nitrogen cycle described in this chapter? How do they fit into the detrital food chain?
2. After enjoying a day of fishing on one of the many lakes in the author's home state, you wonder what to do with left-over angleworms. Should you throw them into the nearby woods? Explain your reasoning.
3. Pocket gophers, which tunnel in the soil, eating plant roots and leaving piles of soil above ground, are considered a pest of pastures, lawns, and many other settings. Could they be considered desirable to soil? Explain,
4. How should you manage soil to maintain a healthy microbial population?
5. In no-till agriculture, one does not plow or cultivate the soil, and crop residues are left as a mulch on the surface. How might this affect earthworm populations and would it be beneficial? (Hint: think also about biopores.)
6. Is Indian pipe, mentioned in this chapter, autotrophic or heterotrophic? Does this make it an unusual plant?
7. Discuss symbiotic plant-microbe relationships described in the text and state why they are important.
8. Discuss two biological processes in the soil that might contribute to global warming. Where are they most likely to occur? How might they be avoided, if possible?
9. Describe the biological part of the nitrogen cycle.
10. Distinguish the type of mycorrhizae that would be typical of forest trees and those typical of most crops.

ENRICHMENT ACTIVITIES

1. Isolate nematodes from a soil sample and examine them under a microscope. Test kits for this purpose are available.
2. Culture soil microorganisms on an agar medium and observe them under a microscope.

3. Grow several soybeans or other legume plants in sterilized soil. Inoculate one group with the correct Rhizobium but not the other. Compare the plants' growth. When the experiment is over, carefully wash soil off the roots and note the difference between the two root systems.
4. For general web sites on soil biology, with some photography, try:
   <http://www.statlab.iastate.edu/survey/SQI/SoilBiology/soil_bioloby_primer.htm>.
   <http://www.blm.gov/nstc/soil/index.htm>.
5. This site is an extension bulletin called "Soil Biology and Management":
   <http://www.extension.umn.edu/distribution/cropsystems/DC7403.html>.
   What does it suggest as cultural practices to promote healthy microbe populations?
6. This site has information on the process of inoculating legumes: <http://www.rhizobium.umn.edu>. Select FAQ.
7. Here is a site on mycorrhiza <http://mycorrhiza.ag.utk.edu>.

CHAPTER 6

Organic Matter

OBJECTIVES

After completing this chapter, you should be able to:

- explain what organic matter is and how it forms
- describe what organic matter does in the soil
- list several ways to maintain soil organic matter
- discuss the problem of nitrogen tie-up
- define organic soil, listing uses and problems

TERMS TO KNOW

amino acid
carbon-nitrogen ratio
cellulose
colloids
compost
cover crop
fibric
green manure
hemic
hemicellulose
humification
humus
labile
lignin
muck
nitrogen depression period
organic matter
organic soils
peat
protein
recalcitrant
sapric
starch
subsidence

FIGURE 6-1 Some woodlands of the eastern United States were unable to support long-term agriculture and are now devoted to forest products such as lumber and maple syrup. This maple is being tapped for sap to make maple syrup. *(Courtesy of USDA NRCS)*

Early settlers in America cleared the woodlands of the eastern colonies to create their farms, but many of those farms were later abandoned and have since returned to forest (figure 6-1). By the middle of the nineteenth century, pioneering farmers were turning over the prairie sod of the Midwest. Farming in the Midwest continued to expand until the prairie retreated to a few preserved areas.

Why could soils of the grassland support long-term agriculture while some eastern woodland soils could not? One difference is the high organic matter content of prairie soils.

The Nature of Organic Matter

Organic matter is that portion of soil that includes animal and plant remains at various stages of decay. In forests, it comes from fallen leaves, dead tree trunks, and tree roots. In prairies, much of the organic matter comes from grass roots and tops. In farmland, crop residues add to the organic matter.

Chemical Makeup of Organic Matter. Organic matter consists of complex, carbon-containing compounds. Carbon atoms, unlike other elements, naturally form long chains. These long chains provide a framework upon which are attached other elements like hydrogen, oxygen, nitrogen, and sulfur, to make the wide array of organic compounds necessary to life. Soil organic matter begins as material like fallen leaves, which largely consist of compounds such as sugars, starches, cellulose, hemicellulose, protein, fats and waxes, and lignins.

Sugars, starches, cellulose, and hemicellulose, all carbohydrates, make up the bulk of plant dry matter. Most plant sugars are short carbon chains of five or six carbons with many oxygen and hydrogen atoms attached. These sugar molecules can themselves link together to form various long chains to make more complex carbohydrates. One version is **starch,** which we consume in such foods as potatoes or bread, and acts as food storage for the plant. Another is **cellulose,** which forms long fibers in plants and constitutes the main structure of a plant. We cannot digest cellulose, and so it is fiber in our diet. Wood and paper are mostly cellulose. **Hemicelluloses** are a group of several different carbohydrates of complex structure.

Proteins are also long chain structures made up of individual units called **amino acids.** Amino acids, unlike sugars, are rich in nitrogen and sulfur. Enzymes are protein, as is much of muscle tissue in animals. Most of the nitrogen in soil organic matter derives from protein in litter.

Fats and waxes include a variety of long carbon chains with mostly hydrogen atoms attached. Fats can be high-energy compounds in seeds and are important in cell membranes. Waxes usually coat leaves to form a protective layer and may occur in other parts of plants as well.

Lignins can be 10 percent-30 percent of plant tissue and act as a structural component of plants. It glues together cellulose fibers to make wood, coats cellulose to protect it from microbial attack, and lends rigidity to plant tissue. They are large, highly complex molecules of almost random structure, with many rings and branches, and are resistant to decay. Lignins contain no nitrogen.

These compounds vary in how readily they can be attacked by microbes. Sugars, amino acids, and starches make ready food sources, followed in order by protein, hemicellulose, cellulose, fats, waxes, and lignin. Easily decayed materials, like starch, are said to be **labile,** while difficult materials, like lignin, are **recalcitrant.** Recalcitrant materials may contain lots of decay-resistant chemicals like lignins, may be low in the nitrogen needed by microbes, or may actually contain chemicals that are toxic to decay organisms. Pine needles, for instance, resist decay because they are high lignin, low nitrogen, and high in toxins called tannins.

The Process of Decay. Fresh organic matter becomes soil organic matter by the process of decay, essentially the biological oxidation of carbon for energy. That is, microbes use organic matter as a food source, and decay is the result of microbial respiration. Decay follows a series of four overlapping steps we can call solution, fragmentation, decay, and humification. During solution, free amino acids and sugars, as well as potassium and other water-soluble components, quickly dissolve out of litter into nearby soil water. Soil microbes rapidly exploit this food source and colonize the detritus. Soil meso- and macrofauna, like tiny mites and even earthworms, now shred the material, feeding on both the microbes and the organic matter. This fragmentation breaks through protective lignin and wax coatings, and increases the surface area available to attack by bacteria during decay. During decay, labile materials are broken down quickly, recalcitrant ones much more slowly, mostly by fungi. Complex molecules get split into smaller units and become increasingly oxidized, carbon dioxide is produced, and nutrients like nitrogen are liberated in mineralized forms.

Humification follows a different path. As decay proceeds, labile materials disappear while recalcitrant ones, like lignin remains. Chemical reactions occur in the soil in which soil nitrogen from the more labile protein reacts with lignins and other remains of decay to form new compounds that, like lignin, are large, highly complex, and resistant to attack but which are rich in nitrogen. This material is called **humus,** the resistant residue from decay. Humification is a chemical, rather than biological, soil process.

Humus continues to decay, but this is a very slow process; about 1 percent to 3 percent of the humus in a well-drained soil is lost annually. The chemical nature of humus would be difficult to describe here, but it contains about 50 percent carbon and 5 percent nitrogen, as well as some phosphorus. It is dark-colored and in the solid state appears as tiny particles of clay size.

The decay process just described occurs under aerobic conditions, where oxygen acts as the oxidizing agent (that is, the electron acceptor) for microbial respiration. In waterlogged soils, there is no oxygen available, but decay proceeds using other electron acceptors like nitrates (denitrification), ferric iron, and others. This process is much slower and gives rise to a variety of less desirable chemicals like organic acids, methane, and the sulfides that give wet soil its rotten egg smell. It also creates the greyed color of wet subsoils.

Factors Affecting Organic Matter. Five major factors directly affect the amount of organic matter in the soil: vegetation, climate, soil texture, drainage, and tillage.

Prairies generate the most soil organic matter. Root masses in a humid tallgrass prairie sum between 5.8 to 7.6 tons per acre. In a North Dakota mixed prairie, growth each year generates about 1.4 tons of shoots and about 4 tons of roots. Note that in native grasslands, most of the growth is in the soil, where natural turnover of roots enriches soil organic matter.

In contrast, forests generate organic matter as litter on the soil surface. The litter decays into a thin organic layer, the O horizon, on the surface. Insects, worms, and other animals mix the material into the top few inches of soil, making a shallow, humus-rich A horizon. The needles of conifers are especially recalcitrant, so conifer forests have even less organic matter than other woods.

Grass tops also die back each year, while trees do not. This means most of a grass plant returns to the soil each year. The differing growth of grasses and trees causes the following differences in prairies and woodlands and their soils (see figure 6-2):

- There is about twice as much organic matter in grassland soil as in an otherwise similar woodland soil.
- Organic matter extends deeper into prairie soil, because grass roots can decay deep in the soil while organic matter in forest soils comes mainly from the decay of surface litter.

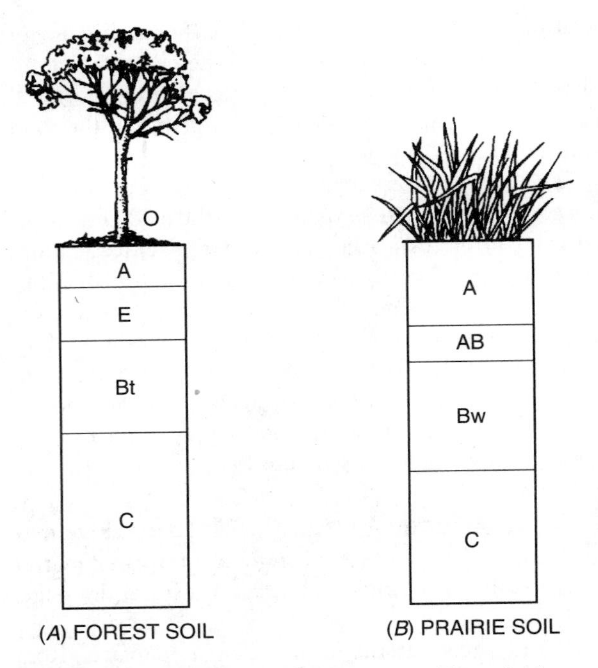

FIGURE 6-2 Typical soil profiles of a prairie and a forest soil. (*A*) Forest soil with a thin O horizon over a thin A horizon. The A horizon is thin because little of the litter mixes deeply into the soil. (*B*) The prairie soil has a deep A horizon because grass roots decay deeper in the soil.

- Most organic matter of the prairie is in the soil. In forests, most of the organic matter resides in standing trees.

Because soils in arid climates support very little vegetation, they are lower in organic matter than either prairie or forest soils. Arid soils, unlike other soils, may gain organic matter under cultivation when irrigated. The gain results from the greater amount of green matter growing under irrigated cultivation.

Temperature and rainfall are key climatic factors that affect soil organic matter. The more rainfall, the greater the total amount of vegetation. Thus, soils in high rainfall areas tend to develop more organic matter than those in drier sites. This is obscured, however, by the tendency of soils in drier climates to host grassland vegetation.

High average temperatures also promote plant growth. However, organic matter decays more rapidly at higher temperatures. Soils in warmer climates tend to contain less organic matter than those in cooler climates. As a simple guide, organic matter is generated faster than decay when temperatures are below 77°F (25°C), and decay almost stops below 41°F soil temperature. In these cooler soils, organic matter can accumulate.

Fine-textured soils tend to have more organic matter than coarse soils like sand. Finer soils grow a large supply of plant materials because they hold water and nutrients well. Because coarse soils are better aerated than fine-textured soils, they have a better supply of oxygen and, as a result, organic matter decay is more rapid in sandy or coarse soils. Fine-textured soils also tend to contain more organic matter because clay protects humus from further decay and because organic matter can be preserved inside anaerobic ped interiors.

Soil drainage has the most dramatic impact on soil organic matter levels. The wetter the soil is, the less oxygen is available to fuel decay and the more organic matter accumulates. The wettest soils usually have a layer of black, decaying material on the surface with a very dark A horizon underneath.

Virgin soils lose organic matter when they begin to be farmed. Organic matter levels drop rapidly at first, but eventually the loss of humus slows and a new balance is reached. The loss occurs partly because erosion washes away some humus along with topsoil. Cropping usually returns less organic matter to the soil to replace the loss than does native vegetation. Interestingly, cultivated crops reverse the root-to-shoot ratio noted previously for grasslands. That is, crops produce far more mass above ground than roots; after harvest, less root mass is left to contribute organic matter to the soil. Most importantly, tillage also stirs oxygen into the soil and raises its average temperature. In one study in the Great Plains, 42 percent of soil carbon was lost after thirty-six years of cultivation.

Functions of Organic Matter

Organic matter improves conditions of all mineral soils for many reasons. Organic matter helps sandy soils by increasing their water- and nutrient-holding capacity. It improves clay soils by loosening them and improving their tilth. This section describes these and other functions of organic matter in the soil.

Nutrient and Water Storage. Organic matter stores nutrients used by plants in two different ways. The first method of storage results from the size of humus particles. Like clay particles, humus particles are extremely small with a relatively large surface area. Particles of this size are called **colloids.** Water and nutrients cling to the large surface area of the colloids.

Organic matter also stores nutrients as a part of its own chemical makeup, released for plant use by decay. Humus contains most of the soil's supply of nitrogen, boron, and molybdenum, about 60 percent of its phosphorus, and 80 percent of soil sulfur. Organic matter acts as a major reservoir of soil nutrients.

Both fresh organic matter and humus absorb water like a sponge, holding about six times their own weight in water. This is extremely important in naturally dry and sandy soils. In fact, the water- and nutrient-holding capacity of organic matter is its major benefit in sandy soils.

While not a nutrient, some humic compounds have been shown to directly stimulate root and plant growth. Preparations of such humic compounds are now offered on the market as plant growth promoters.

Nutrient Availability. Humus not only stores nutrients, but also makes several nutrients more available for plant use. As organic matter decays, it releases mild organic acids, which dissolve soil minerals, freeing them for plant use. Phosphorus in the soil tends to form compounds that do not dissolve in water. These forms cannot move in the soil, nor can plant roots absorb them. Organic acids act on these compounds, making phosphorus more available for plant use.

Some nutrients, such as iron and zinc, react with other soil chemicals to form insoluble compounds. Certain humus molecules form a ring around the metal atom in the process called chelation (key-lay-shun). These **chelates** protect metal atoms from being locked in the soil, helping to keep iron, zinc, and others more available to plants. Copper, on the other hand, is so tightly bound to humus that it is least available in high organic matter soils.

Soil Aggregation. As mentioned earlier, organic matter causes soil particles to clump together to form soil aggregates, and gummy substances produced by soil organisms bind the soil clumps. Better aggregation improves soil tilth and permeability. The soil is easier to work, better aerated, and absorbs water more readily. The better aggregated soils also better resists compaction. This may be the most important way that heavy clay soils respond to organic matter.

Preventing Erosion. Soils kept supplied with organic matter have an improved structure that greatly improves water infiltration. Because water infiltrates soils high in organic matter quickly during rainstorms, less water runs off—water that can remove soil from the field. Data used in the Universal Soil Loss Equation (a tool for predicting erosion rates, described in chapter 18) indicate that increasing a soil's organic matter from 1 percent to 3 percent can reduce erosion by one-third to one-fifth. An equivalent loss of organic matter would increase erosion.

Undesirable Effects. Two undesirable but temporary effects can occur during decay of fresh organic matter. The first effect is that nitrogen is tied up in the bodies of microbes during the decay process. Nitrogen is immobilized and is not available for use by plants. A second effect is that certain plant residues are toxic to other plants. The remains of some plants release chemicals during decay that harm the growth of other plants. Dead quackgrass roots, for instance, until completely decomposed, may slow the growth of crop plants.

Maintaining Soil Matter

It should be a goal of all growers to maintain organic matter at the highest practical level. Equally important is the frequent addition of fresh organic matter. It is new organic matter that provides most nourishment for soil microbes and releases nutrients rapidly. Decomposition of fresh organic matter supplies the greatest amount of gum for improving soil structure. These gummy substances decompose and disappear if not renewed.

The amount of organic matter in soil depends on the balance between organic matter inputs and losses. Inputs include crop residues, manures, mulches, and others. Losses include erosion and decay. Therefore, growers should aim to maximize additions of organic matter while minimizing erosion and slowing decay.

Conservation Tillage. Moldboard plowing (see chapter 16, figure 16-4) drastically loosens soil while burying crop residues. The sudden increase in soil oxygen accelerates the oxidation of humus, and close contact between crop residues, moisture, and oxygen stokes an explosion of rapid decay that consumes new organic matter quickly. At the same time, the bare soil that remains after is exposed to erosion.

Conservation tillage, especially no-till, reduces the influx of oxygen into the soil. Much of crop residue remains on the soil surface to decay slowly. Because the soil is covered with organic litter, erosion is reduced. No-till farming, which disturbs the soil and residue cover the least, has the most powerful effect. Research has shown that no-till in certain cropping systems can increase soil organic matter up to a ton per acre per year for several years, so significant improvements in organic matter content are feasible under conservation tillage.

Conservation tillage concentrates organic matter in the top few inches of soil, especially the top inch. Better soil conditions near the surface improve moisture infiltration, reduce crusting and erosion, and make nutrients available where roots are most numerous.

Chapter 16 covers conventional and conservation tillage.

Crop Residues. Simply leaving crop residues in the soil is an easy way to provide organic matter (figure 6-3). With the exception of root crops, such as carrots, plant roots automatically stay in the field after harvest. The aboveground parts of crops for which only the seed is harvested are also usually left in the field. Nationally, growers harvest about one-third of crop residue for feed, animal bedding, or fuel. While there may be good economic reasons for harvesting or burning crop debris, the practice results in a loss of organic materials for the soil.

Growers can increase the quantity of crop residues being returned to the soil by proper fertilization. A well-fertilized crop produces a greater bulk of vegetation, both roots and tops, resulting in more organic matter for the soil.

Green Manuring and Cover Cropping. Some crops are planted to be turned into the soil rather than for harvest. **Green manure** may be incorporated into four broad systems of production. First is traditional green manure, planted as a main crop for later plowdown. This method is particularly useful to reduce erosion and weed growth on land that has been idled (figure 6-4). Otherwise, the loss of a paying crop for one season seldom justifies the practice for most growers. However, it is a standard and important practice for nursery growers to renew organic matter between crops of trees and shrubs.

Two types of plants can be grown as green manures. Legumes such as clover or vetch are useful because, in addition to the organic matter they leave behind, the nitrogen they fix supplies later crops. To get the most bulk of organic matter at the least cost, or if nitrogen additions are undesirable, grasses like oats or rye may be used. Sudangrass, a tropical grass that grows to six feet, develops the greatest amount of green matter (figure 6-4) and is particularly popular with nursery growers.

Where winter erosion is a problem, a **cover crop** may be planted in the fall after the main crop is harvested. The green cover protects soil during the fall, winter, and early spring when it is most erosion prone. The cover crop is then plowed down the following spring, or even better, simply killed with an herbicide and planted into no-till. Winter rye and other winter crops work quite well for this purpose.

A cover crop may also be planted immediately after harvest of a rapidly maturing crop, like oats following peas in the same season. The crop could be fall-plowed. If left until the following spring, this method also provides winter protection of the soil.

Lastly, a cover crop may be planted between rows of the main crop. This cover is sometimes called a companion crop or living mulch. The cover crop may be planted later in the development of the main crop to reduce yield losses due to competition. Again, planting no-till into the killed material the following spring increases the benefits over plowing it down.

Crop Rotation. Economic factors often cause growers to avoid rotating crops, because it means growing less profitable crops some years. However, rotation, where feasible, improves soil humus. Studies show that continuous cropping of row crops like corn causes the greatest decline in soil organic matter. Grains cause smaller loss, while meadow or hay legumes (e.g., alfalfa and clover) actually increase organic matter levels. Thus, a crop rotation of row crops, small grains, and legume hay is better than continuous cultivation of row crops.

FIGURE 6-3 Crop residues left in the field add organic matter to the soil. Cornstalks, high in lignin, decay slowly.

FIGURE 6-4 This farmer grew sudangrass as a green manure crop that controlled erosion, smothered weeds, and contributed organic matter to the soil.

Organic Matter Additions. The sources of organic matter described so far are all grown in the field. Many growers or gardeners can use other organic materials, including animal manures, organic wastes, sewage sludge, or compost. The availability of these materials depends on the kind of operation and the presence of local organic waste producers.

Manure is an important source of both organic matter and plant nutrients. For the farming operation that both grows crops and feeds animals, manuring recycles nutrients and organic carbon on the farm. Most manure is cycled in this way. The cycle may be broken, however, if the feeding of animals and the growing of feed crops are separated, as in some large feedlot operations. The handling of manure and sewage sludge is covered in chapter 15.

Many industries generate organic wastes that may be locally useful. Forestry by-products, such as sawdust or woodchips, may be available. Meatpacking operations and canneries also produce organic wastes. Yard and municipal wastes and sewage sludge may also be available near cities. These organic materials are usually composted before use. Because these sources can generate large amounts of compost, they are useful to landscapers and others who might want to amend soils. It has been shown, for instance, that tilling a two-inch layer of compost into a clay soil greatly improves the establishment and growth of new sod in the landscape.

Homeowners usually have leaves, grass clippings, or other sources of organic matter in their gardens. Some gardeners compost the leaves, clippings, and even table scraps for their own use (figure 6-5). Garden centers usually sell bagged composted manures and peat moss, both useful for the home garden.

Mulches. Home gardeners often mulch their gardens by spreading straw, sawdust, woodchips, or other materials several inches deep on the ground. As the organic matter decays during the growing season, it enriches the humus content of the top few inches of soil. Besides adding organic matter, mulches have other benefits:

- Thick mulches smother annual weeds. The more aggressive perennial weeds may grow through normal mulches.
- Mulched soil absorbs water much more readily than bare soil, improving soil-water content and reducing erosion.
- Mulches limit water evaporation from the soil surface, improving the water content of the soil.
- Organic mulches reduce the range of soil temperatures. For example, during a hot, sunny day, mulched soil remains cooler than bare soil, and warmer on cool nights.

The financial returns on typical farm crops like corn do not justify mulching. Conservation tillage, however, does leave a mulch of crop residue on the soil surface. Many growers mulch high-value crops like berries (figure 6-6) or nursery stock. Mulches almost always make striking improvements in the yields of blueberries, strawberries, raspberries, and tree fruits.

Increasing Soil Water. Any of the methods of water conservation or irrigation described in chapters 8 and 9 increases the bulk of organic matter growing in the field. The more green matter, the greater the amount of organic matter returned to the soil.

FIGURE 6-5 A handful of composted leaves.

FIGURE 6-6 The mulch applied to these strawberries adds organic matter, conserves moisture, and smothers weeds. It also keeps the fruit clean.

Maximum Cropping. As much as possible, soil should be covered with crops. The soils will be cooler, and more organic matter will be produced. Where possible, double-cropping, or the production of two crops a year, is desirable. Avoid fallowing as much as possible.

Nitrogen Tie-Up and Composting

Soil flora need both carbon and nitrogen in their diet to grow and multiply. When fresh organic matter is added to the soil—whether it be crop residues, green manure, or mulch—the number of organisms rises because of the new food. These organisms may compete with crop plants for nitrogen, causing a slowing of crop growth.

The organic matter of greatest concern contains a lot of carbon compared with nitrogen. This can be measured by the **carbon-nitrogen ratio** (C:N ratio) of the material. The C:N ratio of well-rotted manure is about 20:1, meaning that there are 20 parts of carbon for each part of nitrogen. Figure 6-7 shows the C:N ratio for soil and several common organic materials. Matter with a low C:N ratio is nitrogen-rich; material with a high C:N ratio is nitrogen-poor.

Material	C:N Ratio
Soil humus	10
Garden soil	12–15
Young alfalfa	12
Compost	15–20
Rotted manure	20
Clover residue	23
Corn stover	60
Straw or leaves	60
Sawdust	400

FIGURE 6-7 The carbon-nitrogen (C:N) ratios of several materials. The lower the ratio, the richer its nitrogen content.

Let us see what happens when a large amount of fresh, nitrogen-poor material begins to decay in the soil. In response to the new food source, the population of decay organisms rises rapidly (figure 6-8). Bacteria themselves have a C:N ratio in the range of 4:1 to 5:1, and thus need to incorporate a lot of nitrogen into their bodies. The flora feed on both the carbon and nitrogen in the material, but with high C:N materials, the nitrogen is quickly used up. To make up the difference, the flora draw nitrogen from the soil. In the initial stages of decay, then, soil nitrogen is rapidly tied up or immobilized.

During the period when nitrogen is being immobilized, there is a temporary loss of free nitrogen, and crops growing on the soil will suffer a nitrogen shortage. Crop growth may slow, and crops may exhibit nitrogen-shortage symptoms. This period of decay is the **nitrogen depression period.**

After a time, most of the food is used up and the process reverses. The microorganism population declines and, when microorganisms die, the nitrogen stored in their bodies is released to the soil. In other words, nitrogen is being mineralized. When decay is complete, the net gain in soil nitrogen can be measured—the original nitrogen plus that in the new organic matter.

Nitrogen tie-up can be viewed in terms of the balance between nitrogen immobilization (which makes it unavailable) and mineralization (which makes it available). Both processes occur at the same time but not at the same rate. The balance depends on the stage of decay and how much nitrogen the flora must pull out of the soil. Materials with a C:N ratio of more than 30:1 favor immobilization; those with a ratio of less than 20:1 favor mineralization. At rates between 20:1 and 30:1, the two processes balance.

Phosphorus, sulfur, and some other nutrients undergo a similar process when organic matter is added to soil. Nitrogen tie-up, however, produces the most noticeable effect.

Nitrogen tie-up occurs largely when low nitrogen materials are turned into the soil. In the soil, rapid decay quickly depletes nitrogen in the root zone. Organic materials lying on the soil surface—mulches and crop residues—decay more slowly, and most decaying organisms on the surface do not have access to nitrogen deeper in the soil. Fungi can be an exception, because hyphae may extend from the soil into surface litter. Mulches tie up nitrogen much less than do incorporated material.

Nitrogen tie-up can be avoided in several ways. One way is to plant crops after the nitrogen depression period is over—when the decay of previous crop residues is mostly complete. Another is to fertilize with enough

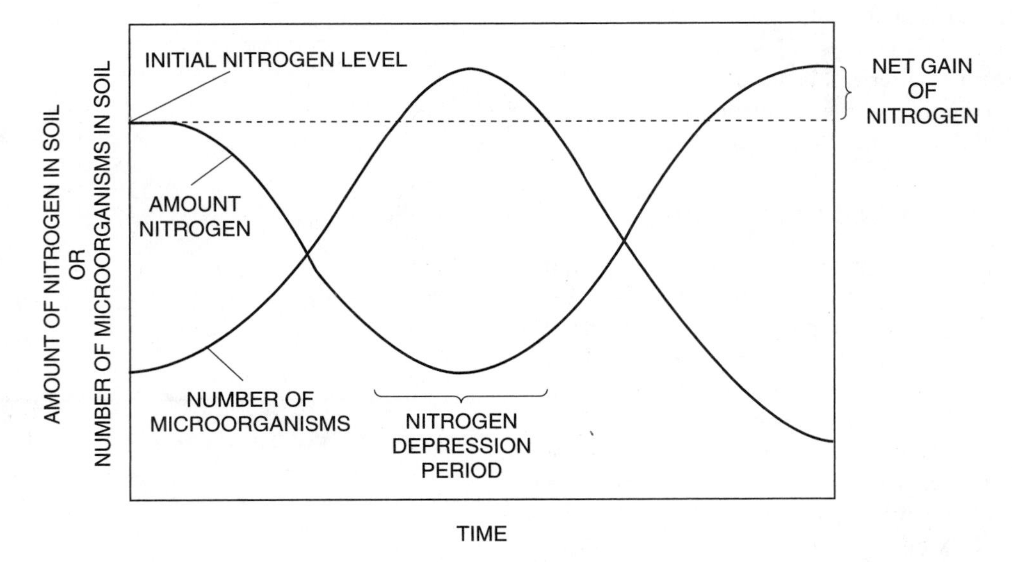

FIGURE 6-8 When a nitrogen-poor material begins to decay, the rising number of microbes robs the soil of nitrogen. During the nitrogen depression period, nitrogen is tied up (immobilized) faster than it is released (mineralized). As decay is completed, the number of organisms declines and nitrogen is released.

nitrogen to provide for the needs of the microorganisms. For instance, if sawdust is being used to amend soil, some additional nitrogen fertilizer can be added as well. Composting lowers the C:N ratio of organic materials before adding them to the soil.

Composting. Gardeners **compost** organic material by storing it in a pile while providing proper conditions for decay. During decay of the compost, the C:N ratio narrows until it reaches about 15:1. The compost can then be added to soil without fear of nitrogen tie-up.

Common composting materials include leaves, grass clippings, wood chips, and even household table scraps. Foods that may attract rodents or dogs, like meats and fats, should not be composted. Materials with a high C:N ratio decompose slowly, so many composters add nitrogen fertilizer to speed up the process. Others include nitrogen-rich materials like manure in the pile. They may also add other fertilizers or minerals to enrich the compost.

Figure 6-9 shows the composition of a common home compost pile. The gardener makes a bin out of snow fencing or chicken wire. While mixing components probably promotes decay, it is more convenient to layer materials like leaves, grass clippings, manures, and so on. Periodically, the pile may be turned by hand to aerate it and water is added. Turning the pile takes materials from the outside of the pile to the center, where most decay occurs. Stirring also adds oxygen to the pile to speed up the decay rate. The compost should be ready in a couple of months. See chapter 15 for details on commercial composting.

Organic Soils

So far, we have discussed the organic matter of soils whose traits are set by their mineral particles. These mineral soils contain only a small percentage of organic matter. Soils containing more than 20 percent to 30 percent organic matter are called **organic soils.** These soils are much different than mineral soils. In organic soils, soil traits are set by the organic matter. Approximately 1 out of every 200 acres of American soil is organic. The five states with the most organic soil are Alaska, Minnesota, Michigan, Florida, and Wisconsin.

Organic soils form in marshes, bogs, and swamps. As aquatic vegetation, such as reeds or cattails, dies each season, it sinks to the bottom. Lacking the air needed for

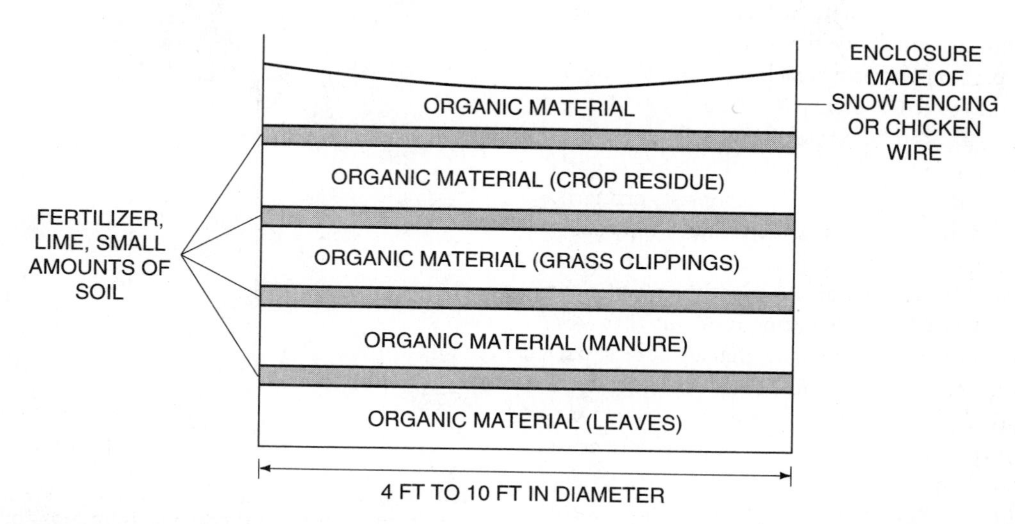

FIGURE 6-9 In a typical home compost pile, various organic materials are layered. The pile may be turned several times to promote decay.

FIGURE 6-10 This organic soil profile shows a twelve- to eighteen-inch deposit of peat on a mineral base. The vegetation is typical of a bog. Note the sphagnum moss near the ruler. *(Courtesy of USDA NRCS)*

rapid decay and oxidation, material builds up on the floor of the wetland. Eventually wetlands may be completely filled in by organic deposits reaching depths ranging from a foot to as much as eighty feet (figure 6-10). Figure 6-11 shows a common organic soil profile.

Organic soils in which plant remains are only slightly decayed are called **peat.** If the deposit consists primarily of fully decayed materials, it is called **muck.** Squeezing a handful of fresh, wet soil can often tell them apart. Water that may be brown, but not muddy, squeezes out of a handful of peat. Muddy water runs out of muck. Further, peat contains plant remains that are at least partially identifiable; muck does not. Soil scientists term slightly decayed plant remains **fibric,** moderately decomposed ones **hemic,** and mostly decayed materials **sapric.**

The nature of organic soils varies not only by the amount of decay, but also by the types of plants the peat derives from. Sphagnum peat forms from sphagnum moss (figure 6-12); it is very acid. Hypnum peat contains

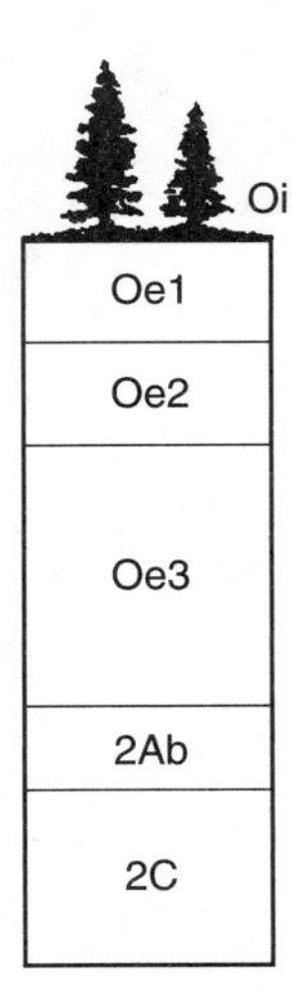

FIGURE 6-11 A profile of a typical organic soil formed in what is now a black spruce/sphagnum moss bog. The growth of different types of vegetation at different periods resulted in the several layers of organic material shown. The lower horizons are old horizons buried by vegetation.

FIGURE 6-12 Sphagnum moss—the main plant of peat bogs in the north. Sphagnum peat is acidic and an important part of horticultural potting mixes.

FIGURE 6-13 Organic soils are favored by sod producers. Turf growth is lush and the sod is lightweight because of the low bulk density of peat. *(Courtesy of Ransomes America Corporation)*

hypnum moss; compared to sphagnum moss, it contains more lime and nitrogen and is less acid. Reed-sedge peat forms in marshes full of reeds and sedges; it is less acid and more decomposed than the other types.

Organic soils are very light, porous, and loose. A cubic foot of fibric peat can weigh as little as a twentieth of a mineral soil. More decomposed types, like sapric muck, weigh more, as do soils that have been cultivated. The heaviest organic soils may weigh about a third of mineral soils.

These soils can soak up great amounts of water. Sphagnum peat can hold ten to twenty times its own weight in water, reed-sedge about five times, and cultivated mucks about twice their weight. Peats are often added to soil to improve its water-holding capacity.

Before fertilization, many organic soils are quite low in plant nutrients. Muck soils are nitrogen-rich, but are generally low in phosphorus and potassium. Sphagnum peat is low in most nutrients. However, once fertilized, organic soils retain nutrients well.

Organic soils are excellent for the production of certain vegetables, including onions, celery, lettuce, carrots, and other root crops. Warm-season vegetables like tomatoes and melons don't perform as well. Mint, hay, and turfgrass seed are also favorite crops for peats. Peats are especially valued for sod production (figure 6-13), because of easy harvest and light weight. About 700,000 acres of organic soils are planted to these specialty crops, for a gross value of about one billion dollars annually.

Unfortunately, peat presents some interesting challenges. After peatland is cleared of brush and drained, exposure to air speeds decay. Soil begins to disappear, changing to carbon dioxide. Added to compaction and wind erosion, the land sinks, a process called **subsidence.** The warmer the climate, the more rapid the loss. In some parts of Florida, soils may lose as much as two and a half inches of depth annually.

In addition, organic soils are flammable, and peat fires are notoriously hard to put out. Being loose and light, wind carries off organic particles easily. Since peats are at low elevation, they are often frost pockets, making it difficult to grow tender crops. Lastly, the black soil can get so hot during warm, sunny days young seedlings may be damaged.

The following practices can help avoid these problems:

- Design drainage systems that keep the water table as high as possible. This will reduce subsidence by decay.
- Install sprinkler systems. They can be used to control frost, cool the soil on hot days, wet soil to reduce wind erosion, or even to drench peat fires.
- Use wind erosion control techniques (see chapter 18), but avoid tall windbreaks that reduce air movement. That increases the chance of frost and can increase soil temperature.

Horticultural Peat. Peat, mostly sphagnum peat, can be harvested for two uses: fuel and horticulture. The Irish have long dug peat for fuel; because of fear of ecological damage, no large-scale energy use yet has been found in the United States.

Horticultural peat, however, is widely harvested. Peat is used to amend problem soils, as a mulch, by nurseries and greenhouses in potting mixes, and as shipping material for nursery stock.

To harvest peat, a bog is cleared and drained. The soil is then plowed and disced or otherwise stirred to promote drying of the surface layer. The dried peat can then be picked up and stockpiled and bagged for sale.

Organic Matter and Global Climate

Let's close this chapter with the observation that soil organic matter affects not only ecosystems and growers but also global climate itself. Soil organic matter interests climate researchers because it is one of the earth's largest reservoirs of carbon. When it decays, carbon dioxide, the main greenhouse gas, is released into the atmosphere. When more soil organic matter is created, carbon dioxide is removed from the atmosphere. Thus, the fate of soil carbon affects global climate. If the earth warms, decay rates should speed up, releasing more carbon dioxide and increasing climate change even more. On the other hand, maybe higher carbon dioxide levels will increase plant growth, and more carbon will be returned to the soil. Unfortunately, models suggest the former is more likely.[1]

In the meantime, we can recognize that growers, by affecting the carbon content of their soils, actually have some influence on global climate. The extent of that influence is uncertain.

SUMMARY

Organic matter, the remains of plant and animal material, is made of such compounds as carbohydrates, lignins, and protein. Microorganisms decay organic matter into carbon dioxide and the more resistant residue, humus. During the decay process, microbes can tie up soil nitrogen.

The amount of organic matter in soil depends on vegetation, climate, soil texture, soil drainage, and tillage. The highest organic matter mineral soils are usually virgin prairie soils formed under fairly cool, moist conditions. Forest soils and those of warm climates are lower in organic matter.

Organic matter and humus store many soil nutrients. They also improve soil structure, loosen clay soils, help prevent erosion, and improve the water- and nutrient-holding capacity of coarse or sandy soils. Organic matter comes from crop residues, animal and green manures, compost, and other organic materials. Proper fertilization, conservation tillage, and crop rotation help preserve organic matter.

Organic soils, widely used for vegetable and sod production, form under the low-oxygen conditions of a swamp or bog. Peat is light and porous, and holds water well. Once the swamp or bog is drained, however, peat slowly disappears by oxidation and wind erosion.

Because soil organic matter is a large reservoir of carbon that can feed carbon dioxide into the atmosphere or withdraw it, soil organic matter is a subject of global climate change research.

[1]e.g. Kirschbaum, M. 1995. The temperature dependence of soil organic matter decomposition, and the effect of global warming on soil organic C storage. Soil Biol. Biochem. 27(6):753-760.

REVIEW

1. Name crops commonly grown on organic soils. Why is such a soil particularly well suited to these crops?
2. Compare the effects on the degree of nitrogen tie-up of turning sawdust into the soil, turning alfalfa hay into the soil, and mulching with straw. Speak in terms of mineralization and immobilization.
3. Other things being equal, which would you expect to have the greatest organic matter content, clay loam or sandy loam. Why?
4. Explain the role of small insects in the decay of fresh organic matter.
5. Could no-till agriculture help reduce global warming? Explain your answer.
6. Maple leaves decay more readily than oak leaves, or are more labile. Speculate on reasons why that might be.
7. Describe the full decay process whereby humus is created from fresh leaves on the forest floor. What soil horizons result?
8. How does soil organic matter affect plant growth?
9. Describe ways of using cover crops. Do you think there are benefits besides maintaining soil organic matter levels? One of the suggested Web sites below has answers.
10. What might be undesirable side effects of harvesting horticultural peat? For help, you can check the Internet site of the Irish Peatland Conservation Council at <http:www.ipcc.ie> or numerous other sites that can be found with your browser.

ENRICHMENT ACTIVITIES

1. Observe the effects of nitrogen tie-up by growing corn in pots using two different soil mixes. Grow one group of plants in a normal but unsterilized soil mix. Grow another group in a mix that is half fresh sawdust. Note differences in crop appearance or growth.
2. Make an actual compost pile. Note that as the compost decays, the pile gets smaller. Why?
3. The Soil Science Society of America has a position statement on carbon storage in soils, the effects of management, and possible climatic effects. It is at <http://www.soils.org/carbseq.html>.
4. Two sites on peat bogs can be found at <http://www.ipcc.ie./wptourhome1.html> and <http://www.epa.gov/owow/wetlands/types/bog.html>.
5. There is good information at the extension bulletin "Organic Matter Management" at <http://www.extension.umn.edu/distribution/cropsystems/DC7402.html>. What practices do they recommend?
6. For more about cover crops and their benefits, try <http://agguide.agronomy.psu.edu>. Go to "Crops and Soils Management," then "Cover Crops."

CHAPTER 7

Soil Water

OBJECTIVES

After completing this chapter, you should be able to:

- identify the role of water in plant growth
- define the forces that act on soil water
- classify types of soil water
- discuss how water moves in the soil
- explain how plant roots remove water from the soil
- describe how to measure soil water content

TERMS TO KNOW

adhesion
adhesion water
anaerobic
available water
bar
capillary
capillary rise
capillary water
cohesion
cohesion water
field capacity
gravitational flow
gravitational potential
gravitational water
hydrogen bond
hygroscopic water
matric potential
osmotic potential
permanent wilting point
potentiometer
preferential flow
resistance block
saturated flow
saturation
soil-moisture tension
soil-water potential
temporary wilting point
unsaturated flow
volumetric water content
water content

How Plants Use Water

On the average, crop plants consume 500 to 700 pounds of water to produce a single pound of dry plant matter; a single corn plant may absorb as much as 50 gallons of water during the growing season. Obviously, most crop plants require very large amounts of water, and water deficiency commonly limits crop growth and productivity. Water availability also partly defines what plants occupy which natural ecosystems. Water is vital to growers—and ecosystems—because of the several functions it serves in plant growth:

- Plant cells are largely made up of water. Plant tissue is 50 percent to 90 percent water, depending upon the type of tissue.
- When plant cells are full of water, the plant is stiff (turgid) because of water pressure in plant tissue. This keeps stems upright and leaves expanded to receive sunlight.
- Photosynthesis uses water as a building block in the manufacture of carbohydrates.
- Transpiration, or evaporation of water from the leaf, helps cool the plant.
- Plant nutrients are dissolved in soil water and move toward roots through the water. Water is thus important in making nutrients available to plants.
- Water carries materials such as nutrients and carbohydrates throughout the plant.
- Water is the solvent in which chemical reactions occur in the plant.
- Moist soil has lower strength than dry soil, easing root growth.
- Moisture is required for microbial activity.

(*A*)

(*B*)

FIGURE 7-1 The temporary wilting point. (*A*) This azalea in dry soil has wilted on a hot day. (*B*) After watering, a moist soil and cooling by wetting leaves have reversed the water balance.

Effect of Water Stress. Water stress is caused by a shortage of water in plant tissue. Stress occurs even at moisture levels that do not cause wilting because as the soil dries, it becomes increasingly difficult for a plant to absorb moisture. As the plant becomes deficient in water, guard cells begin to close the stomata, slowing down the exchange of oxygen and carbon dioxide. Reduced exchange of the two gases slows photosynthesis and plant growth is inhibited.

As the soil dries further, the plant becomes even more water deficient. The plant begins to lose water faster than it can be absorbed and it temporarily wilts. At this **temporary wilting point,** the plant recovers when conditions improve. Wetter soil, cooler temperatures, a more humid atmosphere, shade, or less wind (figure 7-1) help the plant recover. Although the plant recovers, episodes of water stress reduce growth and crop yields. With further drying, soil becomes too dry for the plant to access any water, and the **permanent wilting point** is reached. Now the plant will not recover even if conditions improve.

Plants suffering from chronic water stress exhibit a variety of symptoms, including small, poorly colored leaves, loss of leaf turgor, and reduced growth. Old leaves often turn yellow and drop off. Some plants show specific symptoms of water stress. For example, the leaves of corn plants curl when they need water, and the leaves of sugar maple trees scorch.

Seed germination is very sensitive to water shortage. While seeds efficiently absorb moisture through the seed coat, the emerging seedling is easily injured by dry soil.

Effect of Excess Water. Excess moisture displaces air from soil pores. While small amounts of oxygen are dissolved in water, it is quickly used up by soil organisms. While oxygen can diffuse through water-filled pores, it does so about ten thousand times more slowly than through air-filled pores, so wet soil rapidly becomes oxygen deficient, or **anaerobic.**

Plant roots require oxygen for respiration. Roots lacking oxygen fail to take up water and nutrients properly, so plants growing in wet soil often exhibit symptoms of nutrient deficiency or even wilt. Fungal diseases attack the damaged roots, causing root rots. Carbon dioxide and toxic materials often build up in the soil, further damaging the roots.

Plant species vary in their tolerance of soil saturation. Wetland plants possess adaptations to soil wetness, while some crops, like tomato, may die within a few hours of "wet feet." The common landscape yew is so sensitive to wet soil that it should never be placed where there is any question about drainage.

Forces on Soil Water

A number of forces influence the way water behaves in the soil. The most obvious is gravitational force, which pulls water down through the soil. Other forces, called adhesion and cohesion, work against gravity to hold water in the soil. **Adhesion** is the attraction of soil water to soil particles, while **cohesion** is the attraction of water molecules to other water molecules.

Adhesion and cohesion result from the shape of the water molecule and the way an electron is shared in the oxygen-hydrogen bond. Examine the two water molecules pictured in figure 7-2. Hydrogen consists of one

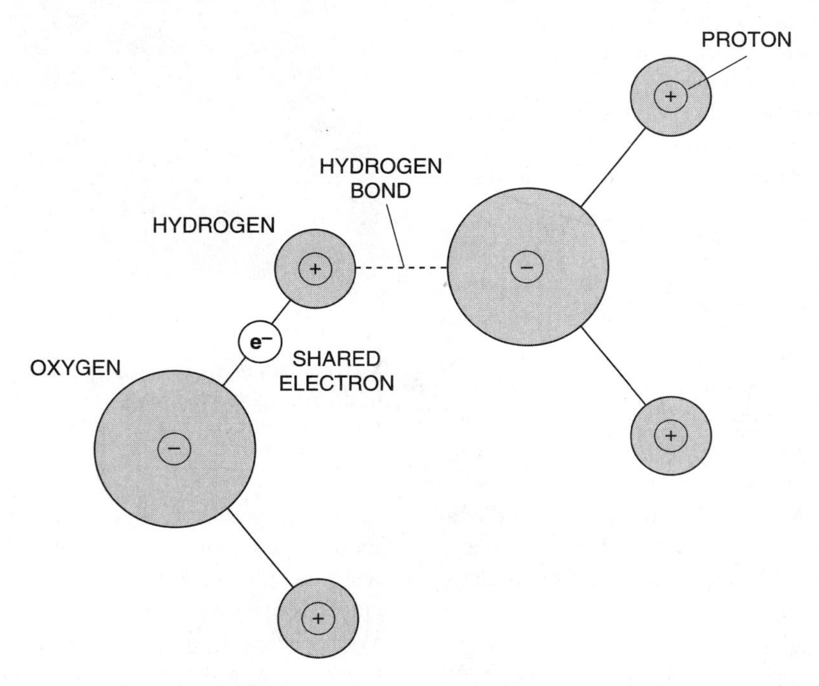

FIGURE 7-2 Two water molecules attract because the electrical charge is unequally distributed. The "plus" side of one molecule attracts the "minus" side of another, forming a hydrogen bond.

proton and one electron. Each hydrogen shares its one electron with the oxygen atom. Each shared electron tends to sit between the oxygen and hydrogen atoms, leaving the positively charged protons positioned on one side of the molecule. As a result, that side has a slightly positive charge, while the oxygen side assumes a slightly negative charge. The water molecule is then like a bar magnet—positive on one end, negative on the other. We call such a molecule a polar molecule, one with a charge separation between the two ends. Like bar magnets, the opposite ends of water molecules attract. The bond between the hydrogen of one water molecule and the oxygen of another, called a **hydrogen bond,** accounts for cohesion.

Hydrogen bonding also accounts for adhesion. The main chemical in soil minerals is silica (quartz is pure silica). Silica, with the chemical formula SiO_2, has oxygen atoms on its surface that can form hydrogen bonds with soil water.

Together, adhesion and cohesion create a film of water around soil particles. The film has two parts. A thin inner film is held tightly by adhesion. The **adhesion water** is held so tightly it cannot move. A thicker outer film of water is held in place by cohesion to the inner film. **Cohesion water,** or **capillary water,** is held loosely, can move in the soil, and can be absorbed by plants. Thus, plants use cohesion water that is clinging loosely to soil particles.

Capillarity. Soil water exists in small spaces in soil as a film around soil particles. The small pores can act as capillaries. A **capillary** is a very thin tube in which a liquid can move against the force of gravity, as shown in figure 7-3. Water is attracted to the glass tube by adhesion so a thin film flows up the side of the tube, while cohesion drags more water along. The liquid rises to the point where gravity balances the adhesive and cohesive forces. The narrower the tube, the higher the water column can rise.

Capillary action, the additive effect of adhesion and cohesion, holds soil water in small pores against the force of gravity. The fact that soil water can move in directions other than straight down is also due to capillary action. The smaller the pores, the greater that movement can be.

Soil-Water Potential. Plants obtain moisture by drawing off water from films surrounding soil particles. The difficulty of the process depends on the strength of the force attracting water molecules to soil particles. Until recently, this force was measured by the **soil-moisture tension** (SMT), which stated how much "suction" is required to pull the water away from the soil particles.

Currently, **soil-water potential** is the concept used to measure soil forces. Soil-water potential, which is just another way to express the forces measured by SMT, is defined as the work water can do when it moves from its present state to a pool of pure water in a defined reference state. More simply, one can think of potential as the tendency of water to flow or move freely in the soil; the higher the water potential, the more freely it can move. The more freely it can move, the more available it is to plants. Potential is a measure of the water's potential energy, its ability to "do work" by movement. As is the rule of energy behavior, water tends to move toward a lower energy state, or to decrease its potential. This rule dictates the behavior of water in the soil.

Consider raindrops falling on a dry soil. This water is capable of great movement, so has high potential. That water can lose potential by getting "stuck," or adsorbed, to a soil particle, which will limit its ability to move. It will also lose some potential energy by moving downward to a lower elevation through the soil.

An important point about the soil-water potential concept is that what matters to a plant is not so much the

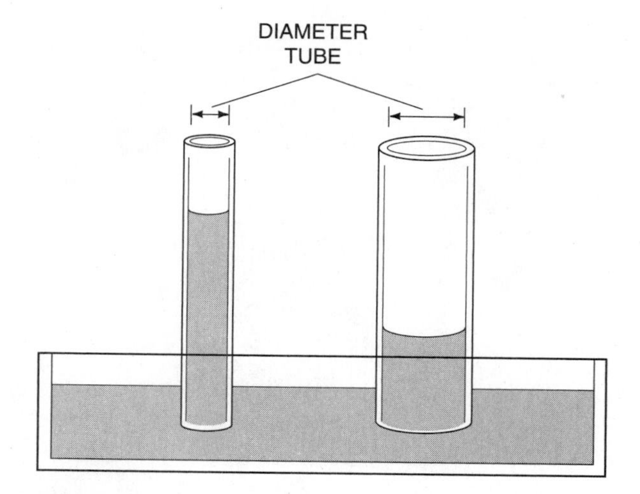

FIGURE 7-3 Capillary action is shown by the movement of water upward against gravity in a capillary tube. The thinner the tube, the higher the water column rises. Soil pores act as capillaries.

amount of water in the soil as it is the *energy,* or potential, of that water. The potential is a measure of the work the plant will have to do to extract water from the soil. While the amount of water in the soil partly defines its potential, so do other factors like pore size and soil salt content. The consequences of this point will become apparent in this chapter.

Now consider a soil-water molecule located far from a soil particle (figure 7-4). At this distance, forces attracting the molecule to the particle (adhesion and cohesion) are weak, and the water molecule can move relatively freely. It has a high water potential. Compare this to a water molecule very close to a soil particle, perhaps in a tiny pore. The attractive force at this distance is very strong, and the water molecule is fixed tightly in place. It has low potential. In small pores, all the water is near a soil particle and is thus at relatively low potential; in larger pores, some water can be farther away and therefore be at higher potential. *The lower the soil-water potential, the more tightly water is attracted to soil particles and the less freely it can move.*

Soil-water potential consists of the sum of several separate forces. Three of these forces are sufficiently important to discuss here. In most soils, the main force is the one just described, the **matric potential,** the potential that results from the attraction of water to soil particles. The matric potential is always a negative value because of the definition of potential. Adsorbed water has less ability to do work than free water in a pool, which is defined as zero potential. Rather than being able to do work, work must be applied to adsorbed water to move it.

A second force is the **gravitational potential.** Soil water is elevated above the water table and so carries potential energy from gravity. To achieve a lower energy state, water simply percolates through the soil to a lower elevation. Gravitational potential is usually a positive value.

The third force, **osmotic potential,** is most important in soils of high salt content. Because water molecules are polar, they are attracted to charged salt ions. For instance, the positive end of a water molecule will be attracted to a negatively charged ion. This attraction lowers water potential. Another consequence is that ions in the soil solution are enveloped in a cloud of water molecules that are weakly attached by electrical attraction. In soils of low salt content, osmotic potential makes a minor contribution to total soil water potential. Osmotic potential is always a negative number, because the reference state is pure water without dissolved ions.

Total water potential can be expressed by the formula:

$$\Psi_{soil} = \Psi_g + \Psi_m + \Psi_o$$

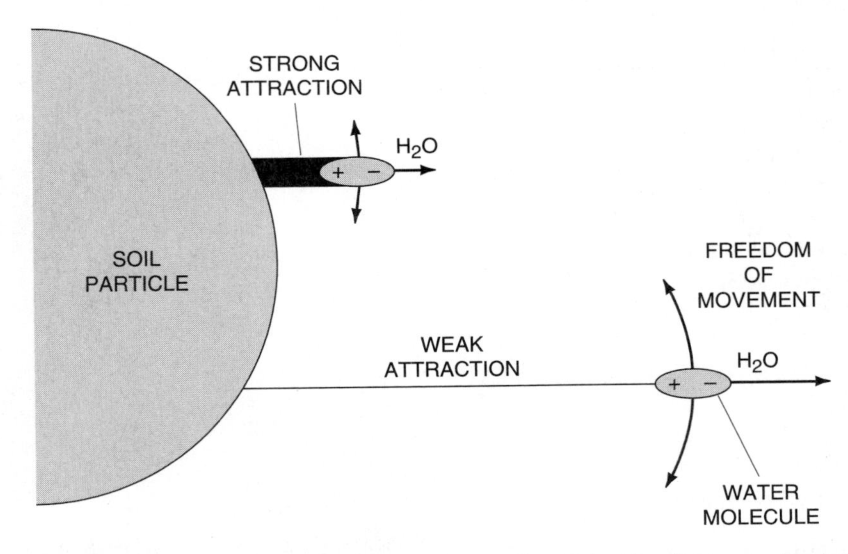

FIGURE 7-4 Energy level, or potential, of water molecules near a soil particle. The closer a molecule is to a soil particle, the more restricted its movement, the lower its potential, and the more tightly bound it is.

This says that total water potential is the sum of the gravitational, matric, and osmotic potentials. These values can be measured and expressed in several different units. The official unit of water potential, acceptable for scientific publications, is the megaPascal (MPa). Still in common usage is the older term **bar,** which is equivalent to 0.1 MPa and is slightly less than one atmosphere pressure (14.7 pounds per square inch).

Soil-water potential is usually a negative number because its largest part, the matric potential, is a negative value. The larger the absolute value of the negative number, the lower the water potential. Figure 7-5 shows a water film with bars of soil potential. The force varies through the film—the closer to the particle, the larger the negative potential, and the more work must be done to pull the water away. The potential for water in the soil, as a whole, is expressed as the most weakly held water in the soil.

Types of Soil Water

Consider what happens after a heavy rain. At first, soil pores are filled with water. This is called **saturation;** see figure 7-6B. At saturation, matric potential is essentially zero, and gravitational potential dominates total water potential. In larger pores, some water is far enough away from the nearest surface that its gravitational potential exceeds matric potential. That is, the force of gravity can overcome the weak attraction of that water to soil particles, and it will flow downward rather than be held in place. The extra water, called **gravitational water,** drains through the soil profile, usually within twenty-four to forty-eight hours in a well-drained soil.

Eventually drainage ceases. The soil moisture level at that point is called **field capacity.** At field capacity, water films are thin enough to hold against gravity, and matric potential now balances gravitational potential. Soil water potential is about 1/3 bar (figure 7-6C). Air fills the interiors of large pores, while micropores remain water-filled, and thick films of cohesion water surround each soil particle. Plant growth is most rapid at this ideal moisture level, because there is enough soil air, yet water is held loosely at high potential.

When drainage stops, water removal by plants and evaporation continues to deplete cohesion water, shrinking the soil-water films. As water films thin, the remaining water clings more tightly, being held at lower

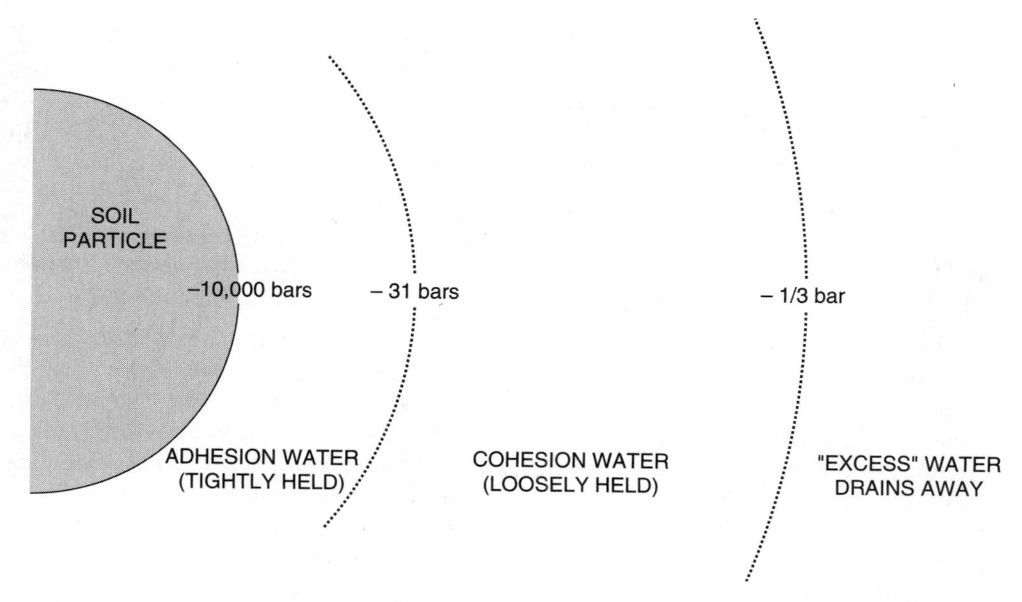

FIGURE 7-5 A water film, showing bars of potential. A plant root must work to overcome that potential.

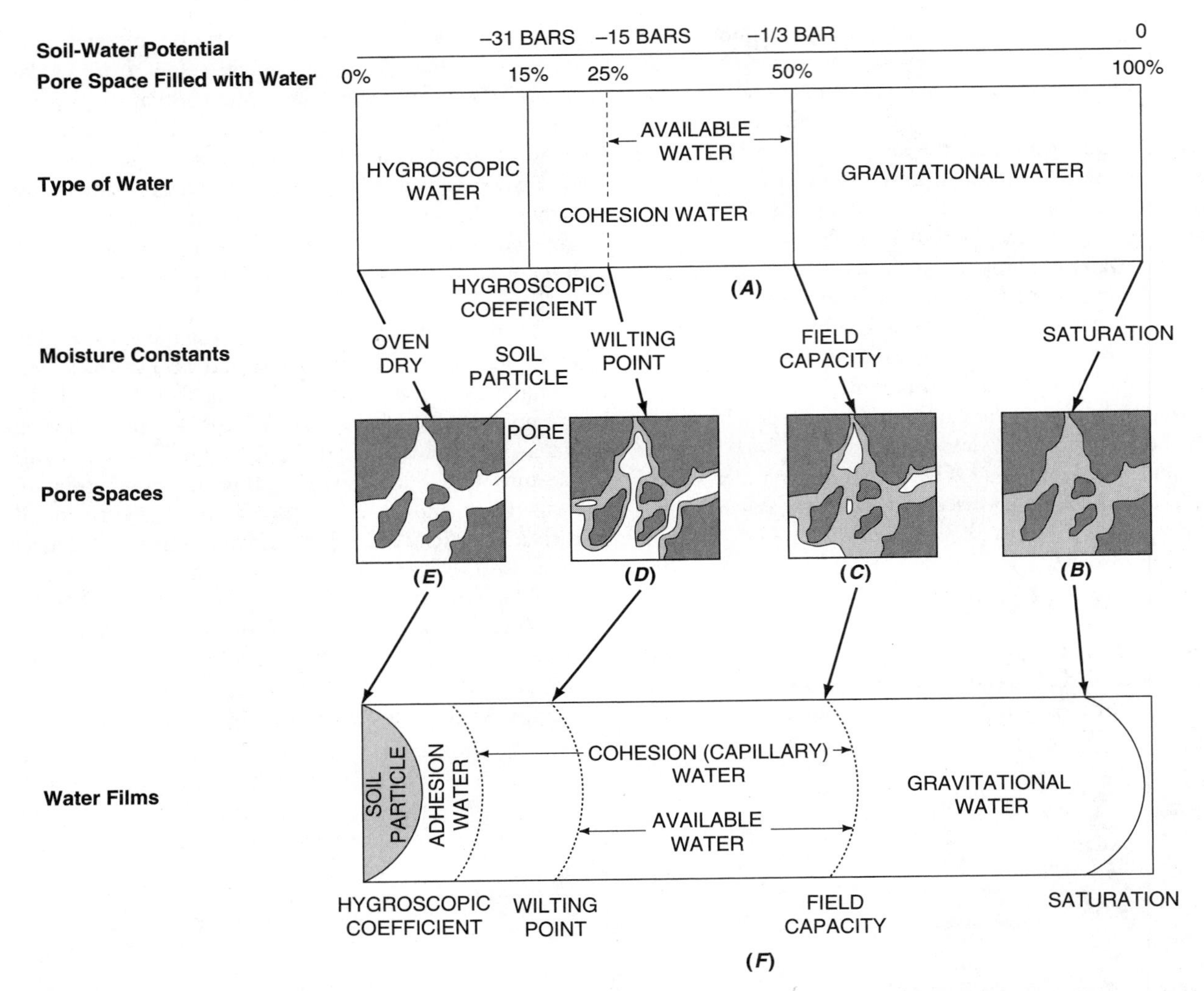

FIGURE 7-6 In *A*, types of soil water are shown, plus moisture constants, corresponding soil-water potential in bars, and average water-filled pore space. Parts *B, C, D,* and *E* illustrate soil pores at each constant, and *F* illustrates the water films.

potential (larger negative value). It becomes increasingly difficult for plant roots to absorb water. Eventually, at the permanent wilting point, most cohesion water is gone and the plant can no longer overcome the soil-water potential (figure 7-6D). The plant wilts and dies. The potential at this point varies according to plants and conditions, but it is usually about −31 bars. At this moisture level, water films are about ten molecules thick.

Beyond the wilting point, some cohesion water remains but is unavailable to plants. The capillary water may also evaporate, leaving only the thin film of adhesion water. This point is called the hygroscopic coefficient, the point at which the soil is air dry. **Hygroscopic water,** as it is called, is held to particles so tightly, between −31 and −10,000 bars, that it can only be removed by drying soil in an oven. In fact, the strength of the soil-water potential is so great that if oven-dry soil is exposed to air, it will bind water vapor from the air until the soil moistens to the hygroscopic coefficient.

Available Water. **Available water** is that part of soil water that can be absorbed by plant roots. Gravitational water is largely unavailable, because it moves out of the reach of plant roots. If the excess water is unable to drain away, roots become short of oxygen and fail to function. Adhesion water cannot be removed by roots, so it is also unavailable to the plant. Only some cohesion water can be used by plants. Available water is defined as lying between the field capacity and the wilting point or between approximately −1/3 and −15 bars. In loamy soil, available water amounts to about 25 percent of the water held at saturation.

Water Retention and Movement

Both the retention of water and the movement of water in the soil are governed by the energy relations just described. We can begin by looking at water retention.

Water Retention. How much water can a particular soil retain and make available to plants? Actually, these are two separate questions, because only the portion of soil water between field capacity and wilting point is plant available. Both the total water-holding capacity and the available water-holding capacity are based mainly on soil texture.

Sand grains are large, so the internal surface area of a sandy soil is quite low. Thus, there is little surface to hold water films. In addition, pores are large enough that much of the volume of each pore is too far from a surface to retain water against gravity. The opposite is true of clay soils—they have small pores and a large internal surface area. Thus, soils high in sand have a low total water-holding capacity, while soils high in clay have a large water-holding capacity.

Not all this water is available to plants, however. In a soil high in clay, clay particles are crowded together tightly, leaving tiny pores. Any water molecule occupying a pore space will be close to a clay surface; therefore, it will be tightly bound at low water potential. Sand is the opposite. With large pores, much of the water can be fairly distant from a grain; therefore, it is held at high potential.

This leads to two rules. First, water in fine soils is held at low potential and water in coarse soils is held at high potential, so it is easier for plants to remove water in coarse than in fine soils. Second, because most water in high-clay soils is held at low potential, much is unavailable to plants. In contrast, most water in a sandy soil is available.

Silt and very fine sand are a special case. They are small enough that there is a high surface area to hold water. Pores are small enough to hold a large amount of cohesion water but large enough that much is held loosely at high water potentials. Thus, soils high in silt hold large amounts of plant-available water.

To hold the largest amounts of plant-available water, soil needs a mixture of large and small pores with many of the medium-sized pores caused by silt and very fine sand. Figures 7-7 and 7-8 show the effects of texture on soil-water retention. There are several important points to note in the figures:

- Fine sandy loam holds more water than regular sandy loam, reflecting the influence of very fine sand.
- Clay has the highest total water-holding capacity, but holds no more available water than a sandy loam.
- Medium-textured soil has the highest available water-holding capacity, especially silt loam.

Organic matter also influences water retention, though the effect is in dispute. It may be that organic matter increases capacity at both the field capacity and the permanent wilting point—meaning no net change in available water. A recent review[1] suggests that organic

Texture	Field Capacity (in./ft)	Permanent Wilting Point (in./ft)	Available Water (in./ft)
Fine sand	1.4	0.4	1.1
Sandy loam	1.9	0.6	1.4
Fine sandy loam	2.5	0.8	1.8
Loam	3.1	1.2	1.95
Silt loam	3.4	1.4	2.03
Clay loam	3.7	1.8	1.95
Clay	3.9	2.5	1.4

FIGURE 7-7 Water retention of several soil textures. *(Adapted from Water: The Yearbook of Agriculture, USDA, 1955)*

[1] *Hudson, B. (1994). "Soil Organic Matter and Available Water Holding Capacity,"* Journal of Soil and Water Conservation, *49, 189–193.*

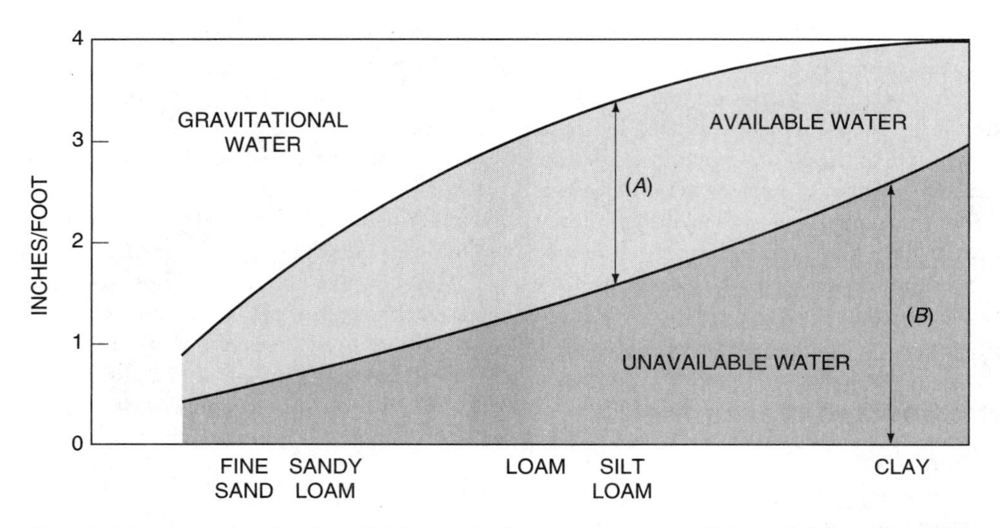

FIGURE 7-8 Water-holding capacity of soils at field capacity (upper curve) and wilting point (lower curve). Available water lies between the two curves. (*A*) Silt loam holds the most available water. (*B*) While clay hold the most total water, most is unavailable. *(Adapted from Water: The Yearbook of Agriculture, USDA, 1955)*

matter does increase the amount of available water. For instance, a silt loam with 4 percent organic matter has twice the capacity of one with 1 percent organic matter.

Water Movement. Horticulturists suggest that trees be watered by letting a hose trickle on the ground. How will water move into the soil?

Figure 7-9 shows penetration of water over time for two soil textures. First, water infiltrates the soil, then it percolates downward through the soil profile. The distance, direction, and speed of travel are set by gravity, matric forces, and hydraulic conductivity.

Directly below the nozzle of the trickling hose is a column of percolating water. This water is gravitational water, moving under the influence of gravity. It is called **gravitational flow.** Gravitational flow only occurs under saturated conditions, when the matric potential is so low it cannot hold water against gravity. Because it occurs under such conditions, it is also called **saturated flow.** Saturated flow resembles the flow of water through water pipes.

In a water pipe, water encounters friction as it flows along the pipe wall, causing a drop in water speed and pressure. The narrower the pipe, the greater the friction losses. Similarly, water flowing through soil suffers friction losses; the smaller the soil pores, the greater the losses. Thus, coarse soils permit more rapid water movement, or have greater hydraulic conductivity. Figure 7-9 reflects this fact.

Because gravity is directed straight down, how does water spread sideways, as it obviously does in figure 7-9? Lateral movement is by capillary flow, the flow of cohesion water through capillary pores in the soil. Recall that water tries to achieve a low energy state. It does this by moving from areas of moist soil (high water potential) to areas of less moist soil (low water potential). Capillary flow can occur in any direction—up, down, or laterally (figure 7-10). In fact, the downward movement of water is aided by capillary flow. Dry soil below pulls water downward from the wetted soil above.

Capillary flow occurs in unsaturated soil, so is called **unsaturated flow.** It depends on unbroken films of water spreading through a series of connected capillary pores. This is like siphoning water—if a bubble gets in the siphon tube, water stops flowing. In sandy soils, large pores contain "air bubbles" that break the continuous water film. Thus, the clay loam soil in figure 7-9 is capable of carrying water farther laterally than the sandy loam.

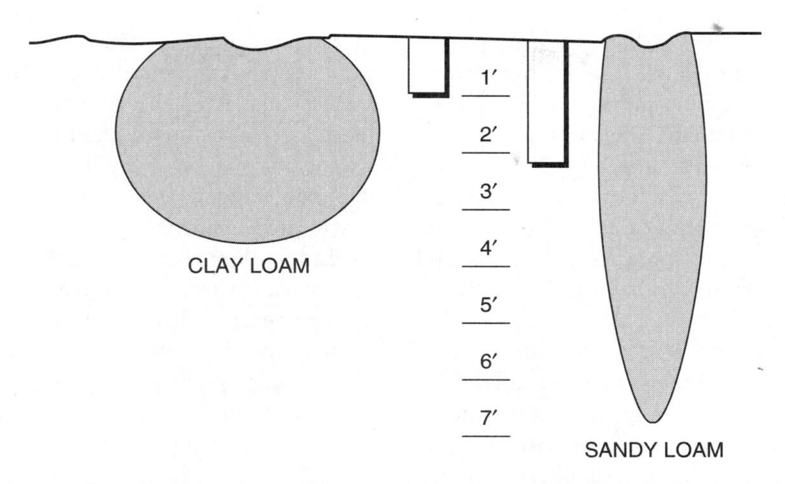

FIGURE 7-9 Wetting pattern in soil. This represents the way water penetrates the soil from a trickling hose. The shaded area shows penetration after twenty-four hours. The bars show how deep two inches of water penetrates for each soil texture. The figure shows that water percolates more slowly and less deeply in the clay loam but moves further laterally.

FIGURE 7-10 Subsurface drip irrigation in a California tomato field. Water trickles out of the device shown. From here, water moves by capillary action in all directions to moisten crop roots. *(Courtesy of USDA ARS)*

As long as the hose in figure 7-9 keeps trickling, the core of saturated soil under the hose supplies free water to keep water films thick where capillary flow is occurring. What if the hose is shut off? The column of saturated soil drains to field capacity, and the soil begins to dry. The water films thin out and become discontinuous, and unsaturated flow slows dramatically. As a result, unless there is a nearby source of saturated soil, capillary flow occurs very slowly over very short distances, often a fraction of an inch per day.

To summarize, saturated flow downward is the movement of gravitational water, usually percolation. It occurs most rapidly in coarse soils with large pores. Unsaturated flow is the movement of water from moist to dry soil (high to low potential) by capillary action, in any direction. Large pores in sand inhibit unsaturated flow.

The Wetting Front. Figure 7-9 shows that water advances into the soil with a definite wetting front—wet behind a distinct line, dry ahead of it. In the dry soil, thin and discontinuous water films have less ability to "pull" water deeper into the soil. Water cannot move forward until films behind the front are so thick that soil particles let go of the water. Behind the front, soil is nearly saturated; immediately ahead of the front, soil is still unwetted. One cannot "half-wet" a soil volume. Either all of a soil is wet, or else part is wet and part is dry. To water crops properly, growers must understand this fact.

Capillary Rise. Water moves upward in the soil as the surface layers dry, moving from areas of high potential to areas of low potential. This upward movement is called **capillary rise.**

Because of capillary rise, soil water can evaporate from the soil surface. However, capillary rise does not continue until the entire soil column dries. When the surface dries, cohesive films become too thin for capillary flow, so upward migration almost halts. This creates a sharp boundary between the dry surface and a moister soil below. The boundary protects against further rapid moisture loss.

Generally, capillary rise does not bring much water from deeper in a soil to plant root systems, because unsaturated flow works solely over a short distance. The plant still extends its roots into lower horizons to draw water. However, where there is a water table near the surface that is a source of saturated soil, capillary rise does carry moisture to plant roots. One method of irrigation makes use of capillary rise. Some potted plants are even watered from below by placing pot bottoms on a saturated fibermat.

In dry climates, or in potted plants, as water rises to the surface it carries dissolved salts with it. These salts are left behind when water evaporates, so they build up in the root zone. Salts may reach levels that injure the plant. The problem is countered by overwatering slightly during irrigation, washing excess salts deeper into the soil or out of the pot.

Effect of Soil Horizons. Figure 7-9 suggests that water flows differently in soils of different textures. Normal soil profiles contain horizons that may differ in texture. What happens when percolating water meets the boundary between two horizons of very different character?

In figure 7-11, water percolating through a fine-textured horizon encounters a coarser soil layer. One might expect that percolation would speed up because of the larger pores of the coarse layer (greater hydraulic conductivity). The soil would become more droughty. In fact, percolation slows down. As the figure shows, water does not enter the coarser layer at first, but instead spreads out along the boundary. Why?

Recall that water is being pulled down by capillarity as well as being pushed down by gravity. The large, non-capillary-size pores of the coarse layer exert much less "pull" on the advancing water than do the fine pores above the boundary, so the clay will not "let go" of the water. The front spreads rather than being pulled across the boundary. When the topsoil is almost saturated, its potential is high enough that sand can pull the water downward.

Commonly, the A horizon is coarser than the B horizon, so a sharp boundary exists between a coarse upper layer and a finer lower layer. As expected, the finer pores pull the water down. However, because water naturally drains more slowly through fine soils with their lower hydraulic conductivity, water will still build up above the fine-textured layer.

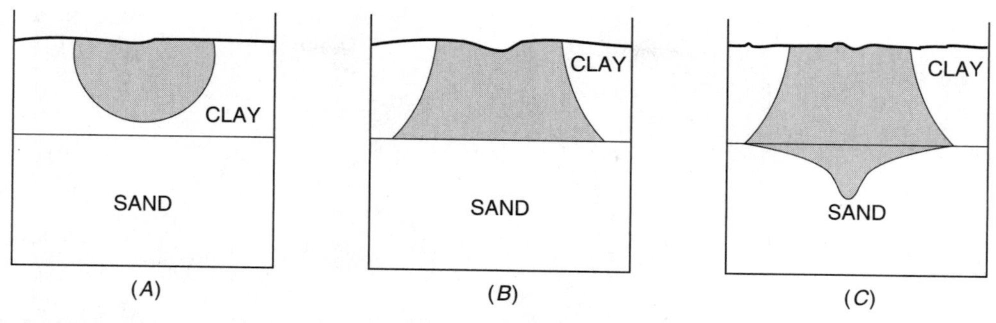

FIGURE 7-11 Effect of textural boundaries. (*A*) Water infiltrates heavy surface soil. (*B*) Wetting front strikes layer of sand and spreads out along boundary. (*C*) When the clay is saturated, water begins to move into the sand.

Any sharp boundary between two layers of dissimilar texture retards drainage. This may be helpful because it can improve the water-holding capacity of a topsoil. In some situations, though, the slowed drainage may keep the soil saturated long enough to injure roots. Landscapers encounter, and sometimes create, these interfaces between different textures. An example is top-dressing a yard with "black dirt" before laying sod, which may be more damaging then helpful (see chapter 17).

Golf greens may be designed to hold moisture by creating the conditions shown in figure 7-11B. A twelve-inch layer of special root zone mix is laid atop a two-inch layer of sand, over a four-inch layer of pea gravel over drain tile. The design causes a wet zone to develop above the sand, so turf roots are watered from below by capillary action. If the root zone gets too wet, excess moisture drains into the pea gravel and is removed from the green.

Preferential Flow. Besides saturated and unsaturated flow, water can move through the soil by **preferential flow.** Under saturated soil conditions, as occurs during heavy rains, water may encounter biopores or other soil channels and quickly enter the channel and flow downward under the force of gravity deeper into the soil profile. Preferential flow can greatly increase infiltration and percolation, reducing runoff and allowing deeper penetration of water. It can also mean deeper penetration of pollutants like pesticides because the water bypasses the filtering action of the soil matrix. This can increase the chance of groundwater contamination. Preferential flow is most prevalent in undisturbed soils or under conservation tillage.

How Roots Gather Water

Uptake of water by plant roots is also governed by soil-water potential. Each root cell has its own water potential; an osmotic potential from chemicals dissolved in cell water; a pressure potential like the pressure in a water balloon; and others. If the water potential inside a root is lower than the surrounding soil water potential, then water flows into the root across the cell membrane. We could say that roots absorb water if:

$$\Psi_{soil} > \Psi_{plant}$$

The larger the difference between soil- and plant-water potential, the more easily and quickly roots will take up water because of a steeper gradient. The smaller the difference, the less steep the gradient, and the more slowly water is taken up. At field capacity, water films are at their thickest. Therefore, soil-moisture potential is high, and water moves easily into a root hair. A root hair pulls a bit of water from this thick film (figure 7-12), so the film becomes thinner at that point and the water potential decreases. This causes capillary flow toward that point from nearby thicker areas of the water film. Therefore, the root hair can continue to draw on the water around it until the water film becomes too thin. The distance in which this capillary flow can occur is less than an inch.

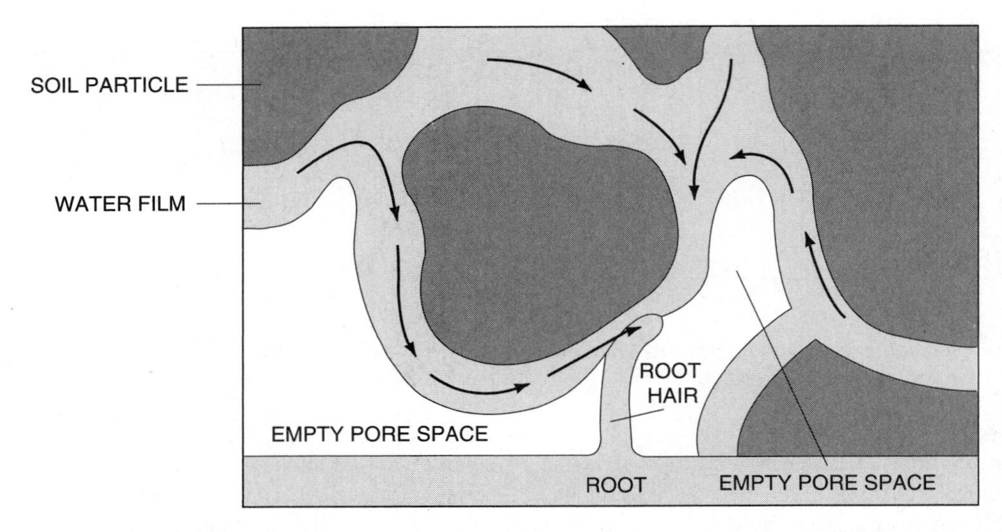

FIGURE 7-12 Water withdrawal by plants. As a root hair removes water from a pore, potential falls at that point, and nearby water flows by capillarity toward the low potential site where the root hair is withdrawing water.

A plant can be compared to an unbroken column of water, beginning in the soil volume that supplies water to a root and ending in the air outside the leaves. Normally, the "driest" end of this column—the lowest water potential—is the air. The "wettest" end—highest potential—is the soil. There is, then, a moisture potential gradient with soil at the high end and air at the low end. Water moves into roots, up the stems, into the leaves, and out into the air, high to low potential.

Another way to visualize the gradient is to picture the atmosphere as drier (less humid) than air spaces in leaves. Water transpires out of the leaf into the drier air, making the leaf drier. That water is replaced by water from the stem, which becomes drier, and so on down the stem into the roots. Water in the roots is replaced from the soil.

As soil dries, it becomes less moist compared to the air; the potential gradient becomes less steep so there is less pressure pulling water into the root. The remaining soil water is more tightly held, so more pressure is needed to remove it. The remaining water films become thinner as well, slowing capillary movement to the root. As the soil dries, then, it becomes increasingly difficult for plants to extract water from the soil. Eventually, the permanent wilting point is reached.

We could also say that as soil dries, soil-water potential approaches plant-water potential, the gradient becomes more shallow, and water flow slows. If the soil becomes dry enough, the two potentials become equal ($\Psi_{soil} = \Psi_{plant}$) and there is no driving force for water absorption. This is the permanent wilting point.

In saline soils, matric potential is joined by the osmotic potential. In this case, both salt ions and particles are holding soil water. As a result, even moist soil can be at quite a low potential, keeping water from entering roots. Thus, saline soils present special water-management problems. If soil salts are concentrated enough—as could happen if one applies too much fertilizer—soil-water potential could even move lower than plant-water potential ($\Psi_{soil} < \Psi_{plant}$), and water would actually be drawn out of roots.

Pattern of Water Removal. Because capillary flow cannot supply all the water needs of a plant, it is very important that roots spread easily, seeking moist soil. Further, roots tend to grow best where oxygen and moisture are highest. Because the oxygen level is highest near the soil surface, plants first use the water near the surface. As soil dries, processes occur in the plant that favors root growth over top growth. Therefore, in response

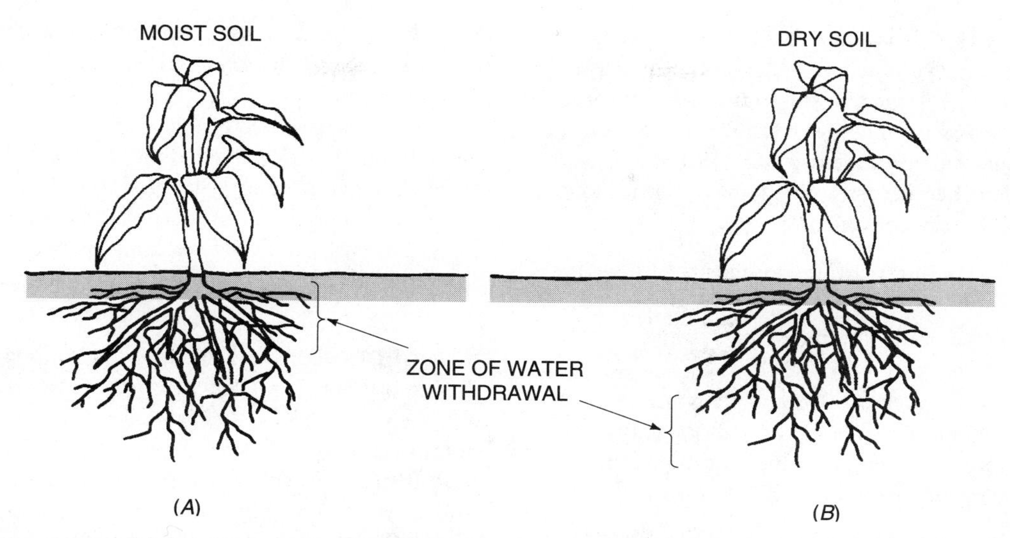

FIGURE 7-13 Water withdrawal from soil. (*A*) When the whole rooting zone of a plant is moist, it draws first from soil near the surface. (*B*) As the surface dries, the plant begins to draw more heavily from deeper soil.

to drying, root systems grow deeper and more densely. As the surface zone is depleted, roots grow more deeply, and absorption shifts downward to zones where the soil is still moist (figure 7-13). If the surface layer is rewetted, moisture absorption shifts back to the surface.

The pattern of water absorption suggests how irrigation water should be applied. If surface soil is kept moist enough, then the plant is well supplied by near-surface water, top growth is favored over root growth, and plants will tend to have more shallow, sparse root systems. The plant is unable to exploit lower soil levels quickly when the surface dries, and is, therefore, vulnerable to drought. Proper irrigation wets most of the rooting depth of the crop and allows for some drying between waterings.

Measuring Soil Water

People who design or use irrigation systems need to be able to measure the amount of water in a soil. They also need terms to name the amount of water present. Four methods are common: gravimetric measurements, potentiometers, resistance blocks, and other devices. At the base of all these are gravimetric measurements.

Gravimetric Measurements. Gravimetric methods directly measure soil **water content** by weight. This gravimetric water content can then be converted to other useful quantities.

WEIGHT BASIS. To measure the water content of a soil sample by weight, the sample is weighed and the weight recorded. The sample is then oven-dried, and the dry weight is noted. The difference between the two weights is the weight of water in the soil. Water content is the amount of moisture divided by the oven-dry weight:

$$\text{water content} = \frac{\text{moist weight} - \text{dry weight}}{\text{dry weight}}$$

Suppose one needs to measure the water content of a soil at field capacity. A sample is taken two days after a heavy rain. If the sample weight were 150 grams when wet and 127 grams when dry, the moisture percentage would be

$$\text{water content} = \frac{150 \text{ grams} - 127 \text{ grams}}{127 \text{ grams}} = .18$$

VOLUME BASIS. It is often more useful to calculate water content on a volume basis. However, it is impractical to measure a volume of water in the soil. This problem can be solved by making a weight determination and converting it to **volumetric water content** using soil and water densities (density being mass per volume). The equation for the conversion is

$$\text{volumetric water content} = \text{gravimetric water content} \times \frac{\text{soil bulk density}}{\text{water density}}$$

The density of water is 1.0 grams per cubic centimeter. In the previous example, if the bulk density of the soil sample were 1.5 grams per cubic centimeter, the percent water by volume would be

$$\text{volumetric water content} = .18 \times \frac{1.5 \text{ gm/cc}}{1.0 \text{ gm/cc}} = .27$$

Thus, if one measures in the metric units of grams per cubic centimeter, the volumetric water content is simply the gravimetric content times the bulk density of the soil.

SOIL DEPTH BASIS. A meteorologist measures rain in inches of water; irrigation is measured in inches as well. Inches of water is a convenient, easily visualized unit that can also be used to measure the amount of water in a soil.

Let's say one could take one cubic foot of soil and squeeze all the water out of it into a one-square-foot cake pan. How many inches of water would be in the pan? This can be calculated simply by the equation:

$$\text{inches water per foot soil} = 12 \text{ inches} \times \text{volumetric water content}$$

In above sample, then:

$$\text{inches water per foot} = 12 \text{ inches} \times .27 = 3.24$$

In the sample, each foot of soil depth contains 3.24 inches of water. If a soil profile were three-feet deep, and each foot was the same, then the total of the entire profile would be 9.72 inches of total water.

Inches per foot is a common measurement used in irrigation. Irrigation also uses the acre-inch, the volume of water that would cover one acre of soil one-inch deep. In the metric system, the measurement equivalent to an inch per foot is a centimeter of water per centimeter of soil.

In practice, it would be a bother to make gravimetric measurements to decide whether or not a soil needs watering. Several device may be used to measure soil water for research purposes or for scheduling irrigation. Here we will discuss the two most common and least expensive devices, the potentiometer and resistance block.

Potentiometers.

From the plant's point of view, the important thing is not how much water is in the soil, but the potential it is being held at. A device called a **potentiometer,** or tensiometer, acts like an artificial root. It measures soil-moisture potential and so gives a "root's eye view" of how much water is available.

A potentiometer is a plastic tube with a vacuum gauge at one end and a porous clay cup at the other end (figure 7-14). The tightly closed tube is filled with water, then buried with the gauge sticking out. The dry soil outside the tube pulls water out through the clay cup, creating a partial vacuum inside the tube. The vacuum creates a force that opposes the pull of matric forces in the soil, and water ceases leaving the tube when the two are equal. At any given moment, the gauge reading the vacuum is then also reading the matric potential of soil water. Potentiometers work best in moist soil at potentials between 0 and −0.8 bar—a narrow range but an important one to plants.

Resistance Blocks.

Another device for measuring soil moisture is the **resistance block** (figure 7-15) or Bouyoucos block. Two electrodes are imbedded in a block of gypsum, fiberglass, or other material. When in the soil, the device measures resistance to electrical flow between the two electrodes. It is more difficult for electricity to flow through a dry block than a wet block, so the reading indicates moisture level. Resistance blocks work well in drier soil than do potentiometers, so work better in drier watering regimes.

Actually, pure water conducts electricity very poorly. It is ions in solution that carry the electrons of electrical flow. Therefore, most resistance blocks sense both water content and salt content of the soil, though the block is designed to buffer salt effects. These blocks should be calibrated for

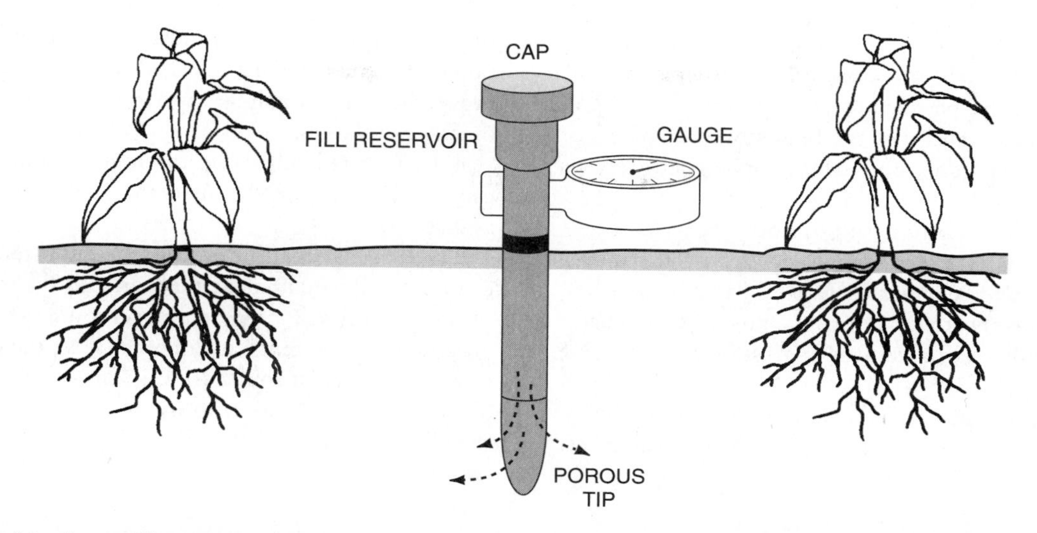

FIGURE 7-14 A potentiometer. As dry soil pulls water out of the porous tip, a partial vacuum is created inside the tube that is measured by the gauge. The vacuum pressure equals the soil-water potential.

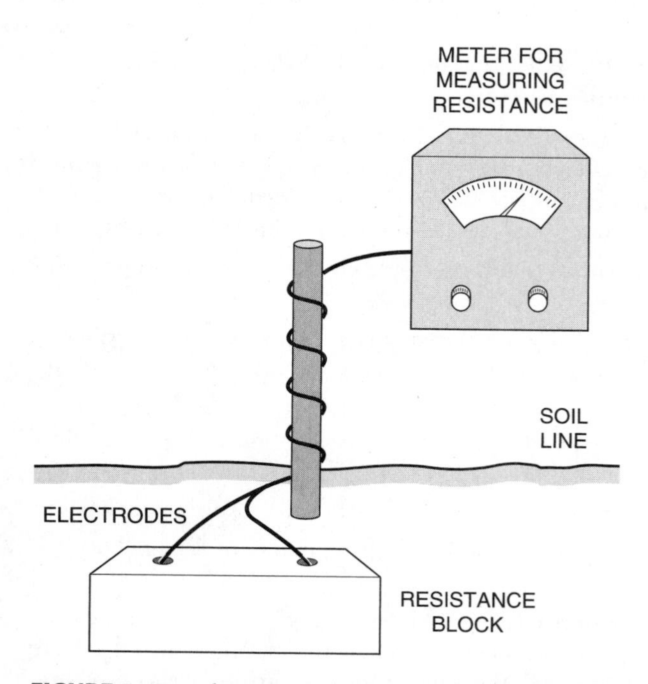

FIGURE 7-15 When using a resistance block, the meter reads resistance to electrical flow between two electrodes buried in the block. The drier the soil, the greater the resistance.

each soil to obtain moisture readings, because of the different salt content and matric potential of different soils.

An inexpensive and convenient variation of the resistance block is a simple device with a meter and two electrode rods that is pushed into the soil. Resistance to electrical flow between the two rods appears on the meter, giving a quick reading of the moisture level. It has been shown to be fairly accurate with practice; however, varying salt levels influence the readings.

SUMMARY

Soil-water potential dictates the behavior of water in the soil. Based on their differing potentials, three types of soil water can be defined. Gravitational water drains away under the force of gravity. Capillary water is loosely held by cohesion and is mostly available to plants. Hygroscopic water is too tightly held by adhesion to be used by plants.

When all soil pores are occupied by water, the soil is saturated. Most plants do not live well in saturated soils because of low soil oxygen. At the field capacity, the last bit of excess water has drained away, and pores are occupied by air and water. Most plants grow best at field capacity. As the soil dries further, plants must work harder to absorb moisture, so water stress increases until the permanent wilting point is reached. When the last water that can be removed by plants or evaporation is gone, the hygroscopic coefficient is reached.

Water moves through soil by both gravity and capillarity. Both forces contribute to downward movement, but only capillarity carries water upward or laterally. Downward movement is most rapid in coarse soils, while lateral movement is most extensive in fine soils.

Water penetrates soil along a wetting front. Below the advancing front, soil remains dry. Whenever a wetting front meets a sharp boundary between soil layers of different texture, percolation slows. This may either improve the water-holding capacity of a soil or cause drainage problems.

Fine-textured soils have the highest water-holding capacity but medium-textured soils retain the greatest available water.

Water is absorbed into plants because transpiration causes low potential inside the plant. As the root draws from a water film, more water moves towards the root by unsaturated flow. Since such flow is slow, the extension of roots into moist soil is critical. The water is used by plants in photosynthesis, to transport materials within the plant, and to keep plant tissues stiff.

REVIEW

1. Of the three water potentials, which one would be most affected by compaction? Explain your reasoning.
2. Which soil texture would have to be watered more often, coarse or fine? Which would need more water each time? Explain your answers.
3. A 112-gram sample of soil, after drying in an oven, weighs 100 grams. Assume a soil bulk density of 1.4 g/cc. Calculate water content by weight, by volume, and inches of water per foot.
4. The soil moisture of a soil was measured at several intervals after a heavy rain. Immediately after the rain, while still saturated, a sample weighed 132 grams. Two days later, just after drainage stopped, the sample weighed 116 grams. When plants wilted in the soil, the soil weighed 108 grams. Finally, the sample was oven dried, for a weight of 100 grams. Assuming a bulk density of 1.4 g/cc, calculate the inches available water per foot for this soil at field capacity.
5. How much water should be taken up by a plant when the air around it is completely saturated with water—100 percent humidity. Explain.
6. Some indoor foliage plants are potted in self-watering pots that water the soil mass from a reservoir of water in the bottom of the pot. How does this work? Might there be a problem if the soil was allowed to dry excessively before the reservoir was refilled?
7. Distinguish temporary wilting point from permanent wilting point. Which one is a trait of the soil only? Could a plant be taking up water and still wilt?
8. Describe some ways you might improve the available water-holding capacity of sandy soils. Explain your answer.
9. Do you think the air-dry point of a soil is constant, or does it very according to the humidity of the air above the soil? Explain your answer.
10. A case study: Ecologists have found that plants like maple trees engage in "self-watering" when surface soils begin to dry. To explain: leaf stomata close at night, breaking the continuous column between soil water and atmospheric air described in this chapter. Little water is then lost through the leaves. Often, as surface layers of soil dry, deeper soil may still be moist. They have found that at night, deep roots absorb water, which rises to the shallow roots, where it is released into the soil. Ecologists call this hydraulic lift. Explain how this could happen in terms of water potentials. How might this be beneficial to the maple tree or other plants in the vicinity?

ENRICHMENT ACTIVITIES

1. Test the effects of cohesion and adhesion using two glass microscope slides. Hold the two slides flat against each other, then pull (not slide) them apart. Now put a drop of water between them, and repeat. Feel the difference.
2. Grow several tomato plants in pots until each has several leaves. Then divide the plants into three groups, watering each differently. Water one group so often the soil stays wet. Water the second group when the soil surface dries. Water the third group only when the plants wilt. Observe the differences in growth.
3. A fellow Delmar author, Mark Coyne, suggests this exercise in his book on soil microbiology. It uses potato sticks to demonstrate osmotic potential and water absorption. Cut a potato into identical thin, long sticks. Place some into distilled water, others into a strong solution of table salt, and wrap some in cellophane and refrigerate; this last one is the control. After a couple of hours compare the length of the sticks. Those soaked in distilled water should be longest, those in a salt solution shortest. Can you explain the results in terms of osmotic potential of potato cells (caused by chemicals dissolved in the cell water) versus the solution they were soaking in? What would this say about water uptake in soils that are high in salts?
4. A nifty hydraulic properties calculator based on the soil texture triangle is at the following site. Click on any proportion of clay and sand on the triangle to get properties for that texture:

   <http://www.bsyse.wsu.edu/saxton/soilwater/>.

   Use this triangle to compare the available water holding capacity and hydraulic conductivities of a sand with 90 percent sand and 10 percent clay, a silt loam with 20 percent sand and 20 percent clay, and a clay with 45 percent sand and 55 percent clay.

CHAPTER 8

Water Conservation

OBJECTIVES

After completing this chapter, you should be able to:

- explain the source of our fresh water supplies
- explain the need for water conservation
- describe ways to make better use of water
- discuss water quality

TERMS TO KNOW

aeration (turf)
antitranspirants
aquifer
arid climate
buffer strips
conservation tillage
consumptive use
contour tillage
evapotranspiration
furrow-diking
groundwater
hydrologic cycle
hydrophilic gel polymers
mulch
nonpoint source
point sources
precipitation
runoff
semiarid climate
soil pitting
strip-cropping
stubble-mulching
subsoiling
surface water
syringing
terrace
water table
water-use efficiency

The Hydrologic Cycle

Of the nine planets occupying our solar system, only Earth is known to contain large amounts of liquid water. Most of this water—97 percent—occupies the oceans.

The **hydrologic cycle** (figure 8-1) is an engine that transports water from the ocean to land and back again. The engine is fueled by the sun's energy. Air moistened by evaporation from the surface of the ocean passes over the continents, where it is shifted upward by warm air rising from the land mass. When moist air has risen high enough, water vapor condenses into **precipitation** (rain, snow, and hail). Some rainwater runs into streams and lakes. Most of this water finds its way into rivers that finally flow into the ocean. Other water is absorbed into soil to later evaporate or be used by plants. Finally, some rainwater percolates into the **water table,** which is the upper surface of saturated underground material.

Let's look more carefully at the point in the cycle when precipitation lands on the soil. The water proceeds in the cycle by one of four paths:

- Rainfall or snowmelt that cannot be absorbed into soil fast enough runs into low areas, streams, or lakes. This water is called **runoff.**
- Gravitational water percolates below the root zone of plants. Some may enter the water table.
- Some of the remaining water stored in the soil evaporates from the surface back into the atmosphere.
- Most water taken up from soil by plants is transpired into the atmosphere. The total water loss due to transpiration and evaporation is called **evapotranspiration.**

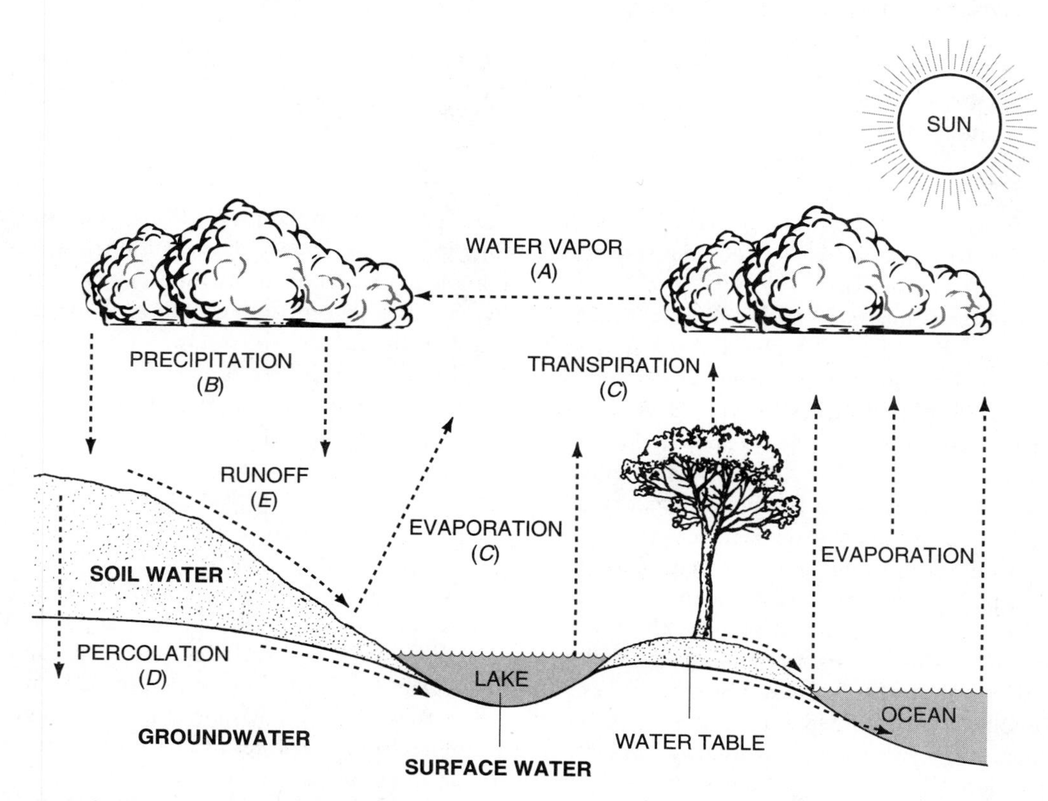

FIGURE 8-1 In the hydrologic cycle, water cycles between the ocean and the land, turning saline ocean water into fresh water available for our use. The letters labeling the cycle refer to the quantities listed in figure 8-2.

	Yearly Amount	Inches/Year	Billion Gallons per Day
(A)	Water vapor	—	40,000
(B)	Precipitation	30	4,200
(C)	Evapotranspiration	21	2,900
(D)	Percolation	3	411
(E)	Runoff	6	822
	Consumptive use	—	106

FIGURE 8-2 Each year in the United States tremendous amounts of water move through the water cycle. *(Source: USDA Appraisal, Part I: Soil, Water, and Related Resources in the United States, 1980)*

FIGURE 8-3 Rangeland in Idaho. Raising animals, like these sheep, is one way to use soil in drier climates. *(Courtesy of USDA ARS)*

Water Resources in the United States

Figure 8-2 shows where the United States' water resources are in reference to the hydrologic cycle. The reservoirs of water we draw on include water vapor in the atmosphere, surface water, and groundwater. Note that enormous amounts of water vapor float over the United States daily. Some work has been done trying to use this water source by "cloud seeding," but until we find practical ways to tap this supply without disturbing weather patterns elsewhere, we rely on natural precipitation to satisfy our water needs.

An average of thirty inches of precipitation falls each year over the continental United States, but the supply of water is unequally distributed. Average annual precipitation ranges from nine inches in Nevada to fifty-five inches in Louisiana. Eastern states and the Pacific Northwest coastal areas tend to have higher rainfall while most western states have a more arid climate and have to depend on irrigation to grow many crops.

More important than simple annual precipitation is the balance of precipitation and evapotranspiration. In the eastern half of the country and parts of the West Coast, more rain falls than is lost to evapotranspiration. Here water does not usually limit crop production, except on excessively well-drained soil or during occasional periods of dryness. In a **semiarid climate,** evapotranspiration slightly exceeds precipitation and moisture limits production. Special dryland farming systems (figure 8-3) or irrigation overcomes the problem. In **arid climates,** where evapotranspiration greatly exceeds precipitation, crop growth depends on irrigation. Despite the importance of irrigation nationally, natural rainfall remains the most important source of water. Seventy-five percent of our food and fiber is grown in "rainfed" fields.

The remaining three sources of water, as shown in figure 8-4, comprise stored water society draws on: saline ocean water, fresh surface water, and groundwater. As figure 8-4 shows, saline ocean water supplies only a small percentage of the water used annually. Fresh surface water and groundwater are critical sources for irrigation and nonagricultural uses.

Surface water occupies lakes, rivers, and ponds and covers approximately sixty million acres of American land. The water is distributed unequally; most surface water is in the Great Lakes and a few other states, especially Alaska, Texas, Minnesota, Florida, and North Carolina. In many western states, surface water for irrigation and other uses depends on rivers fed by snowmelt in mountains; there the annual mountain snowpack is a matter of great concern.

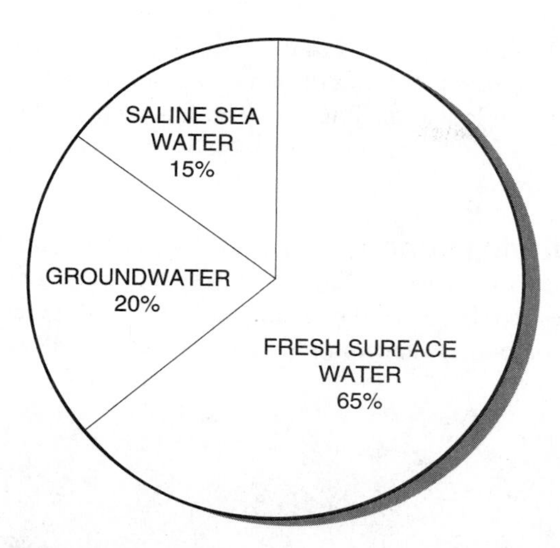

FIGURE 8-4 Water supply sources in the United States. We depend on fresh surface water and groundwater to meet our water needs. *(Source: United States Geological Survey. Circular 1200: Estimated Water Use in the United States in 1995)*

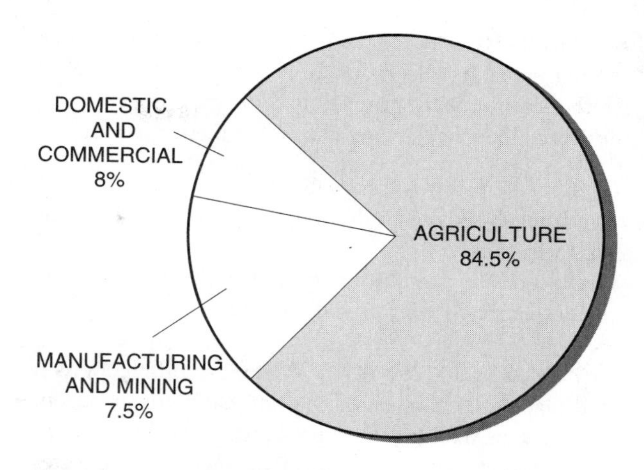

FIGURE 8-5 Agriculture is the largest consumer of water in the United States. *(Source: United States Geological Survey. Circular 1200: Estimated Water Use in the United States in 1995)*

Groundwater is stored in underground formations called **aquifers**. There is far more groundwater than surface water. Experts estimate that some 8,000 trillion to 10,000 trillion gallons of water are contained in mainland United States aquifers. However, this water renews very slowly, averaging about three inches per year (figure 8-2). According to the USDA, about 25 percent of the nation's groundwater supplies are being "mined," withdrawn more rapidly than renewed. This occurs mostly in the Southern Plains and the Southwest, where the water loss per day (in an average year) is fifteen billion gallons of groundwater.

However, improvements are being made in some areas. One of the nation's largest and most important aquifers, the Ogallala, which underlies the High Plains of northern Texas and adjoining areas, has seen drastic drawdowns due to irrigation. In recent years, improved irrigation methods, and the return of some farms to dryland agriculture, has slowed the overdraft.

Reasons for Conservation. Agriculture consumes more than 80 percent of the water used annually in this country (figure 8-5), mostly for irrigation. The nation and its growers benefit from agricultural conservation efforts in three ways: preservation of water resources, increased crop yields, and fewer problems from runoff. For example, schemes that persuade growers to sell some of their irrigation water to expanding municipalities have begun to be implemented in western states. Even worldwide, water use is reaching some limits. Between 1940 and 1990, the worldwide withdrawal of fresh water for human use quadrupled, mostly going to irrigation. Unfortunately, the supply of fresh water is limited by the dynamics of the water cycle.

PRESERVATION OF WATER RESOURCES. It takes forty gallons of water to produce one egg, 150 gallons for one loaf of bread, and 2,500 gallons for one pound of beef. If forecasts are correct, growers will increasingly compete with other users for the nation's water.

As the largest user, agriculture has a special responsibility to conserve, both for national well-being and in its own self-interest.

INCREASED YIELDS. Water conservation methods result in improved soil moisture and better yields. Semiarid areas respond especially well to conservation efforts. For example, about four inches of water is needed to mature a wheat crop. Reports from USDA research indicate that each additional inch of available moisture stored in the soil can raise production from two and a half to six bushels per acre.

Reduced Runoff. Some of the strategies for conserving water involve reducing runoff from farm fields. Methods for reducing runoff do more than improve soil moisture. They have beneficial side effects, such as:

- reduced erosion and topsoil loss
- reduced downstream flooding because less water runs into rivers
- reduced water pollution from fertilizers and pesticides carried off farm fields by runoff

Water-Use Efficiency. Water-use efficiency is a good place to begin a discussion of water conservation. Of the water that lands on a field, little actually becomes part of the dry matter of crop plants. Most of the water is lost to runoff, percolation, or evapotranspiration. The total amount of water needed to produce a unit of dry plant matter, such as a bushel of oats or a nursery tree, is one measure of **water-use efficiency.**

There are three primary means of improving water-use efficiency. One is to capture more of the water from precipitation in the root zone of crop plants. This means improving the infiltration rate and reducing percolation. Having captured more water, the second step is to reduce consumptive use. **Consumptive use** is the sum of the water lost by evapotranspiration and the amount contained in plant tissues. About 1 percent to 10 percent of total consumptive use actually becomes part of the plant; 90 percent to 99 percent of the total is evapotranspiration. The third way to improve water-use efficiency is by improving irrigation systems (chapter 9).

Capturing Water in Soil

To capture into a crop's root zone more of the water landing on a field, a grower can improve infiltration and/or lower deep percolation. Of the two options, improving infiltration is easier. Infiltration rates are, of course, a function of soil texture, structure, organic matter content, degree of compaction, or presence of barriers like hardpans and plowpans. Slope also increases runoff and decreases infiltration.

Structure of the entire soil profile is important, but structure at the soil surface is most critical to infiltration. As noted earlier, the soil surface is sealed by the shattering of surface aggregates by raindrops or puddling. As a result, a crust may be formed that significantly reduces infiltration.

The list of factors influencing infiltration suggests two problems that lower the amount of infiltration. One problem is low water-intake rates and the other is runoff due to slope. Let's look at methods for dealing with slope first.

Capturing Runoff. **Terraces** have long been used to capture runoff water. Terraces consist of a series of low ridges and shallow channels running across the slope, or on the contour, as shown in figure 8-6. When water begins

(A)

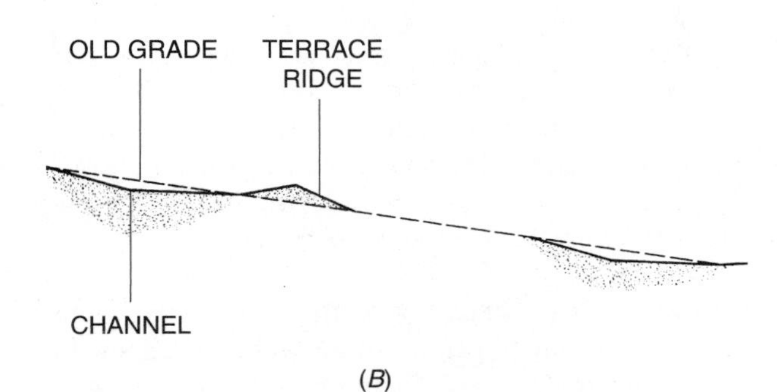

(B)

FIGURE 8-6 (*A*) Measuring water captured on a terrace in the Great Plains. (*B*) Cross-section of flat-channel terrace designed for water conservation. *(Courtesy of USDA ARS)*

running down the slope, it runs into the terraces, where it gathers while it seeps in.

All terraces are built to control runoff. However, in humid areas, the main concern is to control erosion. In drier areas, the primary purpose of terracing is to increase moisture in the soil. To save moisture, terraces are designed to cause ponding of water on the terrace, giving water time to infiltrate. Figure 8-6 shows the cross section of a flat-channel terrace built for this purpose.

Container nurseries and golf courses often alter their topography to capture rainfall and excess irrigation water and direct it into holding ponds for reuse. While a major reason for such a design is to retain pollutants like pesticides and fertilizers on-site, it also conserves water.

Furrow-diking is another means to capture water in drier areas of the country. To furrow-dike, special equipment creates furrows with small ridges, or dikes, across them (figure 8-7). This creates basins that capture and hold water. Somewhat similar is **soil pitting,** a practice that creates tiny pits on rangeland to capture water.

FIGURE 8-7 Furrow dikes in a Texas field. Note that in the undiked furrows on the right, water has already run off. Diked furrows capture water for crop use. *(Courtesy of American Society of Agronomy)*

Contour tillage is practiced by operating all equipment across the slope of the field, on the contour. This practice makes many tiny ridges across the slope. Water ponds behind these ridges, giving it time to infiltrate the soil. In contrast, tilling up and down hills creates actual channels for water flow, contributing to runoff and erosion.

Strip-cropping slows runoff water by alternating bands of different crops across the slope (figure 8-8). One band may be a row crop that leaves most of the soil bare, such as corn or soybeans. The next band would be a close-growing crop (small grains), or a crop that completely covers the soil, such as hay. The close-growing strips slow the water, keeping it from achieving the speed it would on a continuous slope of corn.

Improving Water Intake Rate. A second way to decrease runoff is to improve the rate at which soil absorbs moisture. The soil's physical properties of texture, structure, and permeability set the infiltration rate. Chapter 4 lists ways to improve soil permeability and structure in the section on tilth. Because compacted soil tends to shed water, growers should be particularly careful to avoid severe compaction. The following practices also help soil absorb moisture.

Subsoiling, or deep plowing, shatters plowpans resulting from years of tillage, letting water seep deeper into the soil. At the same time, it allows roots access to deeper soil levels. Power requirements for subsoiling equipment makes this method expensive, especially if it must be repeated later.

Aeration is a term applied by turf managers to the process of punching or coring holes every few inches in turfgrass to break through surface compaction. Improving infiltration is one of several benefits (chapter 16).

Mulch, such as a layer of straw or woodchips on the soil, strikingly improves infiltration by eliminating crusting. Mulch protects the soil surface from the impact of raindrops. Studies show that mulched soil absorbs water two to four times better than bare soil. Gardeners or growers of high-value crops may mulch with straw, leaves, or grass clippings. Landscapers use gravel, woodchips, shredded barks, or other materials.

Conservation tillage, a fairly new method of working soil, preserves both soil and water. Conservation tillage does not bury crop residue; rather, it leaves much of it on the soil surface as a mulch (figure 8-9). Figure 8-10 shows

FIGURE 8-8 This farmer practices strip-cropping by planting alternating strips of corn and hay. Close-growing crops slow the downhill flow of water. *(Courtesy of USDA NRCS)*

FIGURE 8-9 In conservation tillage, the layer of crop debris left on the soil surface acts as a thin mulch. *(Courtesy of USDA NRCS)*

how leaving crop residues on the soil surface improved the infiltration rate in one soil. Conservation tillage may also improve a soil's physical condition.

Capturing Snowfall. In northern areas, capturing winter snowfall is an important way to retain natural precipitation. The key is to keep snow from blowing away from fields. Some growers leave strips of a tall crop standing at right angles to the prevailing winter wind. These **buffer strips,** which act as snow fences, should be about fifty feet apart (figure 8-11). In a North Dakota study, five-feet-tall wheatgrass planted in double rows forty-eight feet apart raised wheat yield from about sixteen to twenty-three bushels per acre. Some strawberry growers in cold climates plant such strips between strawberry rows to capture snow for cold protection.

Crop stubble left standing over winter also captures some snow (figure 1-3). Because the stubble also lowers runoff, more of the snowmelt water is then able to soak into the ground. This practice is called **stubble-mulching.**

Crop Residue (tons/acre)	Runoff (percent)	Infiltration (percent)
0	45	54
1/4	40	60
1/2	25	74
1	0.5	99
2	0.1	99
4	0	100

FIGURE 8-10 Effect of crop residue on water runoff and infiltration on a 5 percent slope. *(Source: Purdue University)*

Reducing Percolation. Water drains very quickly through coarse soils, causing high percolation losses, which are difficult to reduce. The most practical way to reduce percolation losses is to improve the water-holding capacity by maintaining the organic matter level. Since topsoil is usually the most moisture-retentive soil layer, reducing erosion is also important. In fact, the USDA estimates that the loss of moisture-holding capacity due to excess erosion costs American growers $2 billion annually.

The water-holding capacity of soil, especially of sandy soil, may also be improved by incorporating crystals of **hydrophilic gel polymers.** These new materials absorb many times their weight in water, swelling dramatically. Because these gels release water at low potential, moisture

FIGURE 8-11 Wheatgrass buffer strips on a Montana farm. These barriers capture snow that would otherwise blow off the field. *(Courtesy of USDA ARS)*

is retained until the soil dries enough for plants to need access to the water. Gels have been most often used in greenhouses to increase water retention in potted plants, but they have also been used in sand-based sports turf and to improve survival in transplanted tree seedlings. However, when they are buried, pressure from surrounding soil inhibits their swelling, and thus their ability to take up water.[1] High soil salt levels also reduce their effectiveness, so gels may not help in all situations.

Reducing Consumptive Use

Consumptive use is the total water "used" to produce a crop—including evaporation, transpiration, and water that becomes part of the plant. Consumptive use varies dramatically from place to place and crop to crop. It depends on several factors:

- Warm air temperature increases transpiration from leaves and evaporation from the soil.
- Relative humidity also affects evapotranspiration. Dry air increases water losses.
- Wind increases water losses. Normally, a film of humid air surrounds leaves and covers the soil surface. Wind strips away that film of humid air.
- The plant efficiency for using water depends on the health and spread of the root system, the species and variety, and nutritional status.

Reducing Evaporation. The factors that affect evapotranspiration are largely beyond the grower's control. However, there are some methods that can help reduce evaporation. Growers can lower evaporation from soil by covering the soil surface with either vegetation or mulch, shading the soil, and reducing wind velocity at the soil surface. To shade the ground with crops means to grow them so the crop canopy quickly covers the soil.

Mulches have the additional benefit of acting as a barrier to moisture movement. Loose, organic mulches like straw form only a partial barrier, and reduce but do not eliminate evaporation. Many growers of high-value crops, such as berries, nursery stock, or vegetables, mulch not only to preserve moisture but for weed control and other purposes (see chapter 6). For other growers, conservation tillage leaves crop residues as a mulch on the soil surface.

Reducing the number of tillage operations helps control evaporative losses. Each time the soil is worked, moist soil is dragged to the surface where it can dry out. Conservation tillage, use of herbicides for weed control, and combining tillage operations are ways to reduce tillage.

Reducing Transpiration. It is difficult to control transpiration in plants. Three weather conditions cause high transpiration: high temperature, low humidity, and wind. Frequent light sprinkling of plants with water improves conditions by wetting the leaves and raising humidity. This practice is called "crop cooling." Golf course and turf managers may lightly sprinkle turf to reduce heat stress during temperature extremes, a practice they call **syringing.** However, this method can hardly be called a moisture conservation measure because it uses moisture.

Transpiration may be reduced directly by coating leaves with a material that reduces water loss through stomata and leaf surfaces. These **antitranspirants** may be used on golf greens, to reduce transplant stress to landscape plants during planting, and in other special situations. They have been used to reduce drying of evergreen needles during the winter, minimizing winter injury. However, while reducing water loss through the stomata, antitranspirants also reduce movement of carbon dioxide into the leaf, inhibiting photosynthesis. It seems unlikely that antitranspirants would be useful in reducing water loss when active photosynthesis and growth are desired.

Another method of reducing transpiration is to plant windbreaks to cut the wind (figure 8-12). Orchardists often plant tree windbreaks. A few growers protect low-growing crops by planting rows of taller plants across the field. For example, a truck farmer may plant a few rows of sweet corn every fifty feet at right angles to the prevailing summer wind. Between these rows are planted cucumbers or another low-growing crop. The tall sweet corn rows create areas of less wind and so protect the cucumbers from the effects of wind.

Weeds also transpire moisture, so weed control is a basic moisture control measure. Weed control is a central part of summer fallow, a grain-farming technique in semiarid areas of the Great Plains. This will be covered in detail in chapter 16, but it involves leaving the soil

[1] *Volkmar and Chang. (1995).* Canadian Journal of Plant Science, *75:605–611.*

FIGURE 8-12 Windbreaks lower wind speed in the field. This reduces evapotranspiration and water stress. *(Courtesy of USDA NRCS)*

Crop	Transpiration Ratio	Weeds	Transpiration Ratio
Alfalfa	858	Lambsquarter	658
Field pea	747	Russian thistle	314
Oats	635	Pigweed	300
Potato	575	Purslane	281
Cotton	562		
Wheat	505		
Corn	372		
Millet	287		
Sorghum	271		

FIGURE 8-13 Transpiration ratios for several crops and weeds at Akron, Colorado. The ratios are the average over a five-year period.

bare every other year to store moisture for the following season. Weeds are controlled by cultivation or chemicals so they do not pull moisture out of the soil.

Improving Plant-Use Efficiency. This chapter has focused on how to increase the amount of water available for plants in the soil. Another way to conserve moisture is to help plants make better use of soil water. This allows greater production with the same amount of water.

The efficiency of plant water use can be measured by the transpiration ratio. This ratio is the amount of water transpired divided by the amount of dry matter produced. The ratio is affected by climate, the specific plant, and the soil conditions. Figure 8-13 lists sample transpiration ratios. The higher the ratio, the greater is the amount of water needed to produce a pound of dry matter. There is great variation between plants. For example, alfalfa needs three times as much water to produce a pound of material as sorghum.

How can growers improve transpiration efficiency? Windbreaks and crop cooling can help by lowering transpiration. One important method suitable for all growers is to improve the rooting zone of plants. Compaction and plowpans limit the soil volume that a plant can exploit, leading to an increase in the transpiration ratio. Therefore, the practices listed in chapter 3 for improving structure and controlling compaction are important techniques.

Good soil fertility also improves water use by plants. Adequate levels of nitrogen and phosphorus increase the size and depth of root systems and lower the transpiration ratio. For example, work in Texas showed that unfertilized (nitrogen) sorghum produced a 190-pound yield per inch of water used, while sorghum fertilized with 240 pounds/acre of nitrogen produced a 348-pound yield per inch of water used.

Some plants are well-adapted for growing under dry conditions, and the use of such plants can improve water use. New crops can be one answer, but breeders also search for more drought-tolerant varieties of established crops (figure 8-14). Breeders now can also use modern genetic techniques to engineer low-water using and drought-resistant crop varieties.

Urban dwellers can also replace humid climate grasses, shrubs, trees, and flowers with dryland adapted species. This technique, xeriscaping, is growing rapidly in cities of the American Southwest where landscape irrigation strains water resources (chapter 17). Some communities require xeriscaping in new construction.

FIGURE 8-14 Lima beans bred for drought resistance. These beans performed well when the corn in the foreground became shriveled and brown. *(Courtesy of USDA ARS)*

Water Quality

Water quality is as important as water quantity. Like conserving water, agriculture has a special role in the preservation of our nation's water supplies. Fish kills in trout streams of the author's home state, caused by runoff from insecticide-treated corn fields, are an example. So are the widely publicized problems with waterfowl in the Kesterton Reservoir outside Fresno, California. There, selenium-laden drainage water from irrigated fields found its way into the reservoir, causing death and deformity in waterfowl.

Agricultural sources of water pollution are difficult to pin down, compared to a pipe coming out of a factory or sewage plant. Such **point sources** of pollution are relatively easy to identify, and the nation has made good progress in bringing industrial and urban point sources under control. Most agriculture is a **nonpoint source,** like farm fields. An exception is large manure storage facilities and feedlots, which can be identified and are considered point sources. A number of studies have indicated that agriculture now accounts for the bulk of water-quality problems across the nation.

Current agricultural issues include seepage of fertilizers and pesticides into groundwater (figure 8-15), and pollution of surface waters. Contaminants of the latter

FIGURE 8-15 Testing groundwater in Maryland for nitrogen from farm fields. *(Courtesy of USDA ARS)*

include soil particles, pesticides and fertilizers, organic debris, and disease-causing organisms. Some of these problems are discussed in chapter 15, after fertilizers have been presented.

It has often been assumed that the soil matrix filters chemicals from percolating water. Too often, the assumption has turned out to be false: preferential flow can bypass the soil matrix and move pollutants deeper into the soil quickly. Another case is that of very coarse soils with a high water table, especially if irrigated. Examples are wells of such areas as central Wisconsin and Long Island, New York, that have been contaminated by the toxic pesticide aldicarb used on potatoes.

Parts of the United States with limestone bedrock may exhibit Karst topography, or land in which solution of limestone bedrock by water creates caves, sinkholes, and other channels from the surface. Agricultural chemicals traveling in runoff can flow directly into aquifers through these channels.

The ways to avoid water pollution are largely covered in various sections of this text. Here is a brief summary:

- Reduce runoff that can carry contaminants (earlier in this chapter).
- Reduce erosion that can carry contaminated soil (see chapter 18).
- Reduce fertilizer losses in percolating water (see chapter 15).
- Reduce the use of pesticides, especially soil-applied, by crop rotation and other strategies. Much progress is being made by Integrated Pest Management (IPM), the carefully planned use of many pest control methods such as biological or cultural ones.
- Apply and store manures properly (see chapter 15).
- Retain and restore wetlands that can filter pollutants from runoff (see chapter 9).

SUMMARY

Water conservation is important because usable water is not an unlimited resource. Our major water supplies are rainfall, fresh surface water, and groundwater. In many areas of the nation, rainfall is inadequate and stored water is being depleted. As the major user of water, agriculture has an obligation to make more efficient use of water and to conserve water supplies.

Each grower has an interest in conserving water as well. Making the best use of water increases yields. Further, many water conservation measures control erosion.

Many techniques for improving water-use efficiency have been outlined in this chapter. One technique is to capture more of the water that lands on a field. This can be done by terracing, contour tillage, or furrow-diking. Improving the soil infiltration rate by means of preserving soil structure, mulches, conservation tillage, and subsoiling also helps. In some areas, it is helpful to trap snow in the fields.

Consumptive use can be reduced by lowering the amount of evapotranspiration. Methods include reduced tillage, windbreaks, and weed control. A large, healthy root system, possible in soils of good physical condition and adequate fertility, helps the plant to use more of the soil moisture and reduces the transpiration ratio. The use of dryland adopted plants in agriculture and landscape can greatly reduce water use.

Growers and other soil users need to pay attention to their contributions to polluted water. Fertilizers, pesticides, eroded soil, and organic debris can harm environment and health.

REVIEW

1. Describe what paths water can follow after it falls to earth in precipitation.
2. Distinguish point and nonpoint sources of water pollution. What might be pollutants contributed by agriculture? Is agriculture usually a point or nonpoint source?
3. How does improving the infiltration rate improve water use? What are some methods for doing so?
4. Plant available water can be lost from the soil in ways that do not improve crop productivity. Describe these ways and how they might be reduced.
5. What benefits for growers, society, and the environment result from retaining more water on fields and reducing runoff?
6. Using information from chapter 7, explain why we cannot use the vast amount of seawater directly for irrigation purposes.
7. Go the U. S. Geological Survey Web site on water use in the United States at <http://water.usgs.gov/watuse>. Call up the maps on water use by state, and analyze your state's use of water for irrigation and other purposes. How does it compare with other states? Where does most of your water come from?

ENRICHMENT ACTIVITIES

1. Observe the effect of mulching by filling two deep jars with soil to within three inches of the top. Carefully moisten the soil in each jar with the same amount of water. Mulch the top of one jar, leave the other bare, and compare the drying. Weighing each jar before and after a period of drying may also provide interesting results.
2. The Environmental Protection Agency has a fact sheet on managing agricultural nonpoint sources at <http://www.epa.gov/owow/nps/facts>. Note particularly numbers 1 and 6.

3. Work on using genetic engineering to improve water-use efficiency can be found at <http://isotope.bti.cornell.edu>. Note the section on "Introduction to the Science." What does the heading to the main page say about the effects of water stress on global crop yields?
4. For a broader view of water use in the United States go back to the USGS main Web site on water use at <http://water.usgs.gov>.
5. To get an idea of drinking water shortages around the world, check out the website at <http://www.worldwater.org/drinkwat.gif>. What parts of the world have the greatest water shortages?

CHAPTER 9

Drainage and Irrigation

OBJECTIVES

After completing this chapter, you should be able to:

- define drainage and explain its importance
- explain the difference between wetlands and wet soils
- identify methods of artificial drainage
- identify methods of irrigation
- decide when and how much to irrigate
- name water quality problems for irrigation

TERMS TO KNOW

border-strip irrigation
capillary fringe
center-pivot irrigation
drainage
drip irrigation
furrow irrigation
hand-move irrigation
hydric soil
hydrology
leaching requirement
microirrigation
microspray irrigation
perched water tables
saline (soil)
seeps
solid-set irrigation
soluble salts
subirrigation
subsurface drainage
surface drainage
surface irrigation
traveling-gun irrigation
wetlands
wheel-move irrigation

Many acres of American land suffer from one of two moisture problems—either the soil is too wet or the soil is too dry. With proper treatment, however, some of these acres become productive cropland, popular golf courses, and even attractive wetlands.

The Importance of Drainage

Drainage refers to how rapidly excess water leaves soil by runoff or draining through the soil. The term also describes a condition of the soil—how much of the time soil is free of saturation.

In a well-drained soil, excess water leaves the root zone quickly enough that roots do not suffer lack of oxygen. Poorly drained, or "wet," soils remain waterlogged long enough to interfere with plant growth. Three conditions contribute to soil wetness. First, soils may be wet because they are naturally impermeable, like a compacted clay soil. Second, soil may be inundated by flooding or runoff from higher elevations.

The third and most common cause of soil wetness is a high water table—the upper surface of a zone of saturation in the soil. Commonly there are major, regional water tables at some depth. There may also be shallower, local water tables that form above an impermeable soil layer that restricts deeper percolation. These are called **perched water tables,** because they are perched above the regional water table (figure 9-1).

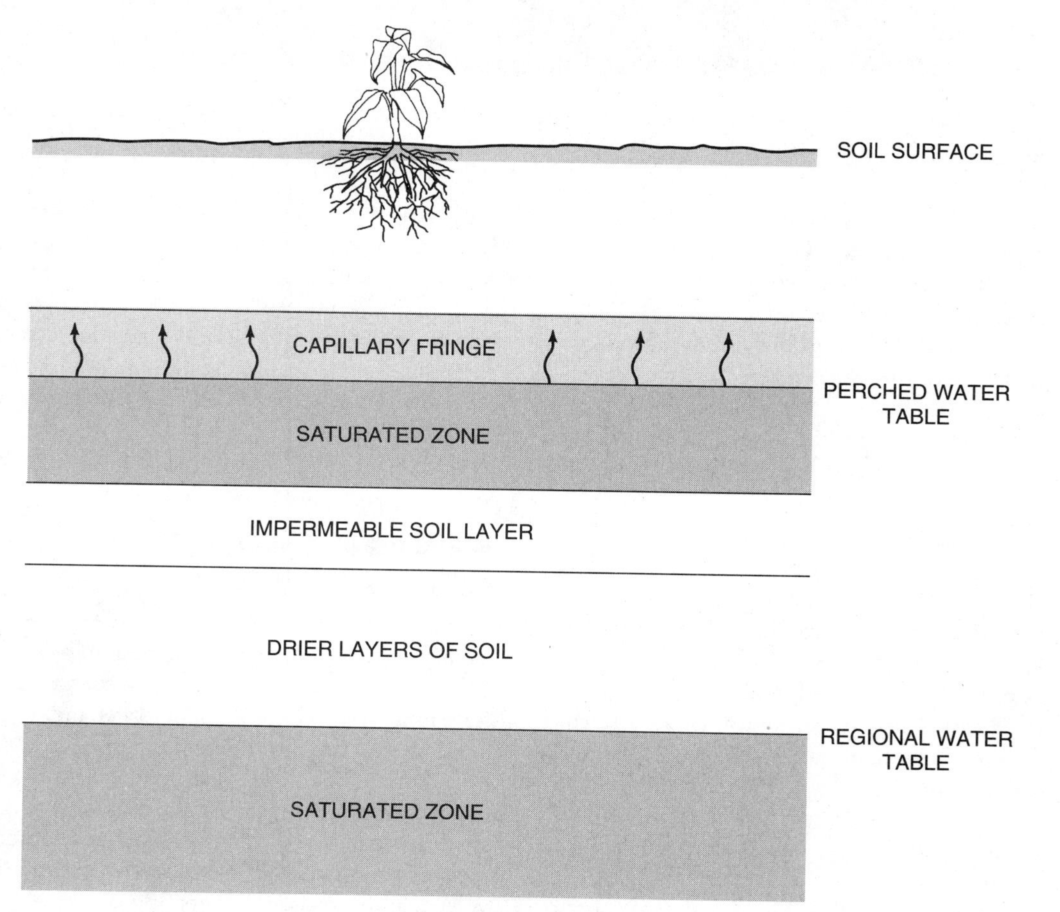

FIGURE 9-1 Water tables in the soil. Regional water tables could be at any depth. Sometimes an impermeable soil layer prevents percolation, creating a perched water table. Above the water table is a zone of wetness called the capillary fringe. Water tables near the surface restrict the rooting of plants.

FIGURE 9-2 Heavy rains on poorly drained soil have halted this haying operation in a by-gone year. *(Courtesy of USDA NRCS)*

Extending above the water table is a wet zone created by capillary rise called the **capillary fringe.** The height of the fringe varies from about six inches in sand to about eighteen inches in fine-textured soil.

Most drainage problems occur in depressions or large level areas that water cannot quickly exit by runoff. Parts of slopes may be wet where a groundwater layer intersects the land surface. Water leaks out into the soil at these locations, called **seeps.**

Effects of Poor Drainage. Where the water table or capillary fringe intrudes into the root zone, soil wetness can create anaerobic conditions that deprive roots of oxygen, as described in chapter 7. For most plants, rapidly growing points like root tips and root hairs are quickly damaged, and root-rotting organisms attack the roots. Water and nutrient absorption suffers, and toxic materials build up in the soil. Rooting will be limited to the aerated zone above the fringe. Thus, soil wetness limits the growth of upland crops such as corn and apple trees or most common landscape plants.

Poor drainage interferes with tillage and other farming operations. Poorly drained soil tends to stay wet later into the spring, delaying planting operations. Slow drying on poorly drained soils also keeps growers out of fields or nurseries after a rain (figure 9-2). Because wet soils warm slowly, they stay cold later into the season and delay planting and seed emergence. Draining these soils lengthens the growing season. In arid and semiarid areas, poor drainage can cause accumulation of salts in the root zone, leading to crop damage and increased susceptibility to erosion.

Wetland and Wet Soils

Wetlands. This text draws an important distinction between wet soils and wetlands. The latter are sufficiently valuable to the nation that their drainage is a questionable practice. As a consequence, federal and state programs offer protection to wetlands (chapter 20), and losses have slowed over the past decade. Nevertheless, the 1997 National Resource Inventory estimates a current average annual loss of almost thirty-three thousand wetland acres, about half to development and a quarter to agriculture. Because of the legal issues involved, and because of competing interests in the use of wetlands, the definition of wetlands is based on both scientific and political factors.

Wetlands are defined as areas that are flooded or saturated by surface or groundwater often enough and long enough during the growing season to support vegetation adapted for life in saturated soils. This condition is currently indicated by three major criteria:

- Wetlands possess **hydric soils,** defined as soils that are saturated, flooded, or ponded long enough during the growing season to develop anaerobic conditions in the upper part. More simply, the water table of a hydric soil is shallow enough that the capillary fringe extends to the surface, or is even wetter, for several weeks during the growing season. Gleying, mottling, and other indicative features in the upper part of the soil are indicators of a hydric soil.
- Wetlands exhibit wetland **hydrology.** That is, the amount, location, and behavior of water in the soil is such that the soil is saturated often and long enough during the growing season to warrant the designation of wetland. Long periods of standing water are not required to meet the definition nor to offer the following benefits.
- Wetlands host identifiable wetland vegetation. National lists of these plants are maintained for purposes of applying this criteria.

Details of applying these criteria are beyond the scope of this text, continue to change and develop, and are the subject of some controversy. A 1995 study by the National Academy of Science, *Wetlands: Characteristics and Boundaries,* concluded that more research needs to be done but supports the criteria devised by soil and wetland scientists.

WATER CONTROL. Wetlands play an important role in the water cycles of the United States. Many wetlands are sites for the recharge of groundwater. Wetlands also store water during snowmelt or heavy rains, reducing the amount of flooding by buffering rivers from an immediate influx of water. When wetlands are replaced by drained fields or suburban streets, large amounts of water enter rivers quickly, rather than being held in wetland reservoirs. Both large, permanent wetlands and small, seasonal wetlands contribute to flood reduction.

WATER QUALITY. Wetlands play a role in improving water quality by trapping and filtering nutrients, sediments, and pollutants. Marsh plants absorb nutrients and even toxic chemicals. Peat on the marsh floor also absorbs pollutants, and high bacterial populations denitrify and remove nitrogen contamination. Thus, wetlands help reduce nutrient and chemical runoff from farms and nurseries. Restored or wholly created wetlands are increasingly being used to cleanse runoff from farm fields, stormwater discharge into lakes, or even as an alternative to septic systems.

WILDLIFE HABITAT. Wetlands function as wildlife habitat. Twenty-six species of game or fur animals are associated with wetlands, especially waterfowl. Even small, seasonal, temporary wetlands are important to many species of waterfowl during spring migration. Hundreds of other nongame animals depend on wetlands for food, protective cover, or sites to raise young. Over 50 percent of threatened or endangered species are associated in some way with wetlands. In addition, wetlands provide spawning areas for many fish and shellfish, particularly coastal marshes that host commercially important species.

RECREATION AND EDUCATION. Wetlands also provide recreational and educational opportunities. Much hunting and fishing depends on wetland habitat, either directly or as nursery habitat for game or gamefish. Science classes often visit wetlands for study. Pollen imbedded in peat layers on the marsh floor are studied to recreate past climatic and vegetative conditions.

Although society benefits from wetland preservation, landowners bear the cost without directly realizing all its benefits. This makes it common to drain or fill wetlands for cropping or development. In many areas of the country, little of the original wetlands remain. Iowa has lost 95 percent, California and Nebraska 90 percent, and the lower Mississippi Valley 80 percent.

Wet Soils. Wet soils have fewer ecological functions compared with wetlands. Nearly 25 percent of all farmland would be too wet to support full production if it were not drained.

The most apparent sign of poor drainage is the presence of standing water several days after a rain. Water-loving plants are often found growing on wet soil. A percolation test serves to infer drainage. In the test, holes are dug and filled with water to wet the soil around the hole. After emptying, the holes are refilled with water, and the time it takes for drainage is noted.

Soil color can be another indication of poor drainage. Typically, poorly drained soils have a very black A horizon and a gray or mottled B horizon. Figure 4-20 presented a chart for determining soil drainage class by color. Soils classed as very poorly drained must be artificially drained for all upland crops, but generally should be preserved as wetlands. Poorly drained soil must be treated for most crops, and even somewhat poorly drained soils often require drainage for best results.

Before draining these soils, however, it is wise to have an expert consider the effect on the local water system. Drainage may lower the water table enough to leave local wetlands dry.

Artificial Drainage

Soils that stay saturated can be artificially drained. Artificially drained soils can be remarkably productive, as they are in much of the Corn Belt. Two types of drainage are available: surface and subsurface drainage.

Surface Drainage. **Surface drainage** systems collect and remove excess water from the soil surface. Surface drainage is best suited to three situations:

- collecting excess surface water on impermeable soil that cannot absorb it readily.
- channeling away water flooding fields from higher elevations.
- collecting irrigation water applied in excess of the soil's ability to absorb it.

Surface systems are inexpensive to install, because they are simply ditches dug through the field (figure 9-3). Water flows off the land into the ditch and is discharged off the field. Random ditches may be dug to drain a few depressions. In large, uniformly wet fields, a series of parallel ditches collect water from the whole area. Some grading may be done to improve the slope of the field to ensure that the water runs into the ditches.

FIGURE 9-3 A drainage ditch on this farm carries off excess water.

The primary function of surface systems is to remove surface water. However, they can also lower the water table in permeable soil if the ditches are dug deep enough to enter the water table. Water leaks into the drainage ditch from the soil. In most situations, however, a subsurface system is needed to remove subsurface water.

Surface drainage may also be improved by land grading. For example, playing fields are often crowned—elevated in the center, gently sloping to the edges—to move water more quickly off the field. Similarly, landscapes are also graded to avoid water problems.

Subsurface Drainage. **Subsurface drainage** collects water that has seeped into the soil and discharges it into surface ditches. A subsurface system consists of buried "pipes" into which water seeps and then flows to an outlet.

Subsurface drains are made of several materials. Common materials in the past were short drain tiles made of fired clay or concrete. The tiles are hollow cylinders one or two feet long. The contractor digs a trench and lays the tiles end-to-end (figure 9-4). The tiles are not sealed together, so water seeps into the line through the joints.

Today, perforated plastic pipe largely replaces tiles in subsurface drainage systems. A long continuous flexible plastic tube with holes (perforations) spaced along its length can be installed much more quickly than tiles (figure 9-5). Plastic is less expensive to install than tile and is especially useful in soils that may break up tile lines by shifting.

Like surface drainage, subsurface drainage design depends on the situation. It cannot be used at all if the soil is so impermeable that water will not flow into the lines. Lines may be installed randomly to drain isolated wet spots. On larger fields, a definite pattern is followed (figure 9-6). The spacing of the lines depends on how rapidly water moves through the soil, which depends on texture and structure. The depth at which the drainage lines are installed depends on the crop (deep-soil rooted crops need the deepest lines) and texture. Figure 9-7 gives a suggested depth and spacing for each of several soil textures.

FIGURE 9-4 Short concrete or clay drain tiles are laid end to end. Water seeps into the line through the joints and is carried off the field. *(Courtesy of USDA NRCS)*

FIGURE 9-5 Perforated plastic drainpipe can be installed more quickly and cheaply than clay or concrete drain tile.

FIGURE 9-6 Aerial view of a recently installed subsurface drainage system: 57,600 feet of tile were installed on 160 acres. *(Courtesy of USDA NRCS)*

Tile lines are also installed around foundations to avoid moisture problems in basements. Some of the most advanced playing fields also include tile lines to return fields to playability quickly after a rainfall.

Both drainage systems present the problem that they carry off whatever is contaminating the water, including pesticides, nutrients, salts, or other chemicals. The contaminants end up downstream where they may cause water quality and environmental problems. For instance, a large dead zone in the Gulf of Mexico has been attributed partially to drainage systems in the Mississippi River basin (chapter 15).

Soil Text	Depth (ft)	Spacing (ft)
Coarse	4.5	300
Medium	4.0	100
Fine	3.0	70
Organic	4.5	300

FIGURE 9-7 Guide to maximum depth and spacing for tile lines.

Irrigation Systems

Irrigation has a long history in world agriculture. We know that irrigation has been practiced for at least four thousand years. The success of civilizations along rivers such as the Nile River in Egypt and the Ganges River of India has been partially attributed to irrigation. In the United States, thousand-year-old irrigation canals can be found along the Gila River in Arizona.

Today, 11 percent, over 500 million acres, of the world's cropland is irrigated, especially in China, the United States, Mexico, and the mideastern nations. About 16 percent of American cropland is irrigated (figure 9-8).

In arid parts of the country water management is an important part of farming. Farmers either irrigate or use dryland farming techniques. Even in less arid regions, irrigation can be profitable on droughty soils or to relieve temporary rainfall shortages.

Irrigation water can be applied in a number of ways. Figure 9-9 lists nine systems and some of their traits. These nine systems can be divided into four categories: subsurface, surface, sprinkler, and microirrigation.

Subsurface Irrigation. **Subirrigation** is watering from below, using capillary rise from a deeper zone of saturated soil. The zone must be high enough that water can rise into the root zone but not so high as to saturate it. In some places, subirrigation occurs naturally. In other areas, pipes can be used in a manner opposite to drainage to produce an artificial water table. Subirrigation on cropland is useful only in special situations. It is, however, increasingly common as a way to water potted plants in greenhouses.

Surface Irrigation. **Surface irrigation** involves flooding the soil surface with water released from canals or piping systems. Surface irrigation is most suitable to

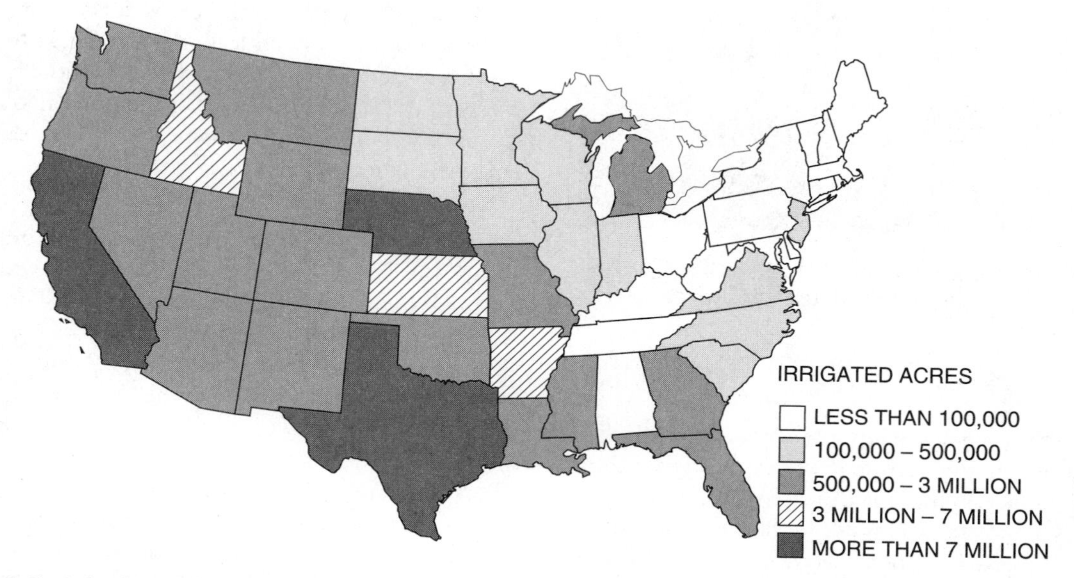

FIGURE 9-8 Irrigation in the contiguous 48 states. Sixteen percent of the nation's cropland is irrigated. *(Source: USDA, 1997 National Resources Inventory)*

Type	Cost to Install	Cost to Operate	Water-Use Efficiency	Flexibility
Subsurface	High	Low	Low	Very low
Surface				
border strip	High	Low	Low	Low
furrow	Medium	Low	Medium	Low
Sprinkler				
hand-move	Low	High	Medium	High
solid-set	High	Low	Medium	High
traveling-gun	Medium	Low	Low	High
center-pivot	Medium	Very low	Medium	Medium
wheel-move	Medium	Low	Medium	High
Microirrigation	Medium	Low	Very high	High

FIGURE 9-9 Several irrigation systems are compared. The flexibility rating is based on how easily the system can be adapted to different soil types and terrain.

FIGURE 9-10 A irrigation canal in the Rio Grande Valley of Texas. Surface irrigation is most used where regional networks of canals exist.

level or slightly sloping land of moderate permeability. In preparing land for surface irrigation, fields are carefully leveled to a slight slope. A system of canals (figure 9-10) uses gravity to carry water to the farm and among the fields. Surface irrigation is especially suited to areas where region-wide canal systems can be built, as in parts of the American West. After the water has reached the fields, it is distributed mainly by one of two ways—border strips or furrows.

BORDER-STRIP IRRIGATION. **Border-strip irrigation** is practiced by covering the entire soil surface of a field with a sheet of water. Each field is divided into smaller parts by low dikes. Each section is flooded in turn from a ditch or pipe running along the head of the field. Because of the large surface area of the water flooding the ground, evaporation causes some waste of water. Runoff and percolation can also be extensive.

FURROW IRRIGATION. **Furrow irrigation** (figure 9-11) distributes water through furrows, with crops planted on the ridge between two furrows. Furrows are suited to row crops like vegetables, corn, or cotton. Evaporation is less of a problem than in border strips because less surface is exposed to the air. Nearly half of all irrigated land in the United States is furrow irrigated.

Sprinkler Irrigation. Sprinkler systems pump water under pressure through pipes to sprinklers that spray water out in a circular pattern (figure 9-12). Sprinkler irrigation can be used where it is not possible to surface irrigate. For example, soil that is too permeable or too impermeable for surface irrigation can be sprinkled. Ground that is not level can be sprinkled without leveling. However, wind can disrupt the sprinkling pattern so irrigation in windy weather may not be uniform.

Sprinkler irrigation equipment can be used for other purposes in addition to watering crops. Some growers apply agricultural chemicals by sprinklers, a technique termed "chemigation," or "fertigation" for fertilizers. Sprinklers can also substitute for rainfall to activate herbicides. One sprinkler system can be used for frost control, which is described later.

FIGURE 9-11 A furrow-irrigated field in Arizona. Here, water is siphoned into the furrows from a canal at the head of the field. *(Source: USDA)*

HAND-MOVE IRRIGATION. **Hand-move irrigation** is the least expensive sprinkler system to install. The system consists of lightweight aluminum pipe that can be easily moved from place to place by a single person (figure 9-13). This system is labor-intensive and the operating cost is high. Hand-move systems are most suitable where irrigation is used to supplement natural rainfall rather than to replace it.

SOLID-SET IRRIGATION. **Solid-set irrigation** uses the same equipment as hand-move setups, except that an entire field is set up at planting. The pipes remain in place until harvest. The larger number of pipes needed to supply all fields increases initial equipment costs but eliminates additional labor during the growing season. Solid-set systems are also commonly installed for both irrigation and for frost protection in high-value crops like strawberries. Turning on sprinklers when frost threatens keeps the crop wet, preventing leaf and flower temperature from dropping below freezing.

FIGURE 9-12 Sprinkler irrigation in a California field. Compare to the layout drawn in figure 9-13.

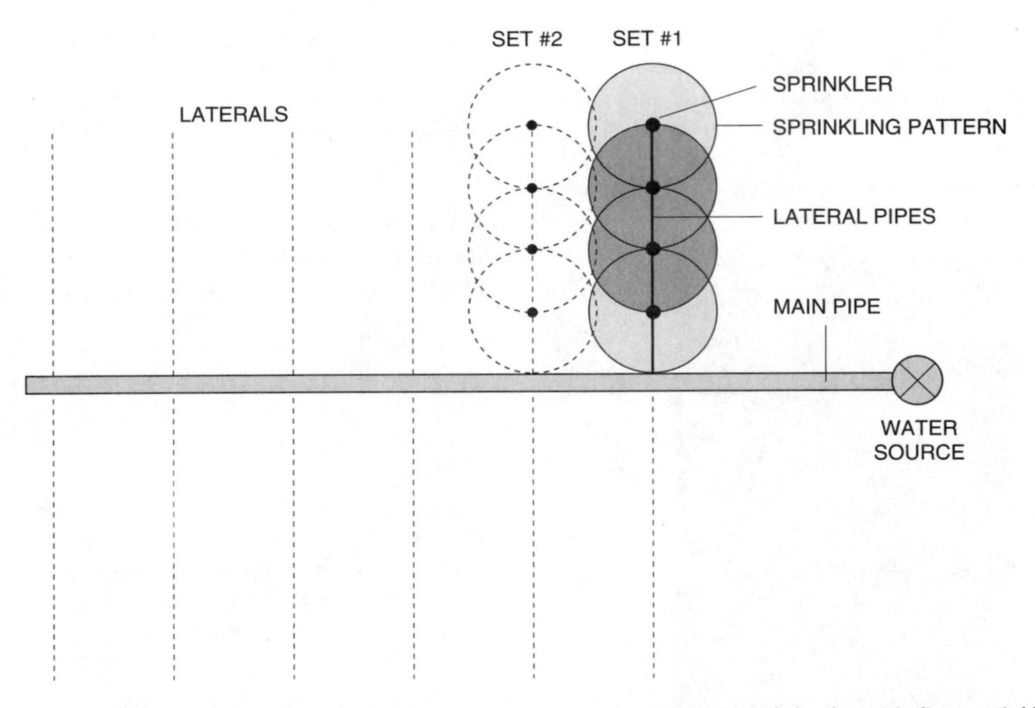

FIGURE 9-13 In a hand-move system, the lateral is moved from one set to the next along the main line. Sprinklers are spaced so their patterns overlap. Solid-set is similar, except all the laterals are set up at once and remain in place.

Turf irrigation systems in lawns, golf courses, or sports fields may also be viewed as solid-set sprinkler systems. In turf systems, pipes are buried and pop-up sprinklers are installed at soil level. Pop-up sprinklers remain out of the way of equipment or people using the turf until water pressure "pops" them up.

TRAVELING-GUN IRRIGATION. **Traveling-gun irrigation** uses one very large sprinkler mounted on a trailer that moves across a field (figure 9-14). The gun sends out a single large stream of water that covers several acres. The gun is very susceptible to wind problems. Because it has a very large nozzle, it has also been used to spray liquid manure and other slurries, a subject covered in chapter 14.

CENTER-PIVOT IRRIGATION. **Center-pivot irrigation** systems are anchored to a center pivot point. The watering line is elevated above the crop by towers mounted on wheels (figure 9-15). As the system operates, the line slowly turns around the pivot point. By the time the circle is complete—60 to 120 hours later—as many as 160 acres have been watered (figure 9-16). Center-pivot systems have the lowest labor requirement of any irrigation method and so are very popular. Their use has spread rapidly in recent years.

WHEEL-MOVE IRRIGATION. **Wheel-move irrigation** consists of a line of sprinklers mounted on wheels. In operation, the line of sprinklers slowly rolls down the field until it reaches the end of its hose. The pattern of wetting for wheel-move irrigation is rectangular rather than the circular pattern of center-pivot irrigation. Some growers prefer the wheel-move system because it irrigates all parts of the field and does not leave unirrigated corners as does center-pivot irrigation.

Microirrigation.

Microirrigation includes two irrigation technologies designed to make more efficient

FIGURE 9-14 Traveling gun irrigation in a prairie plant nursery in Wisconsin.

FIGURE 9-15 A modern center-pivot system in Colorado. *(Courtesy of USDA)*

FIGURE 9-16 The half-mile diameter fields resulting from center-pivot systems are visible from the air in many parts of the world. *(Courtesy of USDA NRCS)*

use of water, improve productivity, and reduce environmental problems associated with standard irrigation methods. Unlike other methods, microirrigation delivers water to small, localized zones where the bulk of the roots are. The two techniques, drip (or trickle) and microspray, deliver water at slow rates right at the crop's root zone with little evaporation or runoff and without watering surrounding soil.

DRIP IRRIGATION. **Drip irrigation** was pioneered in Israel, where water conservation is critical. A trickle system is made of 1/2-inch to 3/4-inch flexible plastic tubing running down a crop row, or buried under the row, with special "emitters" spaced along the pipe (figures 9-17 and 7-10). The soil itself conveys the water to the root system. Water drips out at a single point and wets a soil volume based primarily on capillary movement. Therefore, the wetting pattern and volume wetted depends on soil texture, and so must the spacing of the tricklers. On coarse, sandy soils, tricklers must be spaced very close together. Figure 9-18 shows how trickle irrigation can be designed to wet only the part of the soil occupied by crop roots.

Drip irrigation is also widely used by greenhouse and nursery growers of potted plants. In this application, the emitter is on the end of a long "spaghetti" tube lodged in the pot.

Drip irrigation has several benefits. The system operates at low water volume and pressure so limited water sources and small pumps and pipes can be used. The system is nearly 100 percent water efficient compared with an efficiency of 50 percent to 75 percent for other methods of irrigation. Generally, less water is needed in drip irrigation to produce a better yield compared with surface or sprinkler irrigation.

FIGURE 9-17 Drip irrigation on greenhouse roses. This emitter drips water at carefully controlled slow rates. Water losses to percolation and evaporation are almost eliminated.

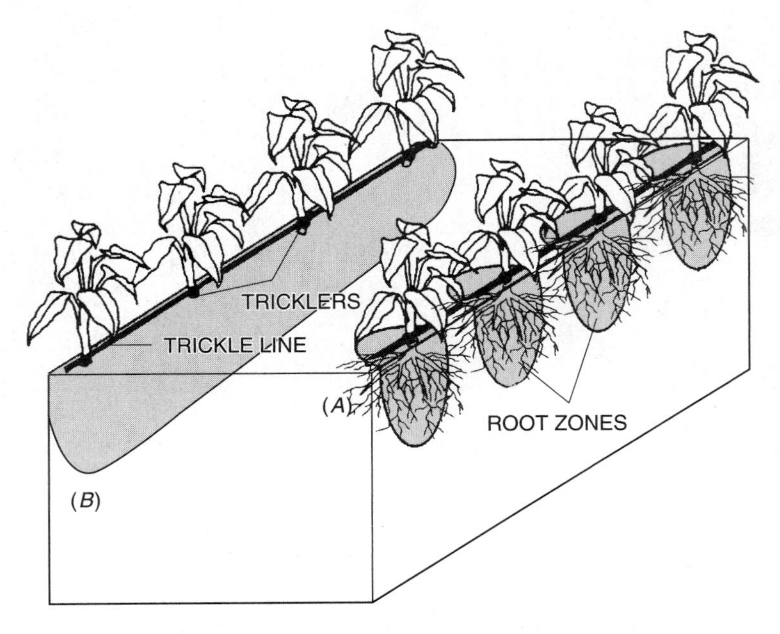

FIGURE 9-18 This figure shows how drip systems water plants. Emitters can be spaced to water individual plants (*A*), or spaced more closely to water a solid strip (*B*). Nearly ideal moisture levels are maintained in part of the root zone, while nearby soil remains unwettered.

FIGURE 9-19 Microspray irrigation watering a landscape tree in Nevada. *(Courtesy of USDA)*

MICROSPRAY IRRIGATION. In **microspray irrigation** small sprayheads replace tricklers (figure 9-19). Like regular sprinklers, water is delivered through the air, but these are located near the ground in spray patterns only a few feet across. Like drippers, they wet only localized areas where roots are most concentrated, and so save water. Application rates and pressure needed, however, are somewhat higher, so larger piping and a larger water supply may be needed to irrigate the same area. Because they wet a larger area, microsprays are particularly useful for irrigating large trees, as in orchards, beds in landscape areas, or in coarse, sandy soils. Their use should be avoided under windy conditions, because wind badly distorts the application pattern and reduces efficiency.

Using Irrigation

The ideal irrigation brings just the soil in the crop's root zone to field capacity. Irrigation engineers—and growers who use the systems they design—decide how much water must be added to reach that moisture level. They must also decide how dry the soil should be before irrigating.

Irrigation practices vary in different parts of the country. In general, irrigation begins when 50 percent to 60 percent of the available water has been used by plants. An exception to this general rule is drip irrigation, which is designed to maintain field capacity.

Deciding When to Irrigate. The moisture level of a soil can be judged using a feel test similar to ribbon testing. A soil sample is taken several inches below the surface. The feel of the soil is then compared with the chart in figure 9-20. Irrigation should begin when the soil reaches the "fair" moisture level. While experience helps, feel tests are less reliable and more work than measuring devices.

Potentiometers (tensiometers) and resistance blocks work very well for deciding when to irrigate. Resistance blocks work over the widest moisture range, but potentiometers are more accurate within a narrower range. Because potentiometers work best within the wet to moderately dry range (0.0 and −0.8 bar), they are best suited to a fairly moist irrigation regime. Resistance blocks are needed where irrigation is infrequent. An irrigation system can be automated by adding a controller to the potentiometer that triggers irrigation when the soil is too dry.

Degree of Moisture	Feel	Amount of Available Moisture
Dry	Powdery dry	0
Low	Crumbly, will not hold together	<25%
Fair	Somewhat crumbly, will hold together	25%–50%
Good	Forms ball, sticks together slightly with pressure	50%–75%
Excellent	Forms pliable ball	75%–100%
Too wet	Can squeeze out water	Over field capacity

FIGURE 9-20 Moisture feel chart. Check the soil below the soil surface. Begin irrigation at the "fair" level. The amount of moisture is expressed as a percentage of the remaining available water.

Some irrigators schedule irrigation by tracking consumptive use. One method measures water lost from a surface of free water, called an evaporation pan. This loss can be correlated to evapotranspiration rates by charts. Knowing how much water a soil holds, and how much has been lost, irrigation can be scheduled.

Modern technology is developing ways to schedule irrigation that greatly increase water-use efficiency. These methods measure actual water stress in the plants, not soil-moisture conditions. For example, devices may be attached to leaves to measure closing of the stomata during moisture stress. Leaf temperature may also be a measure of water stress because reduced transpiration also reduces cooling. A hand-held infrared temperature-sensing device can be pointed at turf or other vegetation to measure water stress.

How Much to Water. It is important to add the right amount of water when irrigating. Too much water causes excess percolation and runoff, resulting in leaching of nutrients, wasted water, possible pollution, and even erosion. Too little water fails to bring soil to the best moisture level. Two factors affect the amount of water to be applied during irrigation: soil texture and rooting depth.

Figure 9-21 is a simplified table showing how much available water soils of different textures hold at field capacity; figure 9-22 gives examples of the average rooting depth of several crops. As an example of how to calculate the amount of water to be applied in one irrigation, let's consider soybeans in a sandy loam soil.

1. A medium coarse soil holds 1.2 inches/foot of water.
2. Soybeans root to a 2-foot depth.
3. Total water available to soybeans = 1.2 inches/foot × 2 feet = 2.4 inches water.
4. Irrigation is turned on when 50 percent of the available water is gone. Thus, 50 percent of the water is to be replaced. Total added = 2.4 inches × 0.5 = 1.2 inches per irrigation.

This calculation tells a grower that if watering is started when the water is 50 percent gone, 1.2 inches of

Soil Texture	Available Water per Foot of Soil
Coarse	0.3–1.1 inches
Medium coarse	1.1–1.8 inches
Medium	2.0–2.9 inches
Medium fine	1.8–2.6 inches
Fine	1.2–2.0 inches

FIGURE 9-21 Available water held in soils of different textures at field capacity. This information is used to design irrigation systems and to calculate how much water is to be applied during irrigation.

Average Rooting Depth (ft)			
1	**2**	**3**	**4**
Beans	Beets	Alfalfa	Nuts
Cabbage	Cane berries	Cotton	Tree fruits
Carrots	Corn	Grapes	Shade trees
Cucumbers	Grains		
Onions	Melons		
Peanuts	Peas		
Strawberries	Potatoes		
Tomatoes	Sweet potato		
Turfgrass	Soybeans		

FIGURE 9-22 Average rooting depth for a variety of crops.

water will bring the root zone of soybeans in medium coarse soil to field capacity. If the rooting zone contains more than one soil layer, each should be calculated separately. Precise information on any given soil is available in soil surveys (chapter 3).

Saving Water. Saving water is an increasingly important task for the grower. In irrigation systems using a pump, saving water also means saving energy. It also helps avoid water pollution from unused irrigation water flowing into streams or seeping underground. The following points are ways irrigation water can be saved:

- Use the most water-efficient system that is practical. Where feasible, trickle irrigation uses the least amount of water.
- In surface systems, level the land carefully. The system should be designed to reuse excess tailwater.
- Make sure that all systems are designed correctly to fit the crops, soil, and terrain. For example, the application rate of a sprinkler system should be no greater than the infiltration rate of the soil. Maintain all systems for efficiency.
- Water should be transported through sealed ditches to avoid seepage, or through pipes, which also stops evaporation.
- All systems should contain devices to measure and control the water flow.
- Use the amount of irrigation water that gives the best return. Using less than the ideal amount may cause some yield loss, but it results in a savings in water. Researchers are developing models for determining the most efficient amounts of irrigation water to be applied.
- Schedule irrigation on actual crop needs, as noted previously, not on a time schedule.
- Stay informed of new developments. For example, in surface irrigation, a fairly new method is surge-irrigation. In this method, water is applied very heavily at brief intervals. In this way, more of the water reaches the end of the furrow before it seeps into the soil near the furrow heads. Many newer center-pivot systems designed with small sprinklers hanging just above crop level lose much less evaporative water.
- Use computers to automate irrigation systems and to make decisions about what crops need to be irrigated when.

Water Quality

Both groundwater and fresh surface water are used for irrigation. The choice depends upon the type of irrigation system used and on the water source that is most practical locally. When obtaining water, the first consideration is the legal availability of the water. Most states have laws controlling access to water, such as water-use permits. Growers using water from federal water projects must also meet federal regulations. The second consideration is the quality of the irrigation water. Water may be contaminated by suspended solids, boron, or soluble salts.

Suspended Solids. Suspended solids are small bits of solid material floating in the water. Groundwater may contain grains of sand or silt. Surface water often has bits of organic matter or small aquatic organisms like algae. Most irrigation systems are not bothered by small amounts of solids, but drip systems can be clogged. All drip systems should include filters to remove suspended solids.

Boron. Tiny amounts of boron are needed for plant growth but slightly larger amounts can be toxic to plants. Some irrigation water has an excessively high boron level, especially for sensitive plants. Most fruits and nuts are sensitive to boron levels, while some crops, such as alfalfa and sugar beet, are relatively tolerant.

Soluble Salts. The most widespread water quality problem is the presence of **soluble salts.** Soluble salts are compounds of sodium, calcium, and magnesium that dissolve in water. These compounds are found in various levels in soil and water. The problems of soluble salts will be examined in detail in chapter 11, but their effect on irrigation will be surveyed briefly here.

When irrigation water evaporates from the soil surface or is removed by plants, salts will be left in the soil. Over time, irrigated fields may accumulate high levels of salts and become **saline.** The problem is most common in the western United States. In the East, because of high amounts of natural rainfall, enough water percolates through the soil to leach salts out of the root zone. Soluble salt problems are especially acute for growers of potted plants because of the large amount of watering.

Soluble salts cause two problems. First, salts cause an increase in the osmotic potential of the soil, causing the plant to work harder to absorb water. Second, one of the cations, sodium, tends to break down soil aggregates. As a result, the soil surface is sealed and crusts form. Relatively salt-tolerant crops include barley, sugar beets, and cotton. Most vegetables, fruits, and alfalfa do not tolerate a saline soil.

Salinity is most severe where land is heavily irrigated with water containing fairly high salt levels. Irrigation water can be classified by salt and sodium hazard levels, as shown in figures 9-23 and 9-24. The units in the table are explained in chapter 11. One answer to the problem of salt buildup from irrigation is to use water low in salts. However, as the demand for water increases, irrigators are usually forced to use increasingly salty water. Thus, growers must learn how to manage salty water.

The key to using salty water is to overirrigate so excess water leaches salts below the root zone. However, if the soil is impermeable or there is a high water table, salts return to the root zone by capillary rise. Drain tiles may be needed to carry salty water off the field.

In irrigating saline land, one must balance the salt coming into the field with the salt going out to avoid a buildup. The saltier the irrigation water, the more excess water must be applied. This excess is called the **leaching requirement.** It is the amount of water to be applied in excess of that needed to wet the root zone of the plant.

Salinity Class/Hazard	Conductivity (Micromhos/cm)	Description
I Low	100–250	Suitable for most crops, little leaching needed
II Medium	250–750	Moderate salt-tolerant crops or moderate leaching needed
III High	750–2,250	Plants with good salt tolerance on drained soil with salinity control
IV Very high	>2,250	Not suitable for irrigation except for occasional use under high salinity control

FIGURE 9-23 Irrigation water salinity classes. *(Source: USDA, Agriculture Handbook #60, 1954)*

Sodium Class/Hazard	Sodium Adsorption Ratio	Description
I Low	0–10	Suitable for irrigation except for crops very sensitive to sodium
II Medium	10–18	Suitable for coarse-textured or organic soils with good drainage
III High	18–26	Soil will need treatment for sodium, or water must be treated to remove sodium
IV Very high	>26	Generally not suitable for irrigation

FIGURE 9-24 Irrigation water classes for sodium hazard. *(Source: Agriculture Handbook #60, 1954)*

SUMMARY

Artificial drainage allows a grower to make a productive field out of soil that is too wet to grow crops. In addition, drainage effectively prolongs the growing season by allowing earlier planting and better growth. While naturally wet soils are candidates for drainage, wetlands carry some legal protection as an important natural resource. Benefits of wetlands include improvement of water quality, reduction of flooding, and retention of wildlife habitat.

Poorly drained soil is deficient in oxygen. The conditions can be indicated by the presence of standing water or subsoil color. Surface drainage carries excess surface water off the field by means of ditches. Subsurface drainage moves excess underground water from the soil through buried drainage lines.

Irrigation is primarily used to supply some or all of the water needs of crop plants. Subsurface irrigation uses capillary rise from a natural or artificial water table to water plants. This method is not widely used in the United States. Surface irrigation floods a field through border strips or furrows. Sprinkler irrigation sprays water over the soil surface through systems like center pivots and solid set. Microirrigation drips water on the soil near crop plants. It is the most efficient system in terms of water use.

A goal of irrigation is to avoid water stress on the plant. The need for irrigation can be judged by feeling the soil, by using tensiometers or resistance blocks to measure soil moisture levels, or by more exacting budgeting methods. By knowing the soil type and rooting depth of the crop, growers can calculate how much water is required.

Water quality is a concern to all irrigators. Some water contains too much boron or suspended solids. Soluble salts are the more common problem, especially in the western United States. Growers can manage salinity by making sure drainage is adequate and by overirrigating to leach salts out of the root zone of crops.

As water supplies dwindle, growers are becoming more concerned about their water use. Over time, irrigation must continue to become more efficient. In addition, more efficient and cost-effective ways must be developed to solve salinity problems.

REVIEW

1. Assume we wish to irrigate turf to a depth of eight inches when 60 percent of available water is gone. Assume also that the root depth is a uniform silt loam texture with an available water holding capacity of 2.5 inches per foot. How much water will we add each time?
2. Describe benefits and drawbacks of artificial drainage.
3. How might topography affect the choice of irrigation systems?
4. Explain why poor drainage harms most crops.
5. Describe why landowners might decide to restore wetlands on their property.
6. Why might an irrigated field need to also have a subsurface drainage system? For what part of the United States might this be most likely?
7. What is a water table and a capillary fringe? How close would a water table have to be to the soil surface to meet the criteria for a wetland?
8. Some garden writers suggest loosening heavy clay soils with organic matter to relieve drainage problems. Under what situations might this help and when would it not?

ENRICHMENT ACTIVITIES

1. Observe the effects of salinity by watering previously established potted tomato plants with saline solutions. Use common table salt in water to create solutions of varying concentrations. Grow some control plants with untreated water, and compare differences in growth over time.
2. Obtain and use tensiometers or resistance blocks.
3. For an interesting historical perspective on wetland drainage, read "The Problem: Drained Areas and Wildlife Habitats" by F. Kenney and W. M'Atee in the 1938 Yearbook of Agriculture *Soils and Men.*
4. Using this EPA Web site, get some picture of the amount of wetlands in your state compared to other states and the percentage loss due to drainage or development. <http://www.epa.gov/owow/wetlands/vital/toc.html>.
5. Here are more details on drainage systems:

   <http://www.extension.umn.edu/distribution/cropsystems/DC7685.html>.

   Find out what the basic system designs are: <http://ohioline.osu.edu/B871/b871_2.html>.

   There is a sequence of pages here about drainage, agriculture, and the environment.
6. Here are more details on center pivot irrigation systems:

   <http://www.ext.colostate.edu/pubs/crops/04704.html>.
7. For photos and descriptions of drippers, microspray heads, or many other components of irrigation systems, numerous commercial sites may be found on the Internet by use of your favorite browser.

CHAPTER 10

Soil Fertility

OBJECTIVES

After completing this chapter, you should be able to:

- name and classify the essential elements
- list four sources of nutrients in the soil
- describe soil colloids
- define cation exchange capacity and related terms
- describe how plants absorb nutrients
- explain other soil fertility factors

TERMS TO KNOW

adsorption
anion exchange
base saturation
beneficial elements
cation exchange
cation exchange capacity
colloid
diffusion
essential elements
exchangeable bases
expanding clays
isomorphous substitution
luxury consumption
macronutrients
mass action
mass flow
micelle
micronutrient
nonexchangeable ions
oxide clays
primary macronutrient
root interception
secondary macronutrient
sesquioxide
silicate clay
soil fertility
soil solution
trace element

Soil fertility is the ability of soil to supply nutrients for plant growth. The soil is a storehouse of plant nutrients, stored in many forms, some very available to plants, some less so. The concept of soil fertility includes not only the quantity of nutrients a soil contains, but also how well they are protected from leaching, how available they are, and how easily roots function. A beginning point for discussing soil fertility is to define the term "plant nutrient."

Plant Nutrients

Plant nutrients are the **essential elements** needed for plant growth. Plants absorb at least ninety different elements; only a few are needed for plant growth. Some are not needed by plants, but, like cobalt, are needed by the animals that eat plants. Many elements are not needed by either plants or animals, and others, like lead, are even toxic. Thus, plants contain many elements not needed for growth. Which elements are essential? The most commonly accepted rules for determining if an element is essential are as follows:

1. A lack of the element stops a plant from completing growth or reproduction.
2. The element is directly involved in plant nutrition, not merely "taking up space" in plant tissues.
3. A shortage of the element can be corrected only by supplying that element.

Based on these rules seventeen, essential elements are identified by most scientists (figure 10-1). (Several others play a role in the nutrition of some plants but cannot yet

Name	Symbol	Ionic Form	Ion Name
Carbon	C	—	—
Hydrogen	H	H+—(not used by plants in this form)	
Oxygen	O	—	—
Primary Macronutrients			
Nitrogen	N	NO_3^-, NH_4^+	Nitrate, ammonium
Phosphorus	P	HPO_4^{-2}, $H_2PO_4^-$	Orthophosphates
Potassium	K	K^+	—
Secondary Macronutrients			
Calcium	Ca	Ca^{+2}	—
Magnesium	Mg	Mg^{+2}	—
Sulfur	S	SO_4^{-2}	Sulfate
Micronutrients			
Boron	B	$B(OH)_3$, $B(OH_4)^-$	boric acid, borate
Copper	Cu	Cu^{+2}	Cuprous
Chlorine	Cl	Cl^-	Chloride
Iron	Fe	Fe^{+2}	Ferrous
Manganese	Mn	Mn^{+2}	Manganous
Molybdenum	Mo	MoO_4^{-2}	Molybdate
Nickel	Ni	Ni^{+2}	
Zinc	Zn	Zn^{+2}	
Others			
Sodium	Na	Na+	
Silicon	Si	SiO_3^{-2}	
Cobalt	Co	Co^{+2}	

FIGURE 10-1 Essential elements and their ionic forms. Nickel has been tentatively identified as an essential element, while Na, Si, and Co are beneficial elements.

be considered true essential elements for all plants. These are identified as "other" in figure 10-1.) Of the seventeen, three account for 95 percent of all plant needs: carbon, oxygen, and hydrogen from air and water. The other fourteen mineral nutrients are obtained from the soil. It is these fourteen elements that are discussed in this chapter.

Plants use six of the fourteen mineral elements in large amounts. These six **macronutrients** are nitrogen, phosphorus, potassium, calcium, magnesium, and sulfur. In the following relationship, the six nutrients are listed in decreasing order from the greatest amount used by most plants (nitrogen) to the smallest amount (sulfur).

$$N \geq K > Ca > Mg \geq P > S$$

Soils are less likely to be deficient in calcium, magnesium, and sulfur than the other three nutrients, and since most soils supply enough calcium, magnesium, and sulfur, soil scientists call them **secondary macronutrients,** or simply, secondary nutrients. The **primary macronutrients,** sometimes called fertilizer elements, are not usually available in large enough amounts for best growth. The three primary nutrients—nitrogen, phosphorus, and potassium—are most often added to soil by fertilization. Note the division into primary and secondary macronutrients is not based on the relative amounts used by plants, but on their importance as fertilizers.

The six macronutrients, except for potassium, are part of the materials that make up the bulk of the plant. Protein, for instance, includes both nitrogen and sulfur. Living tissue also contains very small amounts of certain important chemicals that control life processes, such as enzymes. The other eight essential elements form part of these key materials in plants.

The other eight nutrients, listed in figure 10-1, are labeled **micronutrients** or **trace elements,** because they are used in small amounts. Iron, for example, plays a role in the process that forms chlorophyll. Only a small amount of iron is needed, but too little iron means that chlorophyll fails to form (figure 10-2). The term "micronutrient" does

FIGURE 10-2 A healthy pin oak leaf is shown behind one that exhibits a lack of iron. Iron is a trace element involved in chlorophyll formation. In the iron-deficient leaf in the foreground, little of the dark green chlorophyll forms, resulting in a yellow leaf.

FIGURE 10-3 Micronutrient deficiencies. The plant on the right lacks micronutrients, while the one on the left enjoys complete nutrition. Sometimes micronutrients are called "minor" elements—but obviously a shortage is not minor in its effects.

not, however, mean the elements are unimportant (figure 10-3). Plants will not grow normally without enough of these trace elements.

Figure 10-1 also lists three examples of what might be called **beneficial elements.** These are elements that do not meet the requirements for being essential for all plants, but that are helpful for growth or are needed by some plants. For instance, few plants require silicon to complete growth and reproduction, but its presence strengthens cell walls and reduces insect and disease problems.

The fourteen "micro" and "macro" elements are furnished by soil. Plants absorb these elements in a specific way: as the ions listed in figure 10-1.

Nutrient Ions. Ions, as explained in appendix 1, are charged atoms or molecules. The charge, either positive (a cation) or negative (an anion), results when there is a difference between the number of electrons and protons. One way ions form in soil is when compounds dissolve in water. For example, when the soluble fertilizer potassium nitrate dissolves, the molecule breaks into two ions:

$$KNO_3 \xrightarrow{\text{solution}} K^+ + NO_3^-$$

The concept of nutrients as ions is important. Plant roots *absorb* nutrient ions; soil particles *adsorb* them. Absorb means to take in something, like a sponge absorbing water. Adsorb means to attract a thin layer of molecules to a surface, where they stick. Figure 10-1 lists each nutrient in the ionic form(s) most commonly absorbed by plants. Any special name for the ion is also listed.

Sources of Elements in Soil

Nutrient elements are present in the soil in four forms, as shown in figure 10-4. Together, these four sources perform two functions: to store nutrients and to make them available to plants. We can compare soil to a bank and nutrients to the money deposited there. Money in checking accounts can be obtained simply by writing checks, so checking accounts are a short-term form of money storage with readily available cash. One can also buy savings certificates but can only cash them after some time has passed. Thus, they are a long-term form of money storage, and the cash is not readily available.

The four sources are as follows:

1. Soil minerals. Minerals are the major source of all soil-supplied nutrients except nitrogen. Soil minerals are the longest-term storage. Weathering frees the elements slowly over time, dissolving the minerals into ions. Figure 2-5 lists the nutrients contained in several rocks. By minerals we include here

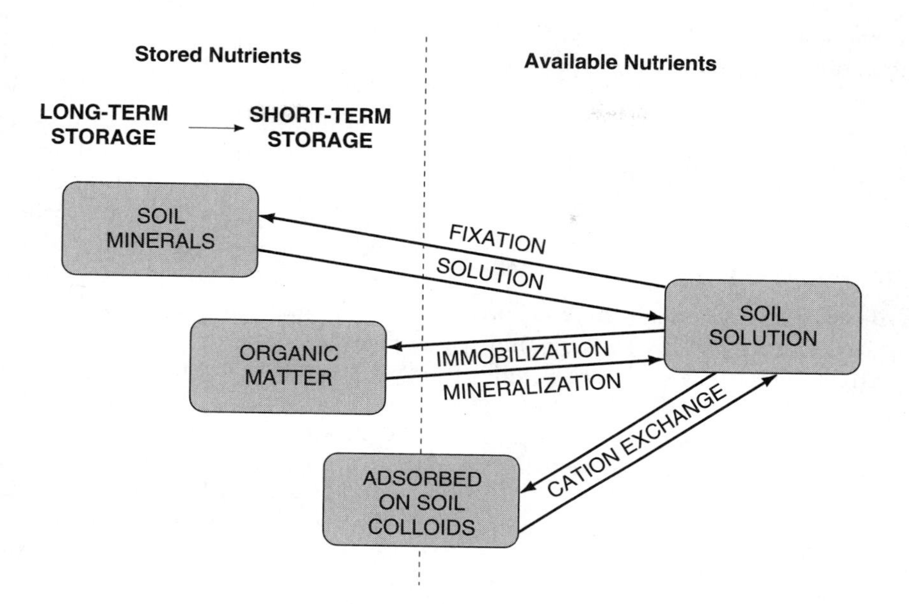

FIGURE 10-4 There are four pools of plant nutrients in the soil. The sources of nutrients that are least available are shown on the left. The sources of nutrients in their most available form are shown on the right. The elements can change forms. When crops remove nutrients, stored nutrients go into solution and become available to plants.

the products of reactions of soil ions that remove them from the soil solution (see enrichment activity 6).

2. **Organic matter.** This supplies large amounts of several elements like nitrogen and the nutrient anions listed in figure 10-1. Organic matter is an intermediate form of storage, since elements are freed for plant use by decay. Some nutrients in fresh organic matter are released fairly quickly while those in humus are released more slowly.

3. **Adsorbed nutrients.** These nutrients are held in the soil because they are attracted to clay and humus particles. Adsorption occurs because clay and humus particles are negatively charged. Many plant nutrients are positively charged and so stick to the soil particles. Adsorbed nutrients are held fairly tightly by particles, but most become readily available to plants.

4. **Dissolved ions.** Dissolved ions are the most readily available form of nutrients. The mixture of ions and soil water is termed the **soil solution.** Plants absorb ions directly from the soil solution. However, these nutrients may be rapidly consumed or leached away by percolating water.

As shown in figure 10-4, nutrients can change form. As plants withdraw nutrients from the soil solution, elements held in reserve can become available to plants. While nutrients are held in reserve, they are protected from leaching or other losses.

Soil Colloids

As shown in figure 10-4, adsorption serves as a source of stored nutrients when plants take nutrients out of the soil solution. Nutrients are adsorbed on soil **colloids,** tiny clay and humus particles that carry a slight electrical charge. This charge is important, because it attracts nutrient ions. The soil contains three types of colloids: silicate clays, oxide clays, and humus.

Silicate Clays. Clay minerals are not simply pieces of silt or sand broken into tinier particles. A clay particle is a tiny crystal of mineral formed in the soil from the weathered products of minerals like feldspar or mica. Feldspar and many others are primary minerals of the earth's crust, while the clay minerals that form from the products of their weathering are called secondary minerals. The formation of these secondary minerals could be shown as:

$$\text{primary mineral} \xrightarrow[\text{solution}]{\text{weathering}} \text{ions in soil}$$

$$\xrightarrow{\text{crystallization}} \text{clay mineral (secondary mineral)}$$

A particle of **silicate clay,** called a **micelle,** is a flat, platelike crystal made of many layers. Each layer, in turn, is made of two or three sheets. The sheets are mainly composed of three elements: silicon (Si), oxygen (O), and aluminum (Al). These three elements combine to form several kinds of sheets, which can combine to form several kinds of clays.

In the soil, silicon combines with oxygen to form the silica sheet. The basic unit of the silica sheet is the silica tetrahedron: a silicon atom surrounded by four oxygen atoms. This forms the shape of a four-sided pyramid or tetrahedron (figure 10-5). Many tetrahedra join together by sharing oxygen atoms to form the silica sheet.

The second important sheet in silicate clays is the alumina sheet. The basic unit of the alumina sheet is the alumina octahedron (figure 10-5). Here an aluminum atom is surrounded by six hydroxyl groups (OH^-) to form an octahedron or eight-sided figure. Octahedra join through the hydroxyl groups to form the alumina sheet.

These sheets can stack atop one another in several ways to form a complete clay crystal. The simplest stacking joins a single alumina sheet to one silica sheet, forming what is called a 1:1 layer (figure 10-6). The alumina sheet sheds some of its hydroxyl groups by sharing oxygens at the "tips" of the attached silica sheet. Note the clear order of the 1:1 layer: hydroxyl groups at the top and oxygen atoms at the bottom. Two layers can now join by hydrogen bonding—the hydroxyl groups of one layer bond to the oxygen atoms of another. Hydrogen bonds hold the layers in the clay crystal together tightly. One clay mineral composed of bonded 1:1 layers is kaolinite.

A second way to stack sheets is to sandwich an alumina sheet between two silica sheets (figure 10-7). Here the alumina octahedra replace all but two of their hydroxyl groups by sharing oxygen atoms with the silica sheets. This is a 2:1 structure. No hydroxyl groups are exposed at the surface, so layers are not cemented by hydrogen bonds. The bonds that hold the layers together are much looser than hydrogen bonds, so most 2:1 clays can "open up."

Types of Silicate Clay. Several types of clays result from the ways in which 1:1 or 2:1 layers bond together. Some clays are highly charged and hold cations well; others are not. Some clays are sticky, some are plastic, and some swell when wet. These traits are the soil consistence factors listed in chapter 4.

Two important traits of silicate clays depend on how easily the layers can be separated. If they loosen easily, then water can enter the micelle between the layers, and the particle will swell when wetted and shrink when

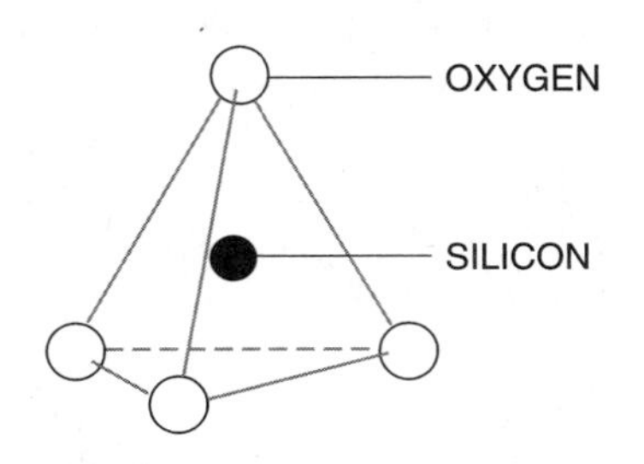

(*A*) **Silica Tetrahedron**

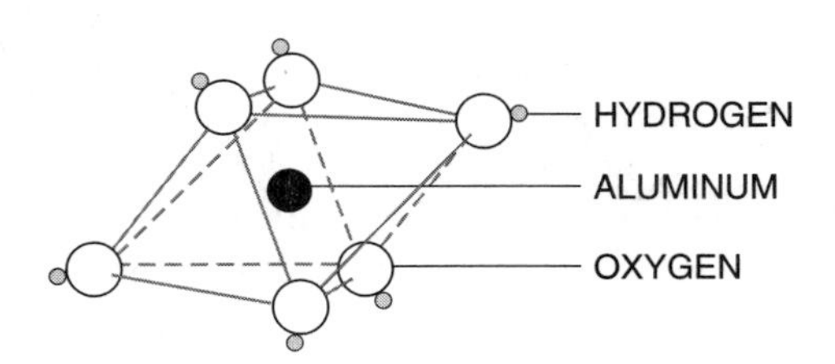

(*B*) **Alumina Octahedron**

FIGURE 10-5 (*A*) In the silica tetrahedron, each silicon atom is located in the center of four oxygen atoms. (*B*) For the alumina octahedron, each aluminum atom is centered among six hydroxyl (OH^-) groups.

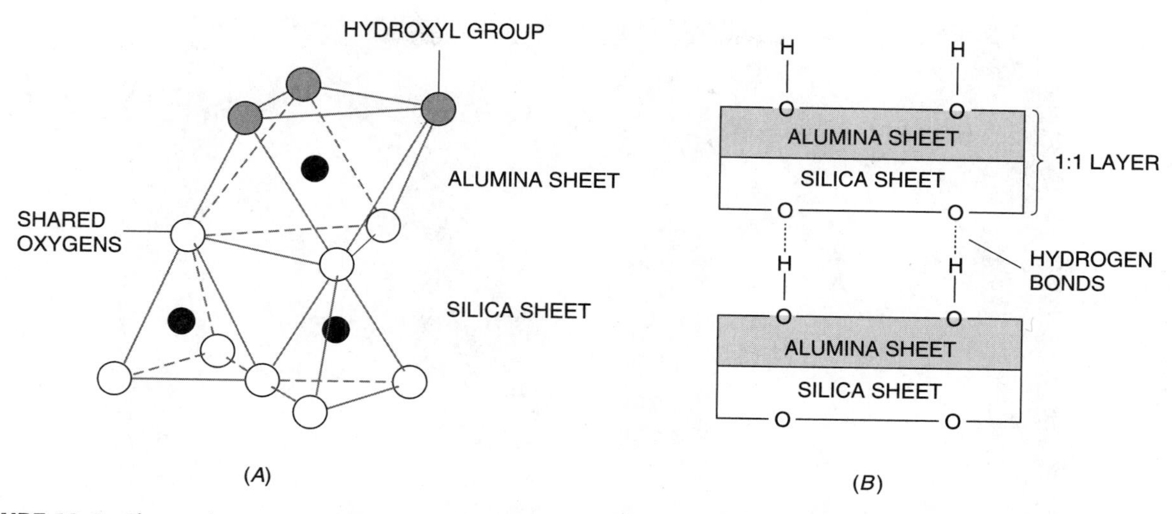

FIGURE 10-6 The 1:1 layer is one way of stacking sheets to form a clay crystal. (*A*) The alumina sheet bonds to the silica sheet by sharing oxygen atoms at the "tip" of each silicon tetrahedron. (*B*) Hydrogen bonding between layers permits several layers to bond together to form a complete micelle.

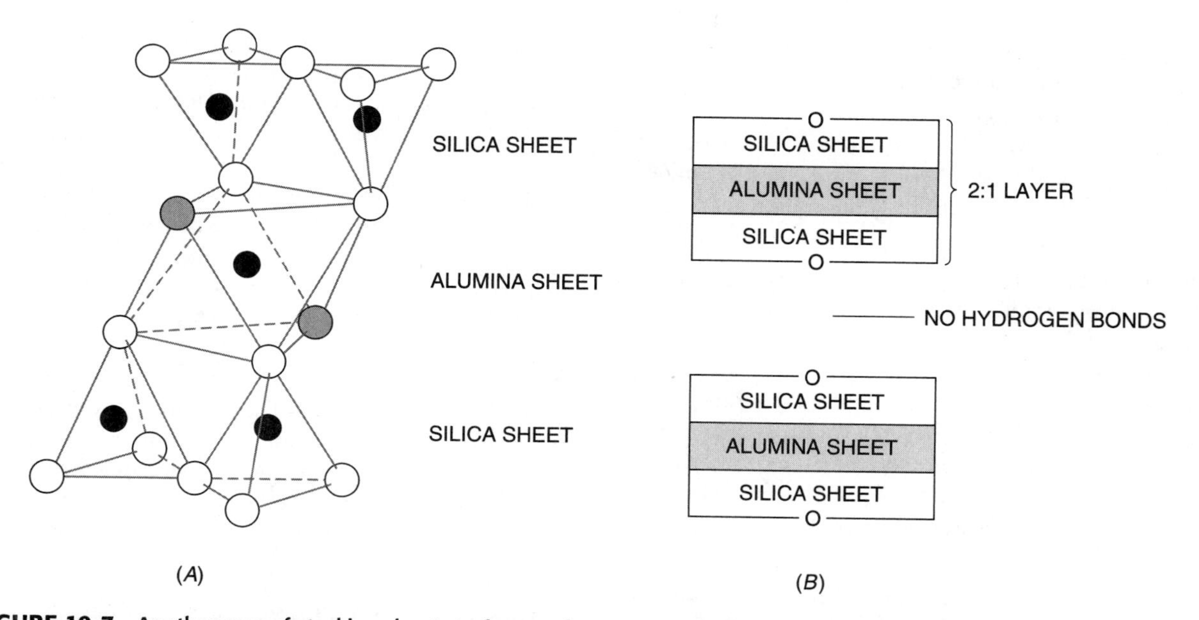

FIGURE 10-7 Another way of stacking sheets to form a clay structure is the 2:1 layer. (*A*) In this structure, sheets are bonded together by shared oxygen atoms. (*B*) Hydroxyl groups are not exposed, so layers are not joined by hydrogen bonds. The layers of many 2:1 clays are less firmly bonded and thus can open up.

FIGURE 10-8 Mud cracks in expanding clay. This Vertisol swells as it gets wetter and shrinks as it dries, forming these deep, wide cracks. *(Courtesy of Howard Hobbs, Minnesota Geological Survey)*

dried (figure 10-8). Such clays are called **expanding clays.** If the layers separate, more surface area is exposed for the adsorption of cations. Thus, clays with loosely bound layers usually hold more nutrients. Figure 10-9 sketches the structure of several types of clays. Figure 10-10 summarizes some of the characteristics of these clays.

MICA CLAYS. Mica clays are 2:1 clays resulting from the weathering of mica minerals. The layers of one mica clay, illite, are firmly bound by bridges of potassium ions. If some of the potassium leaches away, then the layers can open up slightly, so illite is slightly expanding. If all the potassium is lost, a new clay called vermiculite forms. Vermiculite layers are loosely bound by magnesium ions that are surrounded by six water molecules. Vermiculite expands greatly when wetted.

SMECTITE CLAYS. Smectites result either from the weathering of feldspar or from the advanced weathering of vermiculite. They are 2:1 clays, sticky, and highly expanding (figure 10-8). Water fills the space between the layers, so the layers are very loosely held. The bonding force is too slight to hold a large particle together. This means that smectite clays are formed of very small particles. The best known smectite is montmorillinite.

CHLORITE CLAYS. Chlorite layers are tightly bound by a fourth clay sheet. Chlorite is often termed a 2:1:1 clay. The fourth sheet is either another alumina sheet or a sheet of magnesium-oxygen octahedra. This sheet binds the 2:1 layers together fairly tightly.

KAOLINITE CLAYS. These are 1:1 clays. Hydrogen bonds bind the layers tightly, so water cannot get between the layers. Thus, kaolinites swell least of all the clays. They present the smallest surface area for the adsorption of soil cations. The strong bonds allow particle sizes as large as silt. Kaolinite is highly plastic and is used for making pottery.

Oxide Clays. **Oxide clays,** called **sesquioxides,** are tiny particles of iron (Fe_2O_3) and aluminum oxides ($Al(OH)_3$). Oxides are common to old soils in humid

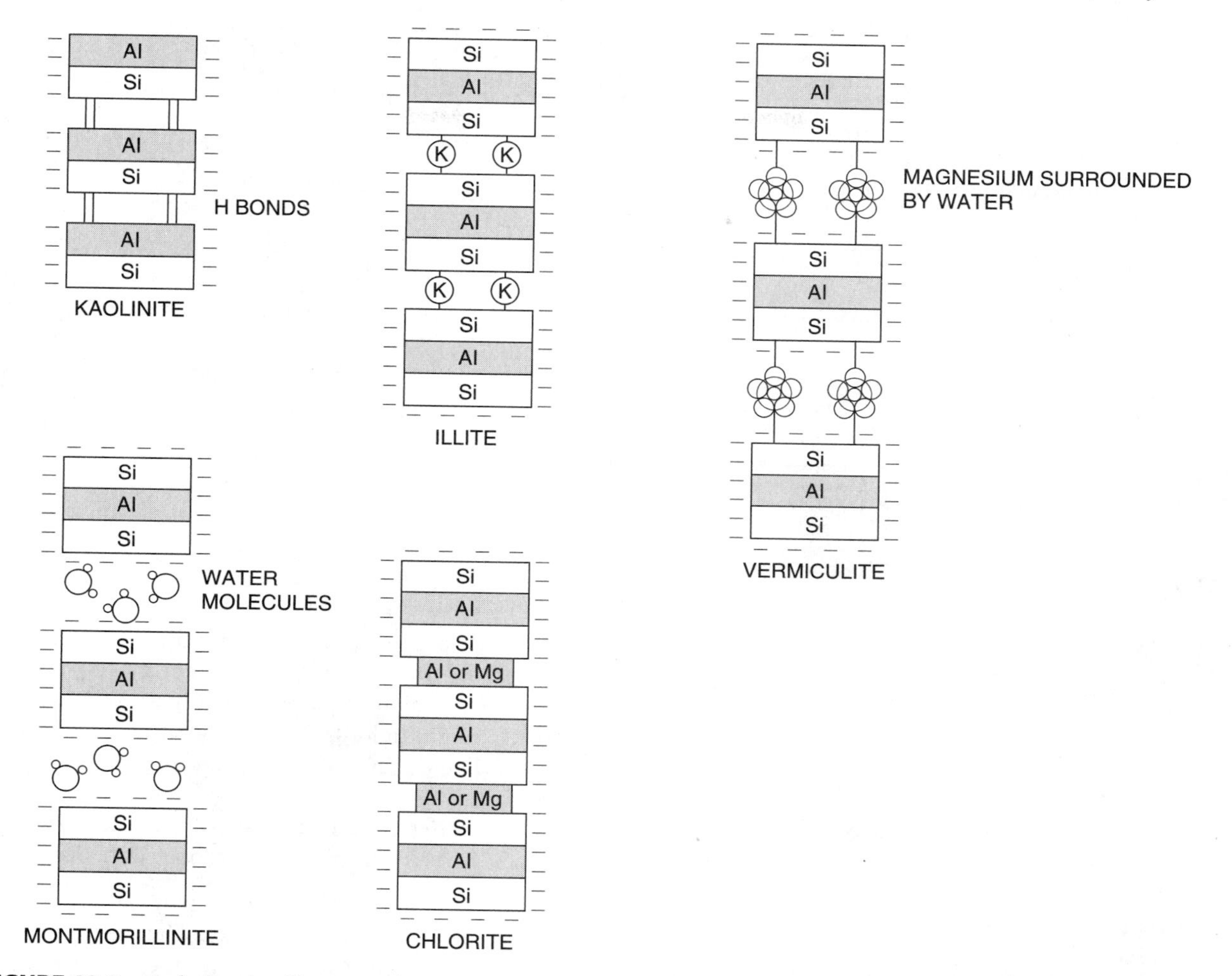

FIGURE 10-9 Each drawing illustrates the micelle structure of different types of silicate clay. The dashes indicate where negative charges, or cation exchange sites, are exposed.

tropical climates. Long periods of weathering leach out silica and some alumina, leaving behind the oxides. Oxide clays tend to aggregate into strong, coated, sand-sized peds that behave much like sand. Oxide clays can form stacked sheets, but do not form the crystalline structure of silicate clays. Oxide clays do not swell, are not sticky, and have limited power to hold nutrients.

Humus. Humus particles are the residues of organic matter decay. They are not crystalline and form irregular, round shapes. They have none of the physical properties of clays, like stickiness or plasticity. However, they have more power to adsorb nutrients than clays. Humus is not stable; it decays over time to carbon dioxide.

Charged Colloids. Colloids usually carry a negative charge that attracts cations from the soil solution. Clay particles gain a negative charge in two ways. First, some hydroxyl groups on the broken end of a

Clay	Wet Consistency	Relative Swelling When Wetted	Interlayer Bonding
Smectites (2:1)	Very sticky	High	Very weak
Vermiculite (2:1)	Sticky	High	Moderate
Illite (2:1)	Slightly sticky	Low	Strong, by K ions
Chlorite (2:1:1)	Nonsticky	None	Strong, by fourth sheet
Kaolinite (1:1)	Plastic, slightly sticky	Very low	Strong H bonds
Sesquioxides	Nonsticky	None	—

FIGURE 10-10 Characteristics of important soil clays.

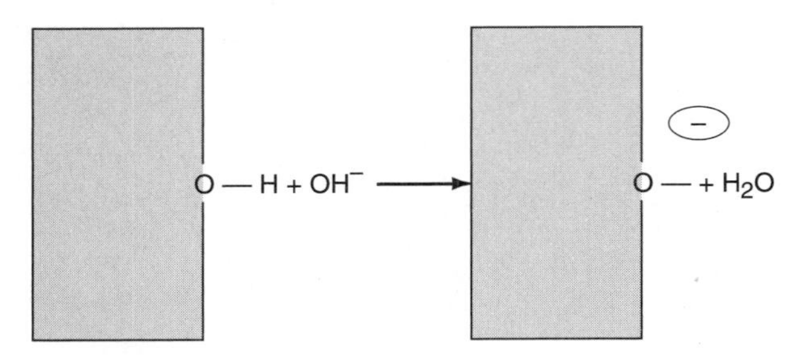

FIGURE 10-11 The loss of a hydrogen ion (proton) from a hydroxyl (OH^-) group on the surface of a clay micelle leaves the oxygen with a negative charge that can attract cations from the soil solution.

clay micelle lose their hydrogen ion (figure 10-11). The hydrogen ion is simply a proton, so this leaves a charge imbalance. The remaining oxygen therefore has a negative charge.

The second process is **isomorphous substitution.** One cation can replace another cation of similar size in a clay sheet. For instance, aluminum (Al^{+3}) can replace a silicon atom (Si^{+4}) in a silica layer. While the cation fits, it has a smaller positive charge. This leaves a "spot" in the crystal that is short of positive charges. These spots acquire a negative charge.

Because a micelle contains a number of these negative "spots," the entire particle maintains a negative charge. The minus charge attracts positively charged cations, so the micelle is surrounded by a swarm of cations. The minus and plus charges balance, for a net zero charge.

Clays differ in the number of sites for negative charges, so their ability to retain cations also differs:

- Kaolinite has a small negative charge, because little isomorphic substitution occurs. The only exchange sites are from hydroxyl groups on the broken ends of a clay micelle.
- Smectites have many negative sites because magnesium ions (Mg^{+2}) substitute for some aluminum (AL^{+3}) in the alumina sheets. In addition, cations can adsorb on sites between the 2:1 layers.
- Vermiculite has an even greater negative charge, because about one in four silicon atoms is replaced by aluminum. Being a swelling clay, some cations can be retained between the 2:1 layers.
- Illite has the same substitutions as vermiculite. However, because the layers are held together by potassium, few cations can be held between the layers. Thus, illite holds fewer cations than vermiculite.
- Sesquioxides have very little negative charge, except for a few hydroxyl groups on the surface.

- Humus has numerous sites. Many organic compounds found in humus have hydroxyl groups as part of their structure. These groups can lose a hydrogen to form negative spots over much of the humus particle surface.

Cation Exchange

The negative charge of soil colloids plays a key role in the way nutrients behave in soil. Because the tiny particle bears a negative charge, it attracts positively charged ions and repels negatively charged ones. The action follows the rule that "opposites attract, likes repel," much like a magnet (figure 10-12).

To explain this further, let's see how it works with a clay particle. Clay particles have platelike shapes and are negatively charged. Figure 10-13 shows a colloid surrounded by soil solution. The negatively charged colloid attracts a swarm of cations from the soil solution. This is called **adsorption.**

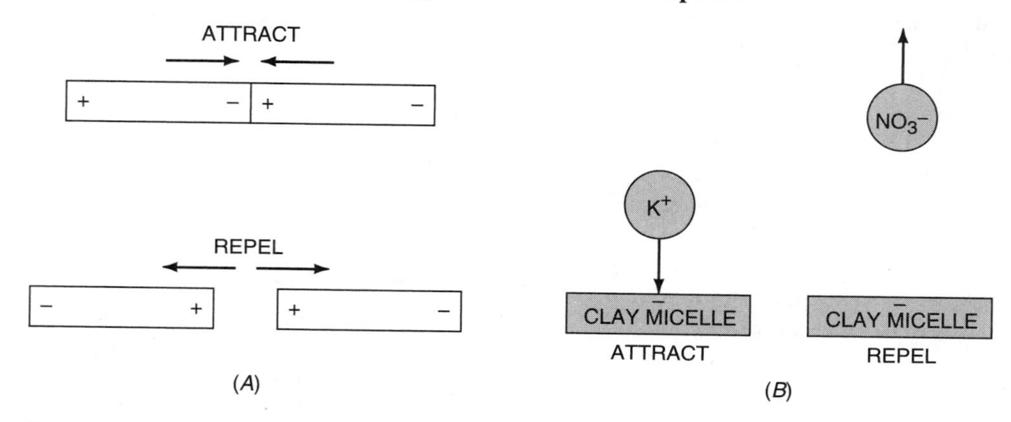

FIGURE 10-12 Opposite charges attract as shown (*A*) in magnets where opposite poles attract and (*B*) in the soil where negatively charged clay particles attract positively charged cations.

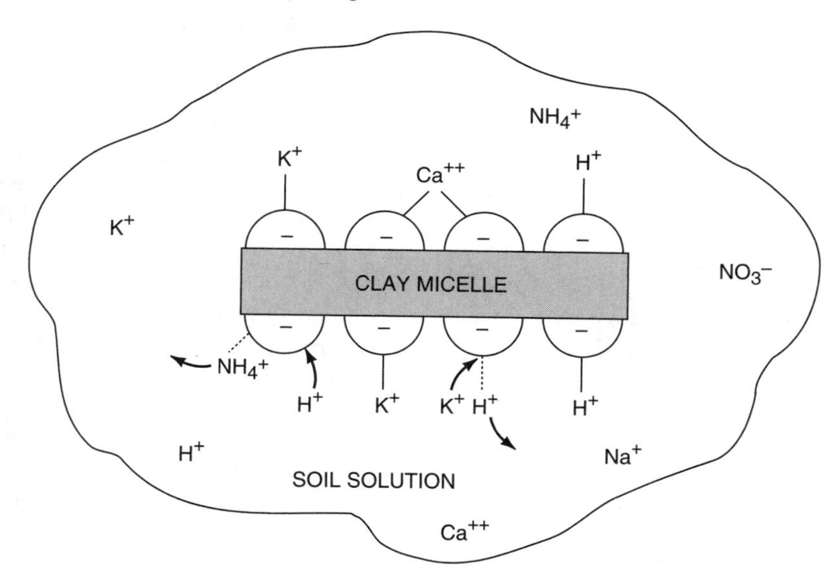

FIGURE 10-13 The colloid surface is negatively charged, so it adsorbs cations. The cations are "exchangeable" in that if one leaves, another takes its place. (*Note:* The semicircles used to represent exchange sites are for illustration only.)

As figure 10-13 shows, cations can move on and off the particles. When one ion leaves, it is replaced by some other cation. We call replacement of one ion for another **cation exchange.** Cations that can be replaced on exchange sites are said to be *exchangeable,* such as exchangeable potassium.

The ability of a soil to hold nutrients is directly related to the number of cations it can attract to soil colloids. This value, determined by the amount of clay, the type of clay, and the amount of humus, is measured by **cation exchange capacity** (CEC), which is expressed in milligram equivalents per 100 grams of soil (mEq/100 g). An equivalent weight is the weight of the number of atoms of an element that is the same as the number of atoms in one gram of hydrogen. CEC may also be expressed in the units centimoles of charges per kilogram of dry soil (cmol/kg), but the CEC values are the same in both systems.

A milliequivalent is used rather than the simple weight of adsorbed cations, because it cancels differences in the weight and charge of the cations on the sites. For instance, one milligram of calcium occupies fewer exchange sites than one milligram of magnesium, simply because it weighs more. Using milliequivalents per 100 grams of soil ensures that the CEC values are the same whatever the cations on the exchange sites.

Figure 10-14 lists cation exchange capacity values for several clays, humus, and soil textures. Note that humus has a much higher CEC than clay. However, clay usually adds more cation exchange capacity to a soil than does humus, because there is so much more clay than humus in most soils. Organic soils, which are mostly organic matter, are an exception. Sandy soils, low in clay, may also gain a large portion of their exchange capacity from humus.

The cation exchange capacity of colloids whose negative charge depends largely on surface hydroxyl (OH^-) groups—oxide clays and humus—is said to be pH dependent. Soil pH is discussed fully in chapter 11; simply stated, the higher the soil pH, the greater the concentration of hydroxyl ions in the soil solution. Examine figure 10-11. If the concentration of hydroxyl ions in the soil is high, then the reaction pictured in that figure is driven to the right (see the discussion of chemical reactions in appendix 1 if necessary), and the number of cation exchange sites increases. Therefore, the higher the pH, the larger the pH-dependent CEC.

Colloid	Cation Exchange Capacity (mEq/100 g soil)
Humus	100–300
Vermiculite	80–150
Montmorillinite	60–100
Illite	25–40
Kaolinite	3–15
Sesquioxides	9–3

Soil Texture *(Temperate Climate Soils)*	
Clay loam	30
Silt loam	27
Loam	24
Sandy loam	17
Loamy Sand	9

FIGURE 10-14 Sample cation exchange capacities of several colloids and soils. The values for soil textures are provided only to show differences in capacity. An individual soil may vary from these values.

Cation Behavior at the Exchange Sites. The negatively charged surface of a clay micelle strongly attracts cations. Cations cluster most densely near the micelle surface, neutralizing the negative charge. The cations can move on the micelle and trade places or exchange with cations in solution. Several factors control the selection of cations that leave the micelle or become adsorbed. Two important factors are (1) the relative bonding strength of each cation and (2) the number of each type of cation.

If two cations are present in the soil in equal numbers, then the one that bonds most tightly to exchange sites will tend to be the one found on the micelle. The most strongly adsorbed cation is aluminum, followed in decreasing order by:

$$Al^{+3} > H^+ > Ca^{+2} \geq Mg^{+2} > K^+ \geq NH_4^+ > Na^+$$

Assume that a soil has equal numbers of calcium and sodium (Na) ions. Calcium tends to take over the exchange sites because it adsorbs more strongly on the micelle. Sodium tends to leach out of the soil solution.

The second controlling factor is **mass action.** Mass action means that the greater the number of an ion in the soil, the more exchange sites it will occupy. As an example, in high lime (calcium carbonate) soils, most exchange sites are occupied by calcium. Mass action is a function of chemical equilibrium. Study the description of equilibrium in appendix 1 for further explanation.

Consider the treatment of high-sodium soils with gypsum (calcium sulfate). In a high-sodium soil, many exchange sites are taken up by sodium (more than 15 percent). When gypsum is added, calcium displaces sodium on the exchange sites. The displaced sodium enters the soil solution and is leached away by heavy watering. Calcium replaces sodium on the exchange sites by means of mass action (there are many calcium ions in solution) and because calcium is adsorbed more strongly than sodium.

Cations that are weakly held, in direct contact with the soil solution, are exchanged fairly easily. These are termed exchangeable cations. Some are held very tightly against the colloid or may be trapped between layers of a clay micelle. These do not normally pass into solution easily and are said to be **nonexchangeable ions.** Even these may be given up slowly if the surrounding solution becomes very low in those ions.

Anion Storage. Several nutrients are available to plants as negatively charged ions, or anions (refer to figure 10-1). The negative charge means that an anion is repelled from a cation exchange site. While these elements, like sulfur, are to a large degree stored as organic forms in humus, an **anion exchange** process stores small amounts of some anions.

Anion exchange sites are the opposite of the cation exchange sites where hydrogen is lost from a hydroxyl group. At the anion exchange site, an extra hydrogen joins the hydroxyl group to produce a net positive charge (figure 10-15). The positive charge can then attract anions. For most soils, anion exchange capacities are quite low. Typical values are a few tenths of a milliequivalent per 100 grams of soil. Anion exchange is greatest in acid soils high in oxide clays.

Applications of the Cation Exchange Capacity. How growers use soil is strongly influenced by cation exchange capacity. High CEC soils, measuring between eleven and fifty units, usually contain a lot of clay. Low CEC soils, measuring below eleven units, usually have a high sand content. Sticky soils are high in the types of clay having the highest cation exchange capacity. Thus, cation exchange capacity is reflected in physical properties of soil, such as texture and consistence.

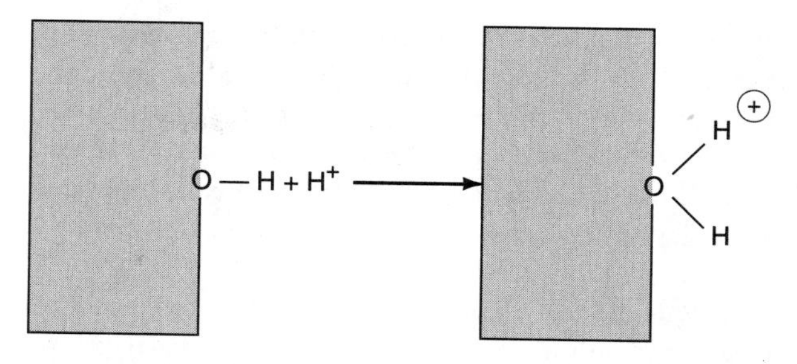

FIGURE 10-15 At anion exchange site, a hydroxyl group on the particle surface picks up a hydrogen ion (proton). The site now has a net positive charge that can attract anions from the soil solution.

Cation exchange capacity is one of the factors that determines how much herbicide should be spread on the soil. Colloids adsorb pesticides as well as nutrients; therefore, clay and humus tend to tie up many chemicals. As a result, we often have to apply higher chemical rates to high CEC, clayey soils than to low CEC, sandy soils.

The amount of lime needed to change soil pH is also a function of the CEC. In the process of liming soil, calcium displaces hydrogen on the cation exchange sites. The more exchange sites in the soil, the more lime is needed. Therefore, growers apply much more lime to correct acidity of fine-textured soils than they do to correct coarse-textured soils.

Cation exchange capacity influences fertilization practices. High CEC soils have greater potential to hold cationic nutrients than low CEC media. Smaller amounts of fertilizer, applied more often, are needed in low CEC soils to prevent leaching losses, while larger amounts may be applied less frequently in high CEC soils. Golf course managers with all-sand greens fertilize lightly often. Greenhouse growers using soilless mixes with low CEC may fertilize lightly with every watering.

The concept of cation exchange suggests that it is easier to improve the cation exchange capacity of a sandy soil by improving the organic matter content than by adding clay. In temperate humid climates, most clays are kaolinite (CEC = 3–15 mEq) and illite (CEC = 25–40 mEq).

In a soil composed of these clays, each percentage clay in the soil adds between 0.03 to 0.5 mEq to the cation exchange capacity of the soil. Each percentage of humus, on the other hand, adds between one and three full mEq. Thus, far less organic matter is required, compared with clay, to raise the exchange capacity of the soil.

Percent Base Saturation. Soil fertility is influenced not only by the cation exchange capacity (how many cations it can store), but also by how much of the CEC is actually filled with plant nutrients. Exchange sites may be filled by members of two groups of cations. One group consists of hydrogen and aluminum, which are not plant nutrients at all. Their primary contribution is to acidify the soil. The other cations are called **exchangeable bases** and include elements such as calcium, magnesium, potassium, and sodium. Except for sodium, the bases are plant nutrients.

The percentage of the cation exchange sites filled with exchangeable bases is called the **base saturation.** It expresses how much of the soil's "potential fertility," the CEC, holds exchangeable bases. For example, if the total CEC of a soil is 10 milliequivalents per 100 grams of soil and bases occupy 6 of the 10, then the base saturation percentage is 60 percent. Most crops grow best at a base saturation of 80 percent or more. These crops require a good supply of nutrients. Some trees that grow on infertile soils do well at a base saturation of around 50 percent.

Nutrient Uptake

This chapter has already noted two factors that affect soil fertility: (1) the amount of storage capacity of a soil (cation exchange capacity), and (2) how much of that storage actually contains nutrients (percent base saturation). A third fertility factor is how easily roots take up nutrients. How do plants take up nutrients from the soil?

Plants absorb nutrients as the ions listed in figure 10-1. Nutrient absorption means that nutrient ions cross the cell membranes of root cells and eventually move to the root's vascular system to be delivered to the rest of the plant. In some situations, some nutrients may be passively absorbed with water entering roots, but otherwise a more active process is required. In fact, roots may have a concentration of some nutrients hundreds of times that of the soil solution. For nutrients to passively "soak in" against such a gradient would be like water running uphill. Roots actively transport nutrient ions through root cell membranes, an active process that uses energy. Because roots produce energy by respiration, conditions that limit respiration, like a waterlogged soil, also limit nutrient uptake. In addition, the active transport of ions across a cell membrane allows some selection—the root can take up some elements more than others.

The soil solution surrounds roots growing through soil pores. Root hairs get ions directly from the soil solution through their own form of cation and anion exchange. If a cation is removed from solution, the root gives up a hydrogen ion (H^+) to replace it in the soil solution (figure 10-16). If an anion is absorbed from solution, the root gives up an anion to replace it. The exchange maintains the electrical balance in the root and in the soil.

Plants take exchangeable bases from solution and replace them with hydrogen ions. Because hydrogen forms stronger bonds on the exchange sites, they replace other cations on the exchange sites (figure 10-16). This exchange renews the number of nutrient cations in solution, allowing plants to continue to draw nutrients from the soil. Over time, however, the exchange increases the number of hydrogen ions bonded to exchange sites and lowers percent base saturation.

When growers fertilize and lime soil, they are reversing the loss of nutrient cations. If, for example, potassium fertilizer is supplied in the form of potassium chloride, potassium will replace some other cations, including hydrogen, by mass action (figure 10-17).

Figure 10-16 raises a question about nutrient uptake. Near the root hair, nutrients should be depleted. How does the root continue to obtain nutrients? Remember, plant roots only absorb nutrients that are in solution at the root surface. Roots continue to grow through the soil mass to find new supplies of nutrients. This growth is especially important because roots most effectively take up nutrients near root tips, where the root hairs are, while older root tissue forms barriers to absorption. New root tips must be constantly being formed for efficient nutrient uptake. These absorbing roots access nutrients by means of root interception, mass flow, and diffusion.

Root interception results directly from the extension of root systems. As new roots grow, they displace a volume of soil that contains nutrients; those nutrients are at the root surface and are readily absorbed. Roots actually

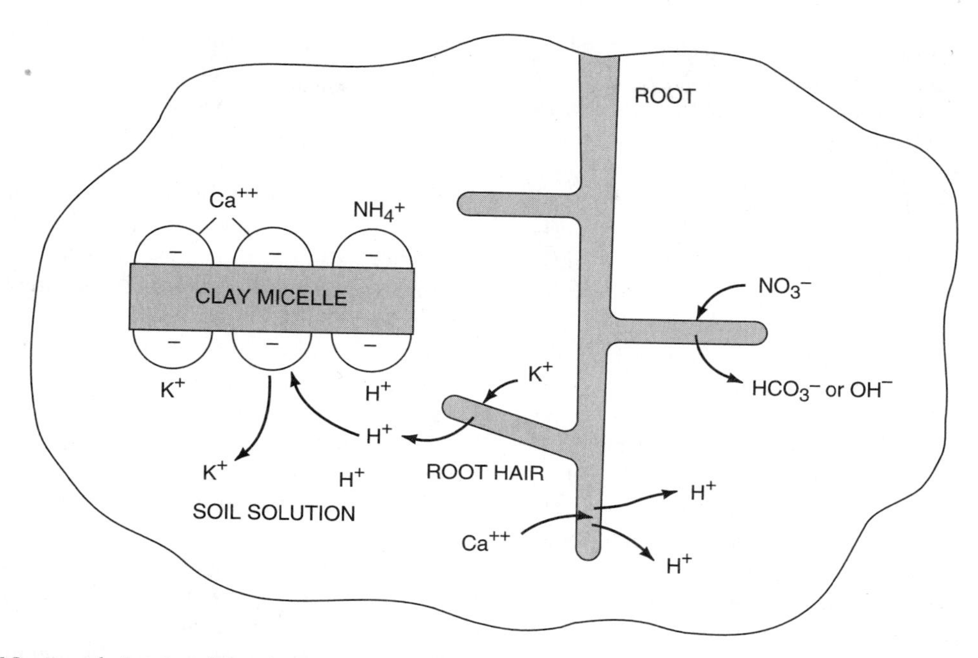

FIGURE 10-16 Root hairs absorb nutrients from the soil solution. To maintain electrical balance, the plant replaces an absorbed ion by releasing another ion to the soil solution. A loss of bases in the soil solution and a buildup of hydrogen ions cause hydrogen/base exchange on soil colloids. (*Note:* Not drawn to scale.)

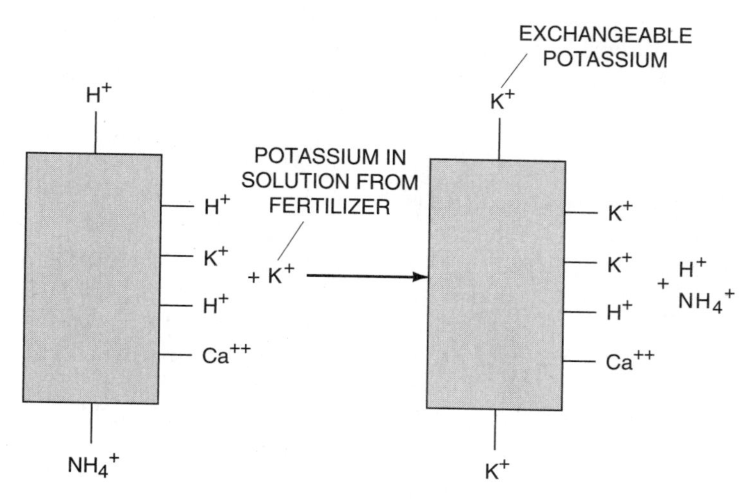

FIGURE 10-17 When a grower limes or fertilizes the soil with a nutrient base, the base replaces other cations on the exchange sites.

closely contact very little soil, however, so other means of obtaining nutrients are needed.

Mass flow carries nutrient ions to roots from nearby soil in water flowing toward roots by capillary action. The driving force for nutrient transport here is simply a water potential gradient. Review figure 7-12, if necessary, to understand how this works. Mass flow obviously depends on water usage by the plant and supplies nutrients most effectively when plants are taking up water rapidly and the soil is moist.

Diffusion also moves ions toward the root from the surrounding soil, but the ions are diffusing *through* soil water rather than being carried *with* it. The driving force is a nutrient concentration gradient.

Consider calcium as an example. Near the root there is less calcium because the root has been removing it. In response, calcium ions move toward the root through soil water (not along with it) to make a new balance. The movement of ions is from areas of greater concentration to areas of lesser concentration, a slower process than that of mass flow.

Since nutrients diffuse through water films, diffusion is particularly sensitive to soil moisture and declines greatly in dry soil.

While all three mechanisms operate at the same time, the relative importance of root interception, mass flow, and diffusion depends on plant species, specific nutrient, soil texture, moisture content, and other factors. For instance, mass flow contributes little to phosphorus and potassium uptake compared to diffusion. In general, diffusion is the most important mechanism, while root interception usually contributes the least to nutrient access.

Factors Affecting Uptake. Several factors affect how well plants take up nutrients. Plant use of nutrients is affected by features of the soil, such as oxygen supply, water supply, and soil temperature. Root distribution in the soil is another factor.

Active uptake consumes energy produced in the roots by respiration; the food to fuel root respiration is produced in the leaves during photosynthesis and delivered to the roots by the plant's vascular system. Anything that interferes with any step of this process also reduces nutrient uptake. For instance, plants growing under low light make less sugar to send to the roots so take up fewer nutrients. Whether it be turf under a tree or a houseplant indoors, one should reduce fertilization compared to plants growing in the sun.

More importantly, since respiration uses oxygen, conditions that limit oxygen supply also limit nutrient uptake. Both poor drainage and soil compaction slow the movement of oxygen into the soil. As a result, these conditions also limit the ability of plants to absorb nutrients. This is another reason for draining wet soils and for avoiding compaction.

Dry soils lower nutrient uptake because the lack of water impedes nutrient flow toward the root hairs by mass flow and diffusion. Phosphorus, for example, moves in the soil largely by diffusion, so phosphorus uptake is sharply reduced in dry soil.

Soil temperature also affects nutrient uptake. The rates of all chemical reactions, including those in soil and plants, depend on temperature. Respiration rates go down in cold soil, so the plant has less energy to take up nutrients. Root growth is also slowed in cold soil, limiting root interception of nutrients. Mineralization of organic matter declines as well, so fewer organic nutrients are made available. For these reasons, cold soil hinders nutrient uptake. Phosphorus and iron deficiencies are common, for example, in spring when soils are cold and wet. Farmers often place an extra amount of phosphorus fertilizer beside planted seeds to overcome deficiencies. Greenhouse growers must be aware of the effect of cold soil, because irrigation water may be cold enough during the winter to induce nutrient shortages. Fall is a preferred time for fertilizing turf and trees in many parts of the country. Air temperature is cool and shoot growth has stopped, but roots remain active in the still warm soil. Fertilizer will be taken up easily at this time and stored in the roots for the coming season, avoiding problems with cold early spring soils.

An increase in the amount of nutrient ions in the soil improves absorption. This factor is obviously one reason growers fertilize their crops. When some elements, like potassium, are present in very high amounts, plants even take up more than they can use. This condition is called **luxury consumption.** However, because the excess is stored in plant cells, it may be used later if something happens to slow uptake by the roots.

Uptake is most rapid, of course, where roots are most numerous. We know roots grow best where air, water, and nutrients are in good supply. Drainage, compaction, and fertilization influence how well roots grow, as does soil or root depth. Plants with deeply growing roots need less fertilization than plants with shallow root systems. Soils that have restricted zones—those with high water tables, plowpans, or bedrock—can cause shallow root systems.

Organisms in the rhizosphere, and their interactions with plant roots, obviously influence nutrient uptake as well. Soilborne pathogens like nematodes or root-rotting fungi damage the ability of roots to take up nutrients, while mycorrhizal infections improve nutrient uptake. Some rhizosphere microbes can make nutrients more available. Insects, on the other hand, may feed on plant roots.

Figure 10-18 summarizes the factors affecting soil fertility and nutrient uptake.

Raises Fertility	Lowers Fertility
High clay content	High sand content
High humus content	Loss of organic matter
Good structure	Compaction
Warm soil	Cold soil
Deep soil	Shallow soil
Moist soil	Dry or wet soil
Good drainage	Excess irrigation or drainage
Fertilization	Erosion
Desirable microbes	Root damaging pests
Near neutral pH	pH too acid or alkaline

FIGURE 10-18 Factors affecting a soil's ability to supply plant nutrients.

SUMMARY

For normal growth, most plants require seventeen essential elements. Carbon, oxygen, and hydrogen come from air and water; the other fourteen elements are absorbed by plants from the soil. Crops take up primary and secondary nutrients in large amounts; trace elements are needed in small amounts.

Four sources of nutrients work together to both store and release nutrients to plants. These sources are soil minerals, organic matter, soil solution, and adsorption by clay and humus. The ability of colloids to adsorb nutrients is based on their large surface area and on negatively charged sites that are part of the colloids' structure. This ability is measured by cation exchange capacity. Percent base saturation is the percentage of the exchange capacity filled with exchangeable bases.

Plants absorb nutrients by transporting ions into root cells. Uptake uses energy. To find new nutrient supplies, roots grow through the soil. In addition, nutrients flow either with or through soil water toward roots. Extreme soil conditions, including soil that is too dry, too wet, too cold, or badly compacted, impair the ability of roots to absorb nutrients.

Nutrient uptake is aided by a deep, well-drained soil. Several farming practices can improve soil fertility. Artificial drainage helps if a soil is poorly drained. Avoiding compaction or subsoiling already compacted soils is useful. Organic matter additions improve the cation exchange capacity and provide nutrients. Organic matter also keeps soil loose to enable sufficient oxygen to be supplied to roots. Proper fertilization also improves soil fertility.

REVIEW

1. What do you think would be the effect on nutrient uptake of removing all the leaves from a plant?
2. Name four soil conditions that inhibit nutrient uptake and describe why they do so.
3. Discuss how we classify the seventeen essential elements and the reasons behind each class.
4. Discuss the four primary pools of essential elements in the soil. Pick one of them and describe possible consequences if that pool were to suddenly disappear.

5. Why might two soils with identical percentages clay have different cation exchange capacities? Explain your answer.
6. Some people call soil a vast "clay factory." Explain this statement.
7. Describe the basic structure of silicate clays, and explain the difference between expanding clays and others.
8. Explain why kaolinite has a lower cation exchange capacity than do smectites.
9. Using information from this chapter, explain why tropical soils are often less fertile than temperate climate soils.
10. Several problems with vegetable and fruit crops are caused by calcium shortages in the developing edible part, and they tend to happen as the soil dries. Why might soil drying contribute to such problems?

ENRICHMENT ACTIVITIES

1. Using molecular model kits, construct models of silica tetrahedra and alumina octahedra. Then try to construct a portion of a clay layer.
2. Examine granules of horticultural vermiculite. Note that it is made of expanded mica sheets, and that it can soak up water between the sheets. Physically and chemically, the granule resembles the structure of a vermiculite micelle.
3. Gentian violet is a positive dye, and eosin is a negative dye. Prepare a water solution of each dye. Pour each solution into a pot of soil until water begins to drain from the bottom. Collect the drainage water. Which dye passed through the soil? Why? Try this on both a very sandy soil and a finer-textured soil to see if there is a difference.
4. This little test can demonstrate solution and chemical fixation of nutrients. Mix a little silver nitrate (used in photo processing) in distilled, deionized water. It will dissolve, like nutrients in solid form dissolving in soil water. Also mix a bit of table salt into some distilled water. Now mix the two solutions together. The white material that results is silver chloride, which is not soluble in water. In the soil, chemical reactions that have the same effect can occur, tying up nutrients like phosphorus or iron in insoluble forms. Do not allow silver nitrate to come into contact with your skin.
5. For more details on clay mineralogy, try this Web site: <http://mineral.galleries.com/minerals/silicate/ clays.htm>.
6. For a nice simple summary of factors affecting nutrient uptake, as it relates to fertilizer use, see this Web site: <http://www.ext.vt.edu/departments/envirohort/factsheets2/fertilizer/jun89pr3.html>.

CHAPTER 11

Soil pH and Salinity

OBJECTIVES

After completing this chapter, you should be able to:

- describe soil pH and its development
- describe how pH affects plant growth
- tell how to lime or acidify soil
- describe saline and sodic soils
- describe methods to treat saline and sodic soils

TERMS TO KNOW

acid (soil)
agricultural lime
alkaline (soil)
basic (soil)
buffer test
buffering
burned lime
calcareous soils
calcitic limestone
calcium carbonate equivalent
chemical guarantee
dolomitic limestone
fluid lime
hydrated lime
marl
physical guarantee
saline soil
saline-sodic soil
sodic soils
sodium adsorption ratio
soil reaction
soluble salt
total neutralizing power

"Sweet" and "sour" are old terms used to describe soil quality, which remind us of a farmer raising a handful of soil to the lips to taste sweetness or sourness. "Sweet" and "sour" are simple terms for soil reaction, or soil pH. What is soil reaction and how does it affect crop growth?

Soil pH

Soil reaction describes the acidity or alkalinity of a soil. Soil users are concerned about soil reaction because it strongly affects plant growth. Reaction is measured by the pH scale, as shown in figure 11-1, which gives sample pH values for common substances. The scale runs from a pH of 0 to a pH of 14.0. Readings between 0 and 7.0 are said to be **acid.** A pH of 1.0 is extremely acid and a pH of 6.0 is slightly acid. Examples of acid materials include vinegar, tomato juice, and lemon juice. These acid foods have a sour taste.

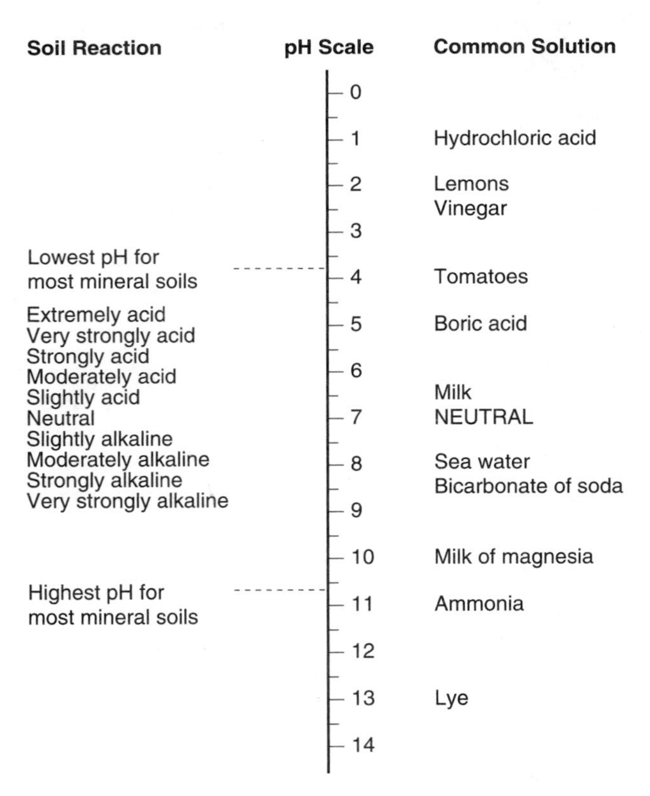

FIGURE 11-1 The pH scale runs from 0 (most acid) to 14.0 (most alkaline). The pH values for several common substances are shown. Note the soil pH range extends from about 3.5 to 10.5.

Readings between 7 and 14 are **alkaline** or **basic.** The larger the number, the stronger the base. Soap is slightly basic, while household ammonia, with a pH of 11, is strongly basic. Bases, or alkaline substances, taste bitter.

The midpoint of the scale, pH 7.0, is the neutral point, which is neither acid nor base. Pure water, which has a neutral pH, can be a model in our discussion of pH. A very few water molecules break up to form a cation and an anion as shown in reaction (*a*):

(*a*) $H_2O \leftrightarrows H+ + OH^-$

water — hydrogen ion — hydroxyl ion

The cation in the reaction is the hydrogen ion (H^+). It makes a solution acid. The anion is the hydroxyl ion (OH^-). It makes a solution basic. In pure water, the number of hydrogen ions equals the number of hydroxyl ions to maintain a balance. Thus, pure water is neither acid nor base. However, substances dissolved in water may change the balance, causing one ion to outnumber the other.

For instance, if pure water is exposed to air, carbon dioxide from the atmosphere dissolves in the water to form carbonic acid. Carbonic acid quickly breaks down to liberate hydrogen ions (*b*):

(*b*) $CO_2 + H_2O \rightarrow H_2CO_3 \rightarrow HCO_3^- + H^+$

Now there is an excess of hydrogen ions, so this dilute solution is acidic. Water in equilibrium with air has a pH of about 5.6. In fact, rainfall itself is slightly acid even without air pollution that creates lower pH "acid rain."

The pH scale indicates how acidic or basic a solution is by giving the concentration of hydrogen ions. The pH scale is a special scale for expressing hydrogen ion concentration as one over the log of the hydrogen ion concentration ($1/\log [H^+]$). The smaller the number on the pH scale, the stronger the acidity of a substance. Each pH point multiplies acidity by a factor of 10. A pH of 5.0 is 10 times more acid than pH 6.0 and 100 times more acid than pH 7.0.

The balance between hydrogen and hydroxyl ions dictates pH. Soil with far more hydrogen ions than hydroxyl ions is very acid. With only a few more hydrogen ions, it is slightly acid. On the basic or alkaline side of the scale, the reverse is true.

Development of Soil pH

Soil does not reach the extreme pH limits shown in figure 11-1—the most acid soil has about a pH 3.5 and the most basic soil is pH 10.5. These are extreme values. Growers more commonly find that soil ranges between pH values of 5.0 and 8.0.

Soil pH results from the interaction of soil minerals, ions in solution, and cation exchange. Different reactions govern at different pH ranges. In the simplest terms, high pH is caused by the reaction of water and the bases calcium, magnesium, and sodium to form hydroxyl ions. Low pH is caused by the percolation of mildly acidic water, which results in the replacement of exchangeable bases by hydrogen ions. To understand the full range of soil pH, it is easiest to start with alkaline soil.

Very basic soils (pH greater than 8.0) are more than 100 percent base-saturated. That is, all exchange sites are filled with bases, and the soil contains particles of mineral carbonates (CO_3) like calcium carbonate ($CaCO_3$). The pH of very alkaline soils results from the reactions of carbonates with water to form hydroxyl ions, according to reactions (*c*) and (*d*):

(*c*) $CaCO_3 + 2H_2O \rightarrow Ca^{+2} + H_2CO_3 + 2OH^-$

(*d*) $Na_2CO_3 + 2H_2O \rightarrow 2Na^+ + H_2CO_3 + 2OH^-$

This reaction with water is called hydrolysis ("hydro" meaning water). The hydrolysis of calcium carbonate in reaction (*c*) results in a pH range of about 8.0–8.5. Soils in this range, which are 100 percent base-saturated and contain enough free calcium carbonate, are called **calcareous soils.** Calcareous soils can be tested with dilute hydrochloric acid—they will fizz from carbon dioxide given off by the reaction between the mineral and the acid (figure 11-2). They result from the weathering of calcareous parent materials like limestone. If the sodium saturation exceeds 15 percent, then the hydrolysis of sodium produces lye (sodium hydroxide, NaOH), which can raise the pH to 10.0 as in reaction (*d*). Such high-sodium soils are termed **sodic soils.**

In humid climates, further weathering and leaching removes excess basic minerals. When these minerals reach a low level, the soil ceases to be calcareous. However, the remaining minerals ensure that the soil is 100 percent base-saturated. This occurs at a pH of about 8.0, though it varies for different soils. At this point, pH begins to be controlled by the hydrolysis of exchangeable bases (rather than free carbonates), as in reaction (*e*):

FIGURE 11-2 Calcareous soil fizzes when treated with dilute hydrochloric acid. *(Courtesy of USDA NRCS)*

(*e*) $\boxed{\text{micelle}}\text{—}Ca^{+2} + 2H_2O \rightarrow$

$$\boxed{\text{micelle}}\!\!<\!\!\begin{array}{l}H\\H\end{array} + Ca^{+2} + 2OH^-$$

This reaction maintains a pH level between about 7.0 and 8.0, depending on soil type. Note that reaction (*e*) replaces calcium with hydrogen. As this reaction proceeds, and rainfall leaches out the calcium, base saturation goes below 100 percent. The same is true of soils of noncalcareous parent materials. Exchangeable hydrogen, if adsorbed, does not acidify the soil. When it does enter the soil solution by cation exchange, however, it makes the soil more acid, reaction (*f*).

(*f*) $\boxed{\text{micelle}}\text{—}H^+ \rightleftarrows H^+ + \boxed{\text{micelle}}\text{—}$

The actual pH of soil with a base saturation below 100 percent depends on the balance between hydroxyl ion production by base hydrolysis, reaction (*e*) and hydrogen ion production by hydrogen exchange, reaction

pH Range	Determining Reaction	Saturation
8.5–10.0	Na_2CO_3 hydrolysis	100 percent base saturation, sodium saturation more than 15 percent (sodic soil)
8.0–8.5	$CaCO_3$ hydrolysis	100 percent base saturation (calcareous soil)
6.0–7.0	Hydrogen exchange	Base saturation below 100 percent, some hydrogen saturation
4.0–6.0	Aluminum hydrolysis	Low base saturation, may be high Al saturation

FIGURE 11-3 Reactions that determine pH ranges.

(*f*). This balance controls pH in slightly acid to slightly alkaline soils.

When pH declines to about 6.0, aluminum begins to leave the structure of silicate clays. Aluminum ions react in several steps with water to form hydrogen ions and aluminum hydroxide compounds. The reactions are summarized in reactions (*g*) and (*h*):

$$(g) \quad \boxed{\text{micelle}}\text{—}Al^{+3} \rightarrow Al^{+3} + \boxed{\text{micelle}}\text{—}$$

$$(h) \quad Al^{+3} + 2H_2O \rightarrow Al\,(OH)_2^{\,+1} + 2H^+$$

Aluminum hydrolysis can lower soil pH to about 4.0. This is the most acidic the majority of upland soils become. Figure 11-3 summarizes the pH ranges and associated conditions.

Causes of Acidity. Relatively young soils—those not exposed to long periods of weathering and leaching—share the pH of their parent materials. Acidic parent materials include granite, sandstone, and shale. These materials are common in New England, the Great Lakes, and the Appalachian states. The soils of many states, including many in the Great Plains, developed from calcareous parent materials like limestone. When young, these soils tend to be neutral to alkaline.

The pH of most soils is controlled by the percolation (or lack of percolation) of acidic water. This percolating water leaches away bases and replaces them on the exchange sites with hydrogen and aluminum ions, reaction (*i*):

$$(i) \quad \boxed{\text{micelle}}\text{—}Ca^{+2} + 2H^+ \rightarrow \boxed{\text{micelle}}\langle^{H^+}_{H^+} + Ca^{+2}$$

Figure 11-4 portrays this reaction graphically. This type of percolation occurs in humid climates, where precipitation exceeds evapotranspiration. In a humid climate, net movement of water over the course of a year is downward, leaching out bases. In semiarid or arid zones, net water movement is upward, since water is being pulled out of the root zone by evaporation or transpiration. With little or no percolation, soils of dry regions tend not to become acidic. They may even become quite alkaline from calcium or sodium being carried upward into the root zone by capillary movement. Figure 11-5 shows that leaching has the greatest effect on the soils of the eastern half of the United States and the Pacific Northwest.

A number of processes produce the hydrogen ions that make soil more acidic. Some processes occur naturally, and others result from human activities. A major natural process that contributes to soil acidity is the respiration of plant roots and other soil organisms. During respiration, organisms give up carbon dioxide which reacts with water to produce carbonic acid, reaction (*j*). Carbonic acid, in turn, breaks down to release hydrogen ions:

$$(j) \quad CO_2 + H_2O \rightarrow H_2CO_3 \rightarrow HCO_3^{\,-} + H^+$$

Thus, plant growth and organic-matter decay both produce hydrogen ions. This is the same reaction that acidifies rainfall, which is a second cause of soil acidity. Crop plants acidify soils in two additional ways. First, when roots take up cation nutrients like potassium, they "give back" an equivalent number of hydrogen ions. Second, growers take calcium and magnesium with each crop harvested. For example, every ton of alfalfa hay is a loss from the soil of thirty pounds of calcium and eight pounds of magnesium. This removal of magnesium and calcium during harvest speeds up the acidification of soil.

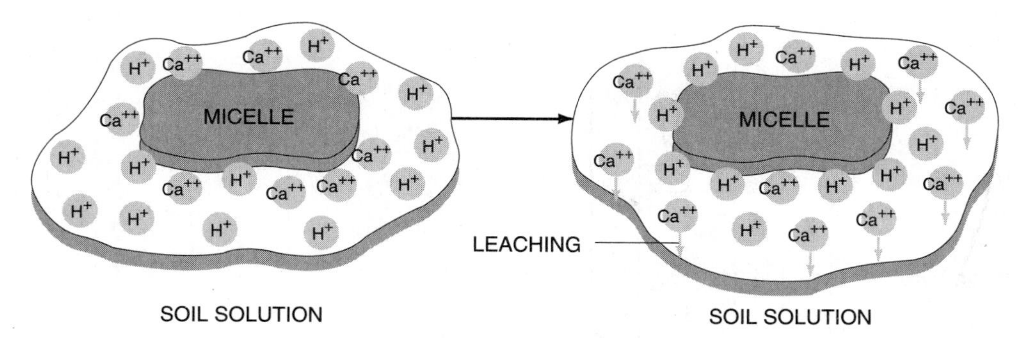

FIGURE 11-4 Percolation of acidic water causes hydrogen ions to replace exchangeable bases on colloids. The calcium and magnesium thus replaced can leach away.

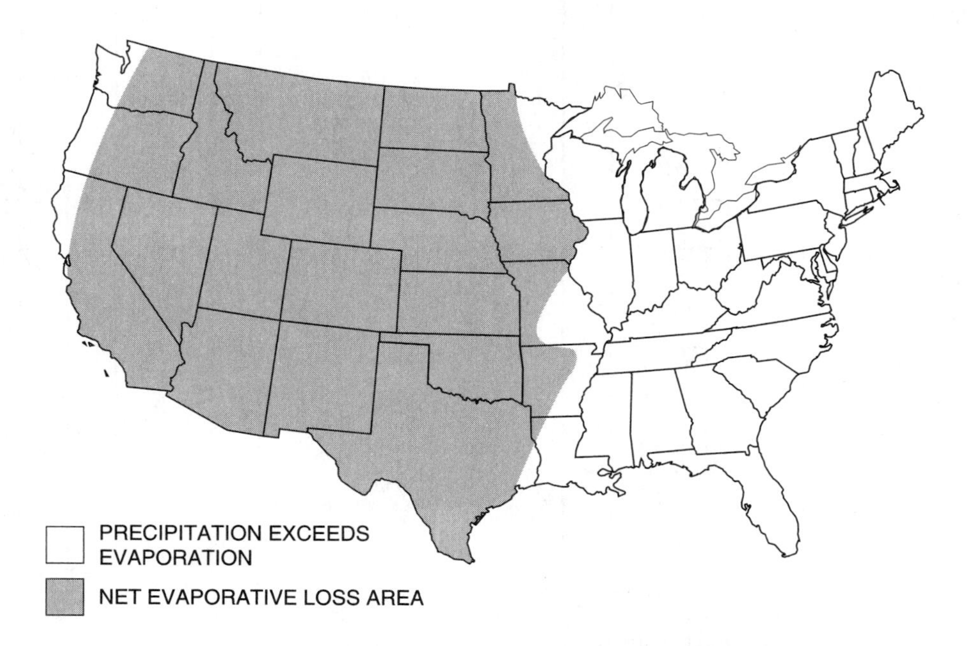

FIGURE 11-5 Soil acidity is greatest where average annual precipitation is greater than average annual evapotranspiration, where there is potential for leaching. Where there is a net evaporative loss, soils tend to be alkaline. *(Source: USDA, Second National Water Assessment, 1975, Appendix C-1, Nationwide Analysis)*

Nitrification also contributes hydrogen ions to the soil. When nitrifying bacteria oxidize ammonium ions (NH_4^+), hydrogen ions result:

(*k*) $2NH_4^+ + 3O_2 \rightarrow 2NO_3^- + 8H^+$

This reaction is most important because of the common use of ammonium fertilizers, otherwise called acid-forming fertilizers.

Effects of pH on Plants

Each crop grows best in a specific pH range. The pH ranges for a selection of crops are shown in figure 11-6. Most plants growing on mineral soils do well at a pH range of 6.0–7.0. For organic soils, most crops prefer a pH of 5.5 to 6.0. An exception is a group of acid-loving plants that includes mostly woody plants like blueberry

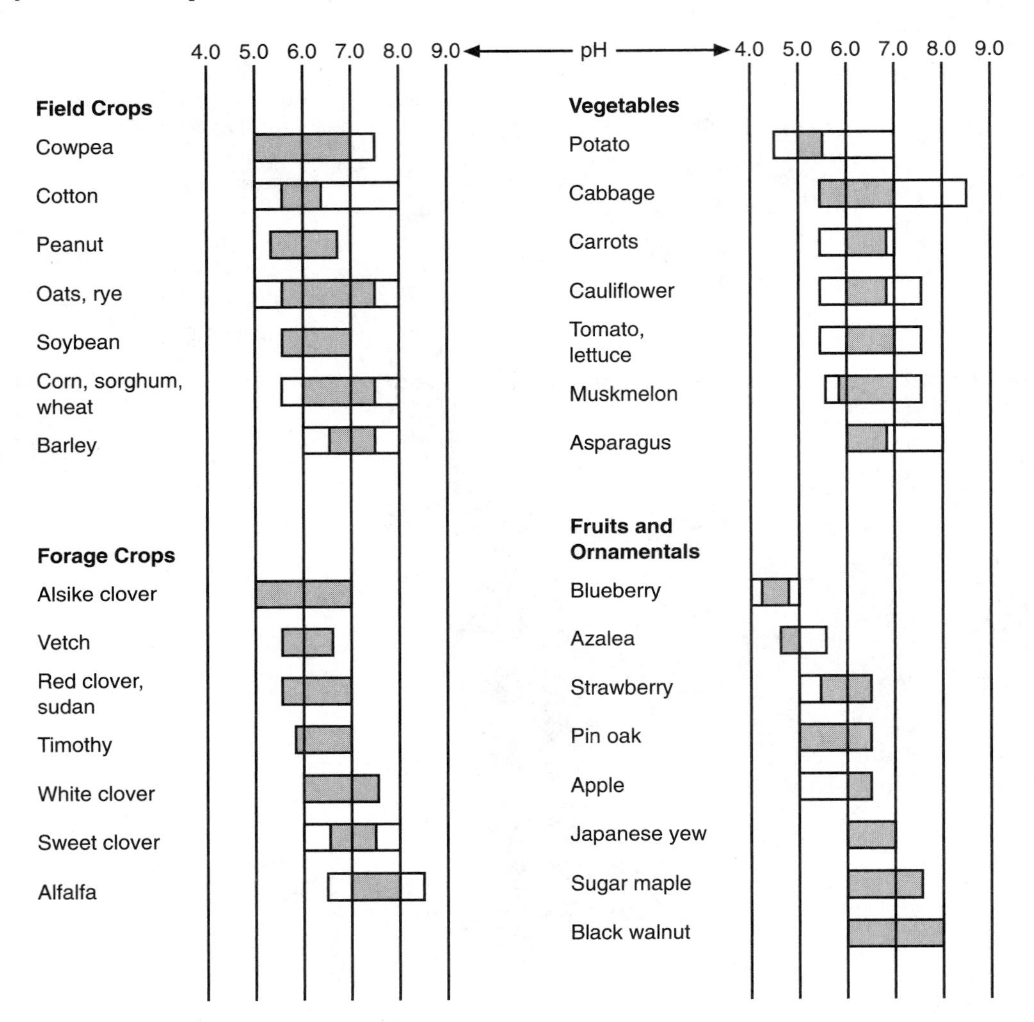

FIGURE 11-6 The horizontal bars represent the pH range of several crops for mineral soils of temperate regions. The shaded area of each bar shows preferred pH, the unshaded areas tolerances.

and azaleas and many evergreens. Alfalfa is one of a few crops that prefer a slightly basic soil.

Except at pH extremes, the actual number of hydrogen or hydroxyl ions does not seem to be the main factor in plant growth. Rather, several soil conditions related to pH are more important to plants. These include (1) the effect of pH on nutrient availability, (2) the buildup of toxic levels of aluminum or other metals, and (3) effects on soil microbes. Which factor has the greatest effect on limiting crop growth varies from soil to soil and from crop to crop.

Effect of pH on Nutrient Availability. Many soil elements change form as a result of reactions in the soil. These reactions, controlled by pH, alter the solubility, and therefore the availability, of nutrients. A good example is phosphorus, which gets tied up with aluminum and iron at low pH, and with calcium at high

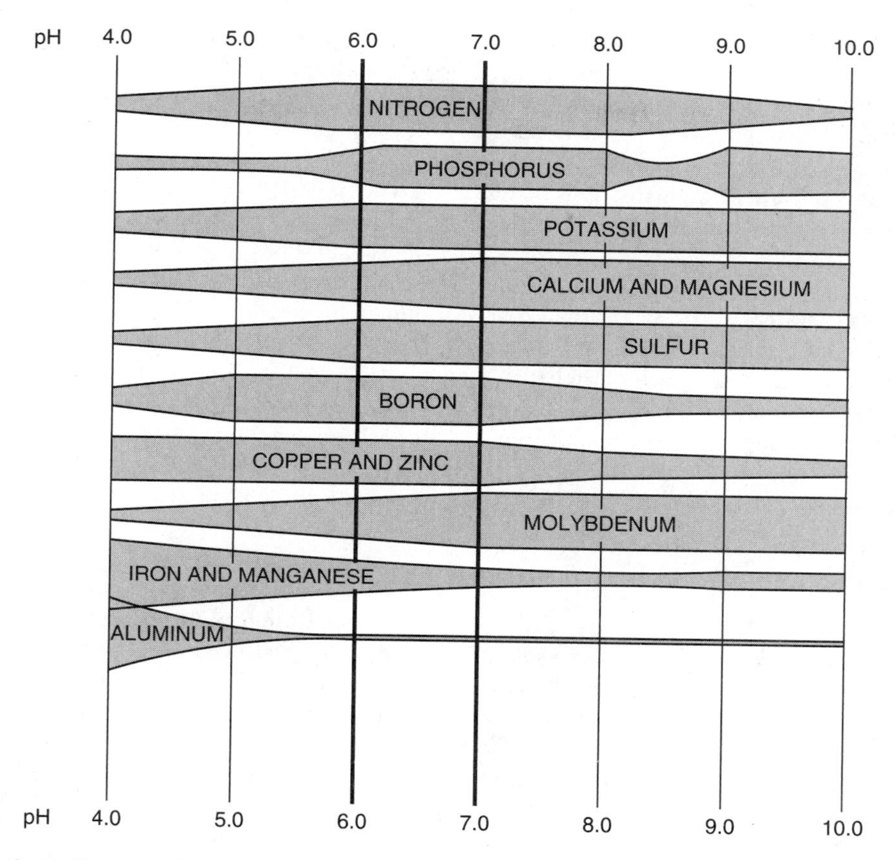

FIGURE 11-7 Soil pH affects nutrient and aluminum availability, here for mineral soils in temperate climates. The thicker the bar, the more available the nutrient. *(Source: Truog, USDA Yearbook of Agriculture, 1943-1947)*

pH. Therefore, phosphorus is most available to plants between a pH of 6.0 and 7.0.

Figure 11-7 shows the availability of nutrients at different pH levels. Note that the major nutrients and molybdenum are most available in near-neutral or higher pH soil. The other trace elements are more available in acid soil. Note that pH in the range of 6.0–7.0 is a good average level for all nutrients. This range is also the best pH range for most crops.

Soils that cannot supply enough of a nutrient may actually contain the element, but the nutrient is tied up because of acid or alkaline soil. Figure 10-2 showed an oak leaf deficient in iron; the deficiency resulted from growing on an alkaline soil.

pH and Element Toxicity. At low pH, particularly below 5.5, aluminum and manganese can reach toxic levels in the soil. Aluminum actually leaves the structure of clay minerals, reaches high concentrations in the soil solution, and occupies most of the cation exchange sites. Aluminum toxicity severely inhibits root growth, especially in acidic subsoils, and restricts the uptake of calcium and magnesium. Aluminum toxicity also increases water stress during dry periods because of poor root growth.

Manganese problems are less common but can be equally toxic under certain conditions. In the greenhouse, iron toxicity may occur on certain crops like geraniums if the pH of the potting mix drops too low.

Figure 11-7 shows that these three elements become highly soluble below pH 5.5.

Aluminum toxicity occurs primarily in mineral soils of temperate climates. Aluminum problems seldom appear in organic soils because these soils have little aluminum. In fact, plants tolerate acid organic soils better than acid mineral soils mainly because of the low aluminum level.

pH and Soil Organisms. Soil organisms grow best in near-neutral soil. In general, acid soil inhibits the growth of most organisms, especially bacteria and earthworms. Thus, acid soil slows many important activities carried on by soil microbes, including nitrogen fixation, nitrification, and organic-matter decay. Rhizobia bacteria, for instance, thrive at near-neutral pH and are sensitive to aluminum.

Liming Soil

The simplest way to ensure proper pH is to choose a crop that matches the present soil pH. Indeed, matching the crop and pH may be the only answer in some cases—growers may find it impractical to lower the pH of calcareous soils or to raise the pH of acid peat soils. Tropical soils may be very difficult to adjust profitably. In many of these situations, it is best to raise crops or select landscape plants that tolerate the existing soil pH. Refer to figure 11-6 and appendix 6 for examples. Breeders are also creating crop varieties tolerant of poor pH conditions, such as high-pH tolerant soybeans.

The other approach is to change the soil pH to match crop needs. Many field crops grow best in slightly acid soil. However, leaching of exchangeable bases, acid fertilizers, and other factors may slowly make soil more acid than is best for good growth. Liming is practiced by growers to counteract soil acidity.

Benefits of Liming. Liming acid soils has long been an important, but sometimes neglected, agricultural practice. Liming improves crop response to fertilizers by improving nutrient uptake, especially phosphorus, reducing aluminum toxicity, and promoting the activities of such desirable organisms as the *Rhizobia* bacteria that fix nitrogen for legumes.

Because calcium is itself a plant nutrient, lime is also a fertilizer, especially for high-calcium crops like alfalfa. Certain limes also supply magnesium, which is important to many acid sandy soils.

Liming Materials. We apply the term **agricultural lime** to ground limestone or other products made from limestone. These materials contain calcium. When lime is mixed into soil, it neutralizes excess acidity. Common liming materials include calcitic limestone, dolomitic limestone, burned lime, and hydrated lime.

Calcitic limestone is nearly pure calcite or calcium carbonate ($CaCO_3$). It forms on the sea floor as deposits of calcium drop out of solution in seawater. Limestone deposits are widespread in the United States. The deposits are mined and ground into agricultural lime (figure 11-8).

Dolomitic limestone is a mixture of calcium carbonate and magnesium carbonate ($CaCO_3$ and $MgCo_3$). Liming with dolomitic lime helps the calcium/magnesium balance in soil. Dolomite is especially helpful in sandy soils, because they often lack sufficient magnesium. Magnesium has the same effect on soil pH as calcium.

Burned lime, or quicklime, is made by heating limestone. Heating drives off carbon dioxide resulting in the lighter calcium oxide:

$$CaCO_3 \rightarrow CaO + CO_2 \text{ (gas)} \quad (l)$$

Because calcium oxide is lighter (has a lower molecular weight), a smaller weight of it has the same effect as a larger weight of ground limestone. Burned lime also reacts more quickly in the soil. However, the material costs more and is hard to handle. Burned lime is caustic and may cake during storage. Burned lime can be used where fast action is needed but is not usually recommended.

Hydrated lime, or slaked lime, is produced by adding water to burned lime, forming hydrated lime, or calcium hydroxide:

$$CaO + H_2O \rightarrow Ca(OH)_2 \quad (m)$$

Like burned lime, hydrated lime is unpleasant and hard to handle, but fast acting. Hydrated lime is used more often than burned lime. Because of processing steps, it is more expensive than regular ground lime, but it may be used where speed of reaction is needed.

FIGURE 11-8 Limestone deposits are mined and ground into agricultural lime. Here, a spreading truck is being loaded in a limestone quarry.

Growers may find other locally useful materials:

- Marl is a soft, chalky freshwater deposit in swamps that receive alkaline runoff water from nearby land. Although marl is difficult to harvest and spread, it may be useful where locally mined.
- Ground seashells, a by-product of shellfish industries, may be used in areas where those industries thrive.
- Lime-rich by-products of several industries may also be locally available.
- Wood ashes can be used by gardeners who burn wood.

It should be noted that gypsum ($CaSO_4$) does not change soil pH, so cannot be used as an agricultural lime.

How Lime Works. Lime neutralizes soil in two ways. First, calcium replaces hydrogen and aluminum ions on exchange sites by mass action. In doing so, liming raises the percent base saturation. Second, lime converts hydrogen ions to water. Let's look at a couple of examples to see how this works.

The simplest reaction is that of hydrated lime (figure 11-9). As hydrated lime dissolves, it releases calcium and hydroxyl ions. Calcium replaces hydrogen and aluminum on the exchange sites, releasing those cations to the soil solution. Aluminum ions undergo complete hydrolysis to form insoluble aluminum hydroxide, with the release of more hydrogen ions. All the hydrogen ions react with the hydroxyl ions from the lime, forming water.

Calcite and dolomite act in a similar fashion, with a couple of additional steps. Hydrogen ions resulting from the other steps react with the carbonate to form carbonic acid, which quickly decomposes to carbon dioxide and water. Figure 11-10 shows this process.

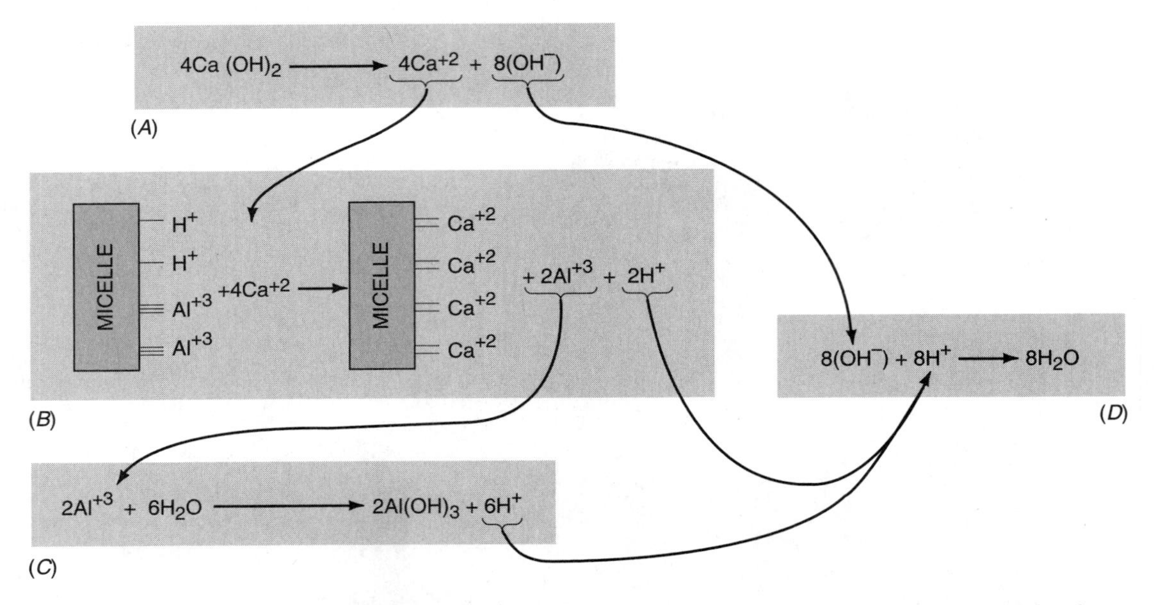

FIGURE 11-9 Hydrated lime neutralizes soil acidity in a sequence of reactions. When calcium hydroxide dissolves (*A*), calcium replaces aluminum and hydrogen on cation exchange sites (*B*). Aluminum is tied up by reacting with water to form insoluble aluminum hydroxide (*C*). Hydrogen ions released by these reactions combine with hydroxyl ions from lime to form water (*D*).

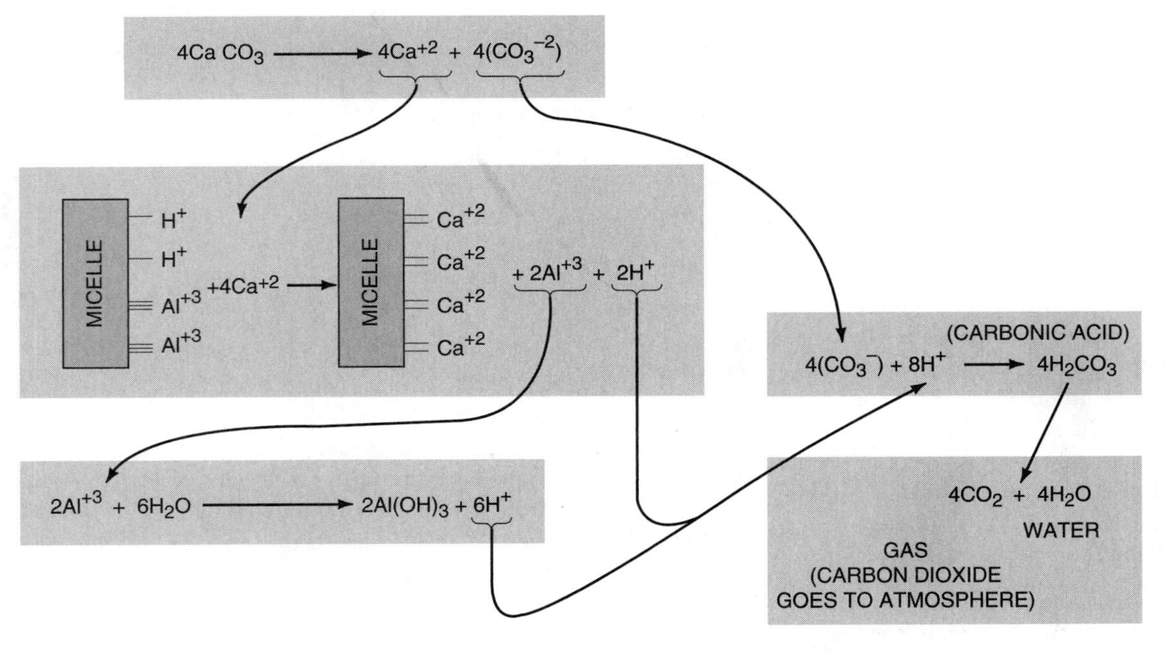

FIGURE 11-10 Ground limestone neutralizes soil acidity in the reactions shown. The cation exchange reaction series is the same as in figure 11-9. Hydrogen ions react with carbonate ions to form unstable carbonic acid, which immediately breaks down to carbon dioxide and water.

Other liming materials undergo similar reactions. The important thing to remember is that calcium (or magnesium) replaces hydrogen and aluminum on cation exchange sites and hydrogen ions are changed to water. The speed of this overall process varies according to the type of material. Hydrated lime dissolves quickly in the soil and reacts quickly. Ground limestone, on the other hand, dissolves more slowly and takes more steps to neutralize acid.

How Much Lime to Apply. Four factors tell the grower how much lime is required: the present pH, the desired pH, the cation exchange capacity of the soil, and the liming material to be used.

By testing the present pH, and by knowing the correct pH for a given crop, a grower or soil testing laboratory can determine how much a pH should change. For example, if alfalfa grows well at a pH of 6.5, and the present pH is 5.0, then the pH must be raised one and a half points. The pH of the soil solution can be measured with methods described in chapter 13. However, pH by itself does not tell how much lime to apply, because it measures only the hydrogen ions in solution, not the potential acidity (hydrogen and aluminum) adsorbed on the colloids. Hydrogen ions in solution can be termed active acidity, while adsorbed hydrogen and aluminum is reserve acidity. While some soil scientists consider these terms out of date, they are useful concepts for this discussion.

Effect of Cation Exchange Capacity on Liming. Picture the soil (figure 11-11) storing hydrogen and aluminum in a large bin (reserve acidity adsorbed on soil colloids) attached to a small one (active acidity in solution). pH measures only the active acidity in the small bin. If we add enough lime to neutralize those hydrogen ions, they are quickly replaced from the large bin. Thus, the soil resists a pH change, a process called **buffering.** Enough lime must be added to draw down both bins before the soil will become less acid.

The size of the large bin depends on the cation exchange capacity (CEC)—the larger the CEC, the more hydrogen a soil can hold, and the more lime it needs. A soil with a CEC of 20 needs twice as much lime as a soil at the same pH with a CEC of 10.

The buffering capacity of a soil depends on the amount of clay in the soil, the type of clay, and the amount of humus. The amount of clay can be estimated by knowing the textural class of a soil. Figure 11-12 suggests how much lime to apply to soils of various textures. The type of clay modifies the effects of texture, being highest for vermiculite and lowest for sesquioxides. The buffering capacity is also increased by the amount of organic matter in the soil.

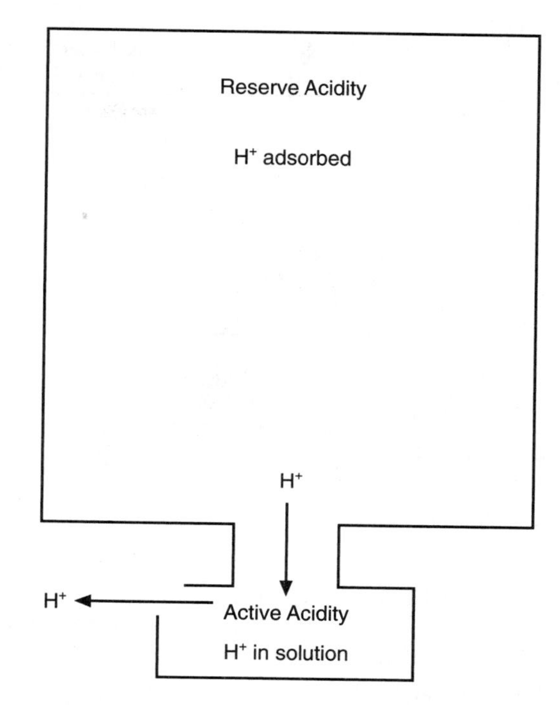

FIGURE 11-11 pH buffering in the soil. As hydrogen ions are removed from the soil solution by liming, they are replaced by ions held on clay and humus.

The lime requirement of an acid soil depends on both pH and buffering capacity. The total lime requirement can be measured directly by a **buffer test,** for example, measuring the reaction of soil to a pH 7.5 buffer solution. The test does not measure actual pH; it is a guide to the amount of lime needed to correct pH.

The pH of a soil sample is first measured to see if liming is needed. Let's say a soil sample has a pH of 5.5. The laboratory technician adds pH 7.5 buffer solution to the soil sample and measures the pH of the new mixture as 6.2. This means the acidity of the soil lowered the pH of the buffer solution from 7.5 to 6.2. The pH of 6.2 is called the "buffer index." Looking at a table of

Textural Class	Change 4.5 to 5.5	Change 5.5 to 6.5
Sand, loamy sand	25	30
Sandy loam	45	55
Loam	60	85
Silt loam	80	105
Clay loam	100	120
Muck	200	225

FIGURE 11-12 Amount of limestone needed to raise pH in a 7-inch soil layer in pounds per 1,000 square feet. These values apply to soils of northern and central states of average organic matter content. *(Source: USDA, Kellogg, Soils: 1957 Handbook of Agriculture)*

Buffer pH	(Tons of Lime/Acre) Mineral Soil	Organic Soil
7.0	0	0
6.8	1.0	0
6.6	2.0	0
6.4	3.0	1.0
6.2	4.0	2.5
6.0	5.5	4.0
5.8	6.5	5.0
5.6	8.0	6.0

FIGURE 11-13 Buffer index tables, such as this simplified one, can be used to determine the amount of lime needed to bring a 6 3/4 inch depth of soil to a pH between 6.5 and 7.0 with 90 percent pure calcitic lime. *(Courtesy of A & L Agriculture Laboratories, Inc.)*

buffer indexes (figure 11-13), the technician reads how much lime is needed to raise soil pH the correct number of points. In this example, the table suggests applying four tons of lime per acre of mineral soil to bring the pH to 6.5–7.0.

Neutralizing Power of Lime. Buffer tests suggest the lime needs of a soil based on an "average" calcite limestone. However, different lime products have different capacities to neutralize acidity. This capacity is called the **total neutralizing power** or **calcium carbonate equivalent.** Neutralizing power is based on comparing an agricultural lime with pure calcium carbonate, or calcite. Two factors affect the comparison: the chemical nature of the lime and the purity of the lime.

Calcium carbonate has a molecular weight of about 100 atomic mass units. Other chemicals weigh more or less, and so have a greater or lesser neutralizing power. For instance, the molecular weight of pure hydrated lime (calcium hydroxide) is 74, less than calcite. Thus, it would take a smaller weight of hydrated lime to obtain the same amount of calcium—hydrated lime has a greater neutralizing power. The calcium carbonate equivalent of hydrated lime is simply 100/74; or 135. This means 100 units of hydrated lime has the same neutralizing power of 135 units of calcite.

The second influence on neutralizing power is purity. For example, calcitic limestone is mostly calcite. However, it also contains other materials, like silt, that have no effect on acidity. A ground limestone that is 90 percent pure is only 90 percent as active as pure calcite. Since calcite has a neutralizing power of 100, the power of the limestone would be 90.

Figure 11-14 gives sample neutralizing values of several agricultural limes. If the lime requirement from a buffer test was based on 85 percent pure limestone, and a grower plans to use a different form of lime, a conversion is needed. The following problem shows how much burned lime with a neutralizing power of 150 is needed to replace three tons per acre of the calcitic lime:

$$\text{rate burned lime} = \text{rate calcitic lime} \times \frac{\text{neutralization value calcitic lime}}{\text{neutralization value burned lime}}$$

$$\text{rate burned lime} = 3 \text{ tons/acre} \times \frac{85}{151} = 1.7 \text{ tons/acre}$$

Most states regulate the purity of agricultural lime to protect the customer. The laws set the **chemical guarantees** for the neutralizing power of lime products used and/or produced in the state. For instance, the average requirement for ground limestone is a calcium carbonate equivalent of about 85.

Form of Lime	Percent Purity	Neutralizing Value
Pure Substances		
Calcium carbonate	100	100
Magnesium carbonate	100	119
Hydrated lime	100	135
Burned lime	100	178
Commonly Available Forms		
Calcitic limestone	85	85
Dolomitic limestone	85	88
Hydrated lime	85	115
Burned lime	85	151
Marl	—	50–70
Basic slag	—	60–90
Wood ashes	—	45–80
Ground seashells	85	85

FIGURE 11-14 Neutralizing values for major sources of lime. The first four values are for pure chemicals; the remaining values are averages for commonly available products.

Lime Fineness. The fineness of the grind affects how rapidly lime acts. The finer the grind, the more rapidly it can neutralize acidity. Lime producers measure the grind by passing lime through screens of a given number of squares per square inch. Figure 11-15 shows the relative efficiency of different grinds.

While finely ground lime acts most rapidly, it is also used up rapidly. A fine powder is also costly and hard to spread evenly. Most labs suggest a medium grind that contains enough "fines" (grains small enough to be dusty) for fast action. All of such a grind would pass an 8-mesh screen, and 25 percent to 50 percent would pass a 100-mesh screen.

Most states regulate the grind of agricultural lime as well as the purity. The laws specify the **physical guarantee** of agricultural limes. Laws differ from state to state. A sample law, for instance, might specify that all of a lime must pass through a 16-mesh screen and that 35 percent must pass through a 100-mesh screen.

Lime Application. Best results are obtained from liming when there is close contact between the grains of lime and the soil. To achieve this, lime should be spread evenly over the field and then mixed well into the soil. Lime-spreading trucks do a good job of spreading the material. The trucks have a V-shaped bed with a spinning disc mounted on the rear. Lime drops out of the truck bed at a controlled rate onto the spinning disc that flings the lime out in a fan-shaped pattern (figure 11-16).

Lime may also be applied in pelletized form, which is more easily applied than ground lime with common application equipment. Pelletized lime is particularly useful when applying over existing vegetation and in small applications like landscapes.

While most lime is spread in a dry form, some is finely ground and mixed with water or a fertilizer solution and sprayed on the field. **Fluid lime** acts more quickly than regular lime and so is useful where fast action is needed. Fluid lime remains active for a shorter time, so it must be reapplied sooner. Fluid lime is becoming popular because it can be applied in a fertilizer solution, combining two operations and saving time. Fluid lime is also popular with growers with short-term use of land, such as a rented field.

After lime is spread, plowing and/or discing mixes the lime into the soil. In established pasture or other situations where plowing is not possible, lime is simply

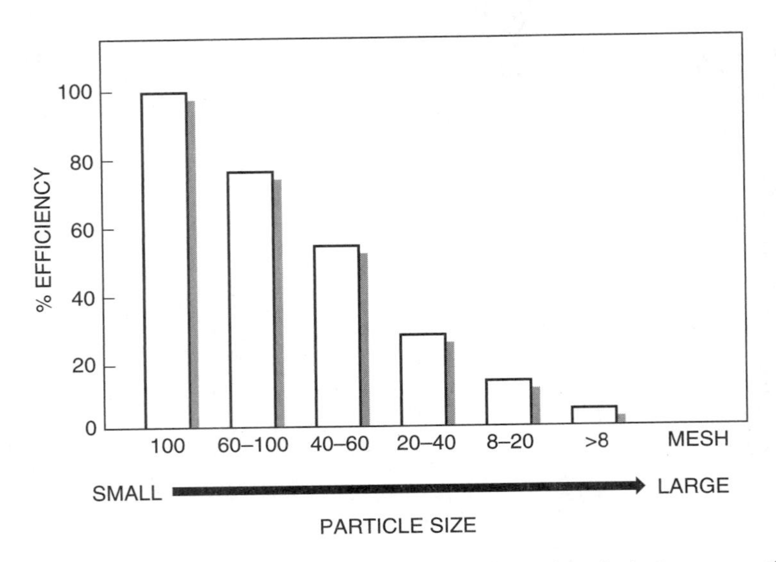

FIGURE 11-15 The neutralizing efficiency of lime depends on its grind (particle size). Coarser grades are compared with 100-mesh grind, which is assigned a value of 100.

FIGURE 11-16 Lime being spread on a farm field.

spread evenly on the soil surface. If it is not mixed into the soil, the lime slowly moves into the soil.

Growers may lime at any time that is convenient. To avoid compaction, however, it is best not to drive the trucks on wet soil. Lime should not be applied with certain forms of nitrogen fertilizer because it can cause nitrogen losses (see chapter 14). The most important consideration is reaction time, because it takes a few months for lime to break down in the soil. If faster action is needed, a grower can use fluid lime, hydrated lime, or more finely ground lime grades.

Acidifying Soil

Crop growth can be inhibited by alkaline soils, common to the areas of the country that are shown in figure 11-5. Here, excess lime or sodium keeps soil pH high, as shown earlier in this chapter in reactions (*c*), (*d*), and (*e*). Overlimed soils may also become alkaline.

As shown in figure 11-7, zinc, manganese, iron, and other trace elements are tied up in basic soils. In addition, free molybdenum can reach toxic levels. Where soil is strongly alkaline, many crops will grow poorly. Even near-neutral soils present a problem to acid-loving plants like azalea and pin oak. Such plants appear to have a very high iron requirement. In the author's home state, new, hardy azaleas are becoming popular, but most local soils are not acidic enough for their needs. Here, landscapers may amend planting beds with acid peat moss prior to planting. However, the pH lowering effects are temporary (the peat decays), and peat amendments are not suitable for large areas.

For longer-lasting pH reduction, and for larger areas, sulfur is preferred. Once applied and mixed into the soil (as described earlier for lime), Thiobacillus bacteria alter sulfur to sulfuric acid:

$$(n) \quad 2S + 3O_2 + 2H_2O \rightarrow 2H_2SO_4 + \text{energy}$$

The sulfuric acid releases hydrogen ions, and the soil becomes more acid.

Sulfur is available in granular and powdered forms and as a flowable liquid. The powdered form acts most rapidly, but is more difficult to handle. Granular sulfur, while slower acting, spreads much more easily with application equipment. Figure 11-17 suggests sulfur application rates.

A number of other chemicals also acidify the soil. These include iron sulfate, $Fe_2(SO_4)_3$, and aluminum sulfate, $Al_2(SO_4)_3$. These materials are less powerful than sulfur and usually more expensive. They should also be used with caution, because toxic levels of aluminum or iron may build up from repeated use. In addition, when increasing pH is a problem, strongly acid-forming fertilizer may be selected. Gypsum, as a neutral salt, is unlikely to lower pH.

Calcareous soils may be very difficult to acidify because there is such a large reserve of lime that must be leached out. The bin example in figure 11-11 pictures this, except one would relabel "reserve acidity" with "reserve alkalinity." Here, free lime buffers soil from pH changes. Nevertheless, in the American Southwest, sulfur is often included in the preparation of flower beds.

Where pH reduction is impractical, fertilization and crop selection are required. Deficiencies may be

To Lower pH by This Amount	Ground Sulfur Pints/100 ft²		Pounds/Acre	
	Sand	*Loam*	*Sand*	*Loam*
0.5	2/3	2	360	1,100
1.0	1⅓	4	725	2,200
1.5	2	5½	1,100	3,000
2.0	2½	8	1,350	4,400
2.5	3	10	1,650	5,400

FIGURE 11-17 Amount of sulfur needed to lower pH for an 8-inch soil layer. *(Source: USDA, Kellogg Soil: 1957 Handbook of Agriculture)*

temporarily corrected by fertilizing with the proper nutrients. Crops should also be selected for tolerance to high pH. In alkaline soils, for instance, alfalfa would outperform soybeans. Varieties of the same species will vary in tolerance, so one soybean variety may grow where another would not. Appendix 6 lists pH preferences of common landscape trees.

Soil Salinity

In humid regions of the United States, acidity is a common problem for growers because percolation leaches calcium, magnesium, and sodium from the soil. Growers in the more arid parts of the nation often have a different but related problem—an accumulation of **soluble salts** of these same bases. A salt is a chemical that results from the reaction of an acid with a base, such as the reaction of hydrochloric acid with sodium hydroxide to form common table salt:

(*o*)
$$\underset{\text{acid}}{HCl} + \underset{\text{base}}{NaOH} \rightarrow \underset{\text{water}}{H_2O} + \underset{\text{salt}}{NaCl}$$

A soluble salt is defined as a salt that is as soluble or more soluble in water than gypsum (calcium sulfate, $CaSO_4$). The soluble salts of greatest concern in the soil are sulfates (SO_4^{-2}), bicarbonates (HCO_3^{-}), and chlorides (Cl^{-}) of the bases calcium, magnesium, and sodium. These salts may come from parent materials, irrigation with salty water, or even deicing salts. Prominent locations for this problem in the United States include the San Joaquin Valley of California, the lower Rio Grande Valley of Texas, and such western and southwest states as Arizona, New Mexico, Utah, and some of the northern Great Plains. Even farmland around estuaries of the East Coast may suffer some salinity problems. Salinity problems affect about 25 percent of the irrigated lands of the United States.

Growers of potted plants—greenhouses, nurseries, and interior landscapers—also experience soluble salt problems. Here high volumes of water, often containing dissolved salts, are needed to supply crop needs. Because fertilizers are salts, fertilization compounds the problem. Chapter 17 discusses this further.

Soil scientists define three types of problem soils based on the types of soluble salts: saline, sodic, and saline-sodic. Figure 11-18 summarizes these three.

Saline Soils. **Saline soils** have high levels of soluble salts except sodium. Soil salinity can be easily measured by passing an electrical current through a solution extracted from a soil sample. The greater the salt content, the more electricity will pass. The amount of electrical flow is called electrical conductivity and is measured by the unit millimhos per centimeter (mmhos/cm). This unit of measure is presently being replaced by siemen per meter, which equals 10 mmhos/cm.

A saline soil is defined as a soil with an electrical conductivity of four or more millimhos per centimeter. However, salinity levels as low as two mmhos/cm can injure very sensitive crops. Most salts are chlorides or sulfates. Less than half of the cations are sodium, and little sodium is adsorbed on soil colloids. Soil pH is 8.5 or less. A white crust may be seen on the soil surface, due to salts migrating to the surface by capillary rise.

The main effect of salinity is to make it more difficult for plants to absorb water from the soil. In nonsaline soils, only the attraction of water for soil particles (matric potential) contributes to total water potential. In saline soil, the water also is attracted to ions in solution (osmotic potential), so less water is available to plants (figure 11-19). Saline soils may also damage plants when chlorine or other ions reach toxic levels in the plant.

Soils can be classified for use based on salinity. Figure 11-20 shows the classification system. Figure 11-21

Salted Soil Class	Conductivity (mmhos/cm)	Exchangeable Sodium (%)	Sodium Adsorption Ratio	Soil pH	Soil Structure
Saline	>4.0	<15	<13	<8.5	Normal
Sodic	<4.0	>15	>13	>8.5	Poor
Saline-sodic	>4.0	>15	>13	<8.5	Normal

FIGURE 11-18 Characteristics of salted soils.

classifies common crops according to their salt tolerance. Appendix 6 lists salt tolerance of common landscape trees.

Sodic Soils. **Sodic soils** are low in the kinds of salts found in saline soils but high in sodium. The exchangeable sodium percentage (or sodium saturation) is 15 or more, and pH is in the range 8.5 to 10.0. Sodium is often measured by the **sodium adsorption ratio** (SAR). The SAR compares the concentration of sodium ions with the concentration of calcium and magnesium ions according to the formula:

$$\text{SAR} = \frac{[\text{Na}^+]}{\sqrt{\dfrac{[\text{Mg}^{+2}] + [\text{Ca}^{+2}]}{2}}}$$

Using this measurement, a sodic soil has an SAR greater than or equal to 13.

Sodic soil has a number of effects on plant growth. The importance of these effects varies according to soil and crop.

- Sodium reacts with water, reaction (*d*), to form lye. The resulting high pH, 8.5 or higher, limits growth of many crops.

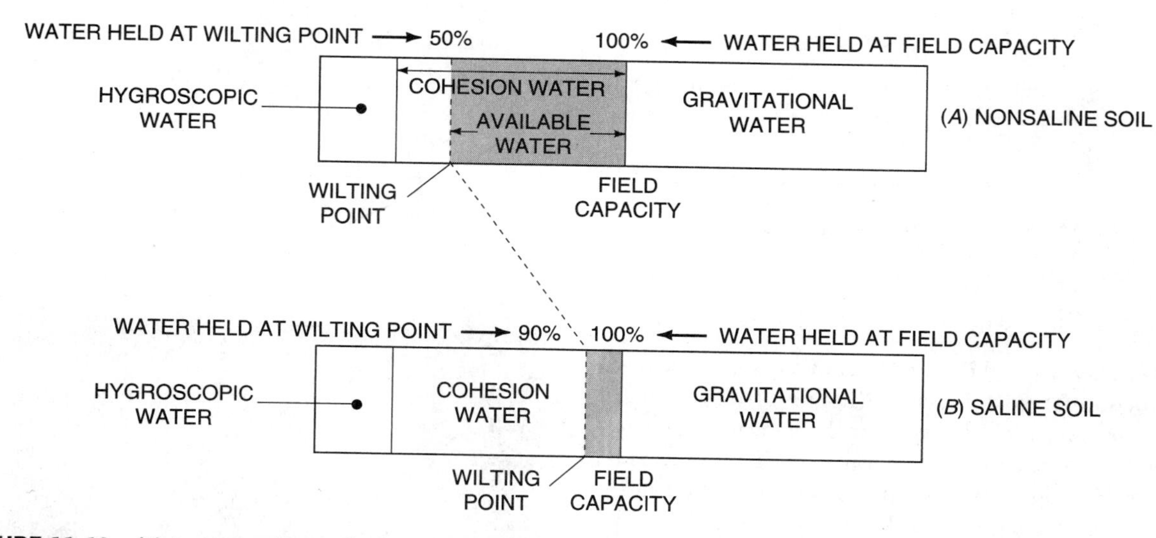

FIGURE 11-19 (*A*) In a nonsaline soil, about half the water held at field capacity is available to plants. (*B*) In saline soil, as little as 10 percent may be available because of osmotic potential.

Class	Salinity (mmhos/cm)	Crop Response
Nonsaline	0–2	Salinity effects unimportant
Slightly saline	2–4	Yields of sensitive crops lowered
Moderately saline	4–8	Yields of many crops lowered
Strongly saline	8–16	Only tolerant crops yield well
Very strongly saline	More than 16	Only most tolerant crops yield well

FIGURE 11-20 Crop responses to soil salinity.

Type of Crop	Tolerant	Medium	Sensitive
Field crops	Barley	Corn	Beans
	Sugar beet	Soybean	Flax
	Cotton	Sorghum	Broadbean
		Wheat	
Forage Crops	Bermuda grass	Alfalfa	Clovers
	Wheatgrass	Orchard grass	
	Tall fescue	Perennial rye	
Vegetables	Beets	Spinach	Lettuce
	Asparagus	Tomato	Bell pepper
		Broccoli	Onion
		Cabbage	Carrot
		Potato	Beans
		Sweet corn	Celery
Fruits	Date palm	Grape	All others
		Fig	
		Olive	

FIGURE 11-21 Tolerances of selected crops to soil salinity.

- For many crops, the main effect of sodium is the destruction of soil structure (figure 11-22). When sodium ions saturate cation exchange sites, the colloids separate and disperse soil aggregates. Tiny soil particles lodge in the soil pores, sealing the soil surface and creating wet "slick spots." Tilth suffers and crusts hard enough to stop seed germination may form. Sodic soils may also show a poorly drained columnar subsoil structure. The effect of sodium is most extreme on fine-textured soils and least extreme on coarse soils.
- Crop plants may take up enough sodium to injure plant tissues. Crops vary in their tolerance to sodium. For the most sensitive crops, like citrus fruits, the nutritional effects of sodium are more important than its effects on structure. For sodium-tolerant crops, poor growth results mainly from soil conditions. Figure 11-23 lists the sodium tolerance of selected crops.

FIGURE 11-22 A sodic soil. Sodium has dispersed soil aggregates, destroying structure. Vegetation is sparse. *(Courtesy of USDA ARS)*

Sensitive Sodium Percentage (Exchangeable [ESP] = 2–20)	Moderately Tolerant (ESP = 20–40)	Tolerant (ESP = 40–60)	Most Tolerant (ESP Above 60)
Deciduous fruit	Clovers	Wheat	Crested wheatgrass
Nuts	Oats	Cotton	Tall wheatgrass
Citrus fruit	Tall fescue	Alfalfa	Rhodesgrass
Avocado	Rice	Barley	
Bean	Dallisgrass	Tomato	
		Beets	

FIGURE 11-23 Tolerance of some crops to exchangeable sodium. Damage to the most sensitive crops is from sodium toxicity. Damage to tolerant crops is due to poor soil conditions. *(Source: USDA Agriculture Information Bulletin No. 216, 1960)*

Saline-Sodic Soils. **Saline-sodic soils** contain high levels of both soluble salts and sodium. The electrical conductivity is greater than 4.0 millimhos per centimeter, the SAR is greater than 13, and pH is less than 8.5. The physical structure of these soils is normal. However, after periods of heavy rain or irrigation with low-salt water, soluble calcium and magnesium may leach out of the soil, leaving behind the sodium salts. The soil may then become sodic, with poor physical structure and drainage.

Reclaiming Salted Soils. The first step in reclamation of salted soils is to decide if the project is practical and will pay for itself. The basic step to reclaiming soil is to leach out salts, so there must be a source of acceptable water. Very fine-textured soils may not allow sufficient drainage. If a decision is made to reclaim the soil, the next step is to ensure good drainage to allow salted water to leave the soil profile.

Many salted soils have problems with drainage, including a high water table, hardpans, or fine soil texture. Subsoiling can help break up hardpans. Soil with a high water table must be drained to a depth of five or six feet so salty water can be removed from the root zone and carried off the field. After proper drainage has been installed, the next steps depend on the type of problem.

Saline soils are most easily reclaimed. Growers flood the soil surface so that percolation leaches salts out of the soil profile. High-quality water works best, but larger amounts of fairly saline water will also work. Treatment water should, however, be low in sodium. Ponding is one way to apply leaching water. In ponding, heavy equipment constructs low earthen dikes to divide the affected land into ponds, which are then flooded. The field may be ponded several times, allowing time for drainage between floodings.

The reclamation of saline soils has been improved by the use of organic mulches. Mulches reduce evaporation of water from the soil surface, increasing the net movement of water downward. In addition, organic matter keeps the soil loose and maintains structure to improve drainage.

Sodic soils cannot usually be reclaimed simply by leaching, because the sealed soil surface inhibits drainage. It is usually necessary to first remove the sodium. This is usually done by treating the soil with gypsum. Granular gypsum may be spread on the soil surface, or finely ground gypsum may be applied through an irrigation system. When gypsum enters the soil, it dissolves and calcium replaces sodium on the cation exchange sites. Sodium sulfate leaches out out of the soil:

(*p*) $$\boxed{\text{micelle}}\genfrac{}{}{0pt}{}{\diagup Na^{+}}{\diagdown Na^{+}} + CaSO_4 \rightarrow$$

$$\boxed{\text{micelle}}\text{—}Ca^{+2} + \underset{\text{(leaches)}}{Na_2SO_4}$$

Gypsum is the least expensive amendment, but other chemicals may be used as well. If soil contains some lime ($CaCO_3$), finely ground sulfur will add calcium indirectly. The sulfur is converted to sulfuric acid by bacteria, reaction (*n*). The hydrogen ions from the sulfuric acid

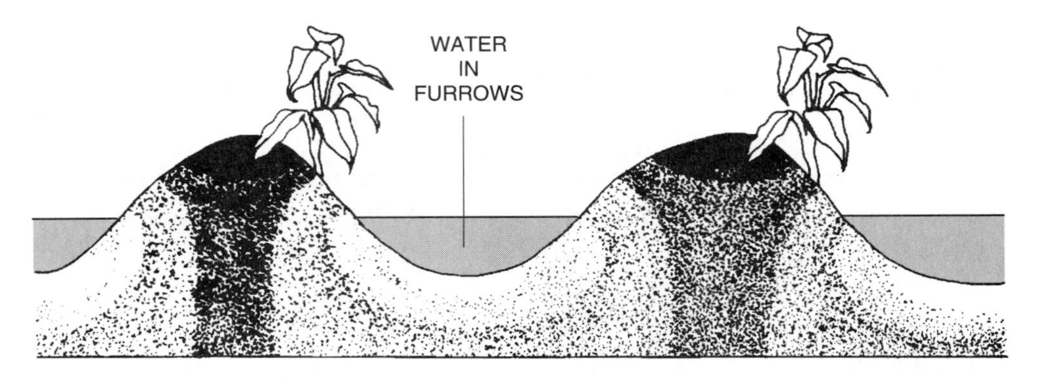

FIGURE 11-24 In furrow-irrigated fields, planting can be done on the shoulders of the furrows. The top of the furrows has the highest salt content, which inhibits seed germination.

can replace sodium on the exchange sites. More importantly, the acid reacts with soil lime to make gypsum:

(*q*) $CaCO_3 + H_2SO_4 \rightarrow CaSO_4 + H_2O + CO_2$ (gas)

The conversion of sulfur to sulfuric acid takes some time, so sulfur treatment is relatively slow. Sulfuric acid can be added directly for faster action. This is more expensive and also more dangerous because sulfuric acid is highly caustic. One USDA research project achieved a similar effect on sodic soils by planting a sorghum-sudan grass hybrid. Its roots released large amounts of carbon dioxide, which reacted with soil water to form carbonic acid, reaction (*j*). The acid dissolved soil lime, freeing calcium which displaced sodium, as in reaction (*p*).

After calcium replaces sodium on cation exchange sites, the soil slowly begins to aggregate. As the soil surface begins to improve, some growers plant salt-tolerant crops like barley. The plant roots and tops, if disced into the soil, help rebuild the soil structure.

A few additional words about gypsum may be added here. Gypsum is often sold to loosen clay soils in gardens and lawns. However, unless the soil is salt-affected, it is likely to have little effect. It may possibly improve soils damaged by road deicing salts.

Saline-sodic soils must be treated to remove sodium. If they are simply leached with low-salt water, calcium and magnesium salts are removed, but sodium remains in the soil, forming a sodic soil. Thus, gypsum treatments are useful. In the initial stages of reclamation, some growers leach these soils with fairly saline water. The calcium and magnesium salts in the water replace some of the sodium on the soil colloids, preventing destruction of soil structure.

Managing Salted Soils. Saline soils, especially irrigated land in arid climates, may be managed to reduce salt problems. One answer, of course, is to grow salt-tolerant crops. This step, however, does not really solve the problem but causes a shift over time to increasingly salt-tolerant crops. A number of practices can be used to help reduce salt problems, as follows:

- Prepare a field properly for irrigation. Proper leveling avoids low spots that collect salts. A grower may also install drainage during field preparation.
- If possible, use high-quality irrigation water. Figures 9-23 and 9-24 list the irrigation water classes.
- Keep the soil moist. Water dilutes soil salts, lowering the effect of osmotic potential. Salts tend to be most damaging in dry soil, when the salts are concentrated and both the salt and matric potential are high.
- Overirrigate enough to leach salts out of crop root zones.
- Return as much organic matter to the soil as is practical, including manures, crop residues, and green manures.
- Avoid overfertilization. Most fertilizers are salts and can compound salinity problems. Chapter 14 lists the salinity of common fertilizers.
- Maintain a good soil-testing program to monitor salinity and avoid overfertilization.
- Plant crops on ridge shoulders in furrow-irrigated fields. Salts tend to concentrate on the top of the ridge (figure 11-24).
- Use drip irrigation; it tends to reduce salt stress because it keeps the soil uniformly moist and moves

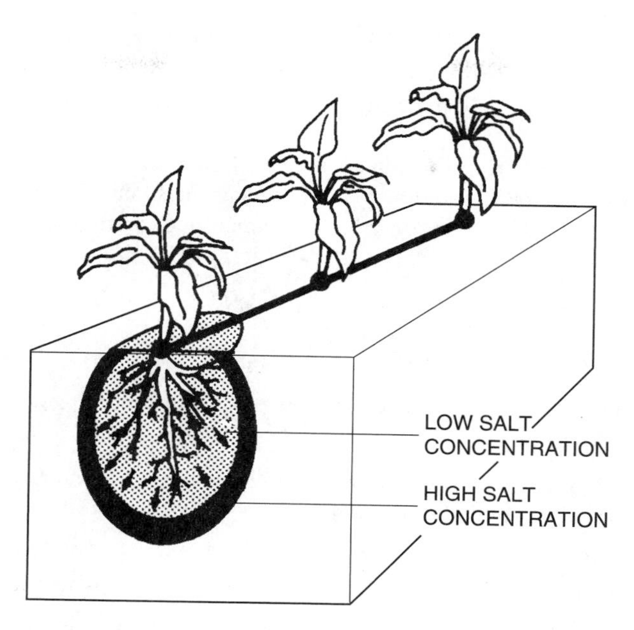

FIGURE 11-25 Drip irrigation can reduce salt stress. As water moves away from the emitter by capillary action, it carries dissolved salts out of the root zone.

salts out of the root zone of the crop plants and into the soil between plants and rows (figure 11-25).

Salted Water Disposal. One difficulty with methods for reclaiming and managing salted soils is that they do not eliminate soluble salts but move them to another place. Salty drainage water reappears in rivers downstream of affected farms, making that water even more salty. Individual growers do have some options to help reduce saline discharges from their fields. These suggestions stress reducing water use and retaining salt safely in the field:

- Improve water delivery systems to reduce seepage and evaporation from canals.
- Use techniques to improve irrigation efficiency, like surge irrigation and careful budgeting to reduce percolation and tailwater losses.
- Where feasible, adopt drip irrigation.
- Practice minimum leaching to carry salts below the root zone but not into the drainage system.
- Reuse salty water on salt-tolerant crops like barley or sugarbeets.

SUMMARY

Soil pH depends on the balance of hydrogen and hydroxyl ions in the soil solution. Alkaline soil, with a pH between 7.0 and 10.0, results from the reaction of calcium and sodium with water to form hydroxyl ions. Acid soil, with a pH between 4.0 and 7.0, results from the leaching of these bases by mildly acidic water and from the release of hydrogen ions by aluminum hydrolysis.

Acid soils affect plant growth by lowering the availability of phosphorus and other nutrients, freeing toxic levels of aluminum, and inhibiting helpful soil organisms. Alkaline soils render several micronutrients unavailable and create many problems associated with salted soils. Most plants grow best between pH 6.0 and 7.0. A few, like potatoes, perform best in acid soil, while a very few, like alfalfa, do best in neutral or mildly alkaline soil.

Acid soils are treated with various forms of agricultural lime, mostly ground limestone. Lime replaces hydrogen and aluminum on the cation exchange sites with calcium and converts hydrogen ions to water. The amount of lime needed depends on the amount of pH change required, the buffering capacity of the soil, and the form of lime. Soils too alkaline for the crop being grown may be treated with sulfur.

Salted soils may be saline, sodic, or saline-sodic. Saline soils, which are high in soluble salts but low in sodium, reduce the water available to plants. Saline soils can be treated by flooding to leach out salts. Sodic soils are high in sodium and exhibit poor physical structure. They are treated with gypsum to displace the sodium. Saline-sodic soils contain both soluble salts and sodium. Care must be taken with these soils to avoid leaching the salts while leaving the sodium. After a salted soil is treated, it must be managed carefully to reduce salt problems.

REVIEW

1. It is said that in humid regions, soil is either acid or in the process of becoming acid. Explain.
2. Describe the effects of pH on plant growth.
3. A buffer test on a sample of mineral soil produces a buffer index of 6.4. How many pounds per acre of an 85 percent pure hydrated lime should be applied to the soil?
4. Explain why fine-textured or organic soils require more lime to neutralize acidity than coarse-textured soils.
5. Name soil conditions for which gypsum is an effective treatment and for which it is probably not. Explain your answer.
6. In excavation of some sites of ancient civilizations, archaeologists note evidence of a decline in agriculture. Early in the history of the site, wheat was grown; later, barley became the dominant crop; then agriculture ceased. Can you explain this?
7. Rising salinity is often a problem for plants grown in containers, as in greenhouses and interior landscapes. What would contribute to that rise? How could it be controlled?
8. How does common calcitic lime neutralize acid soil?
9. Describe four forms of agricultural lime.
10. A case study: assume you want to grow a river birch tree (*Betula nigra)* in your yard. Your soil is a moderately well-drained loam with a pH of 7.0 and a salinity of 2.0 mmhos/cm. Looking at appendix 6 and charts in this chapter, is this soil suitable for river birch as is? Explain your answer for each. If needed, how might you amend the soil, providing numbers?

ENRICHMENT ACTIVITIES

1. Put a few drops of a strong acid on a piece of limestone, and observe the fizz. Can you explain the bubbles? See figure 11-10. How does this relate to calcareous soil?
2. Use pH paper to test a number of household solutions, such as vinegar, lemon juice, tapwater, and ammonia.
3. Check a soil sample for pH by mixing one volume of soil with two volumes of distilled water. Let sit for twenty minutes, then filter with a coffee filter or other filtering device. Measure the pH of the liquid with pH paper or a pH meter.
4. At the beginning of the course, the instructor can mix finely ground lime, sulfur, and gypsum into separate soil samples in pots. Keep warm and moist for several weeks. Then check for pH in class, as described in (3) above.

5. This chapter claims that rainfall is acidic. Check out this map of the rainfall pH's for the United States. Why might it vary? <http://water.usgs.gov/nwc/NWC/pH/html/ph.html>
6. For more information on salted soils, try this Web site: <http://www.ext.colostate.edu/pubs/crops/00503.html>.
7. Here is a site on soil pH and modifying pH (note that it is a PDF file): <http://www.back-to-basics.net/efu/pdfs/ph.pdf>.

CHAPTER 12

Plant Nutrition

OBJECTIVES

After completing this chapter, you should be able to:

- discuss nitrogen nutrition and the nitrogen cycle
- discuss phosphorus nutrition
- discuss potassium nutrition
- answer questions about the secondary nutrients
- answer questions about trace elements

TERMS TO KNOW

ammonia volatilization
ammonification
chelates
chlorosis
enzymes
lodge

Many soil factors, such as texture, structure, and water, affect plant growth. Often these conditions are less than ideal but are not easily or cheaply improved. For instance, growers cannot alter soil texture over large areas. Even irrigation is a costly investment.

The supply of soil nutrient elements, on the other hand, can be more easily controlled. Soil can be tested and fertilized to satisfy crop needs. This chapter takes a detailed look at the essential elements.

Nitrogen

Nitrogen, more than any other element, promotes rapid growth and dark green color. Plants need a lot of nitrogen because it is part of many important compounds, including protein and chlorophyll. Plants respond to nitrogen in the following ways:

- Nitrogen speeds growth. Plants receiving adequate nitrogen have vigorous growth, large leaves, and long stem internodes.
- Plants make large amounts of chlorophyll, a dark green pigment. Thus, leaves are dark green on well-fed plants.
- Protein content of plant tissue will be at its best. The higher protein content makes the plant a better source of forage, feed, and human nutrition.
- Plants use water best when they have ample nitrogen.

Plants with too much nitrogen do not grow properly, however. Here are some problems associated with too much nitrogen:

- Soft, weak, easily injured growth is encouraged. For example, plant stems are weaker and more easily topple, or **lodge,** in the rain. Lodging can turn a good crop into a disaster.
- Soft, high nitrogen growth is more prone to some diseases and insects.
- Overly rapid growth slows maturity and ripening of many crops.
- Too rapid growth also delays the hardening-off process that protects many plants from winter cold. Landscape plants, for instance, may suffer winter damage when nitrogen is applied too liberally.
- Excess nitrogen impairs flavor in several vegetable crops.
- High levels of nitrates may accumulate in some crops, with possible health effects for animals or people consuming them.

About half the nitrogen in a leaf occurs in enzymes involved with photosynthesis, so a well-supplied plant photosynthesizes much more efficiently than a deficient plant. This partially explains why nitrogen stimulates growth. On the other hand, we also know that leaves high in nitrogen also respire—use up the food produced by photosynthesis—more rapidly. In low-light situations, where photosynthesis is light-limited, high nitrogen merely depletes food more quickly. Plants growing under low light, like shady turf or indoor plants, should be fertilized more lightly than plants grown in the sun.

In general, nitrogen promotes vegetative growth—stems, leaves, and roots—more than the reproductive growth of flowers and fruit. Home gardeners see the effect if they overfertilize their tomato plants, promoting lush growth but few fruit. Also, while nitrogen aids growth in both root and shoot, shoot growth tends to be favored. This can be a problem in turf, where high nitrogen fertilization can yield lush growth with an inadequate root system to support it during times of stress.

In the natural world, low nitrogen tends to be the primary nutrient limiting growth in land ecosystems of cooler climates, on young soils, and many marine ecosystems.

The Nitrogen Cycle. Of the essential elements, nitrogen undergoes the most movement and change. The series of gains, losses, and changes is termed the nitrogen cycle. The central portion of the nitrogen cycle operates by the action of soil microorganisms. To review briefly (see chapter 5), nitrogen comes from nitrogen gas (N_2) in the atmosphere, a form unusable to plants. Symbiotic (figure 12-1) or nonsymbiotic bacteria use that nitrogen to form protein for their own bodies or supply it to host plants. When these bacteria, or host plants, die, other microbes mineralize the protein (**ammonification**) to ammonium ions (NH_4^+). These ions can be taken up by plants, but most are converted by bacteria (nitrification) to nitrite ions (NO_2^-) and then to nitrate ions (NO_3^-). Nitrates are taken up by plants or microbes (immobilization) or return to the atmosphere as

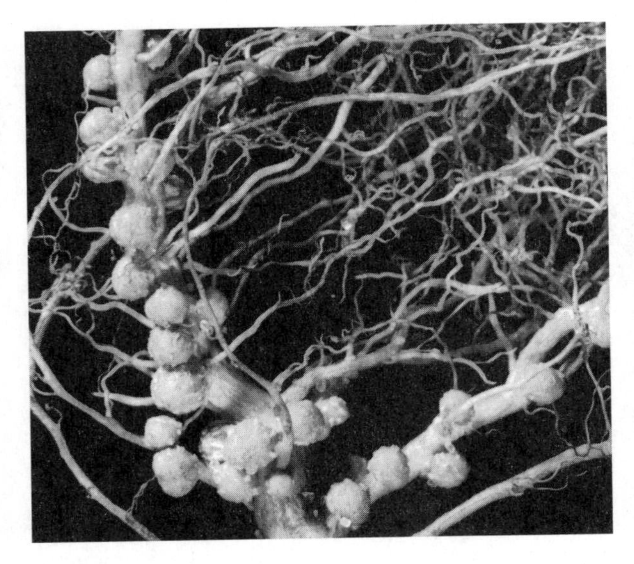

FIGURE 12-1 Symbiotic fixation. These nodules contain *Rhizobium* bacteria that fix nitrogen from the air. *(Courtesy of Crop and Soils Magazine, American Society of Agronomy)*

nitrogen gas through the process of denitrification. The solid lines in the simplified cycle pictured in figure 12-2 summarize this portion of the cycle.

The complete nitrogen cycle includes some nonbiological processes as well, shown in figure 12-2 as broken lines. Two other forms of fixation add usable nitrogen to the soil. First, lightning during storms provides energy to combine gaseous nitrogen and oxygen to form nitrogen dioxide (NO_2). The gas dissolves in water vapor to produce nitric acid (HNO_3). About five to ten pounds per acre of nitrogen fall to earth yearly in rain and snow from this source. Second, large amounts of nitrogen are fixed from the air in fertilizer factories (see chapter 14) and applied to soil by growers.

Two nonbiological losses of nitrogen from soil may be important as well. The nitrate ion is negatively charged and so is not adsorbed by soil colloids. Nor is it held in soil by other means, so nitrates easily leach from the soil. Although ammonia does not leach readily (being adsorbed by soil colloids), it too can be lost by a process called **ammonia volatilization.** Ammonium ions react with hydroxyl ions in the following reaction:

$$NH_4^+ + OH^- \leftrightarrows \underset{\text{ammonia (gas)}}{NH_3} + H_2O$$

The smell of an open bottle of household ammonia (ammonia gas dissolved in water) is a result of this reaction. Normally, this is a balanced reaction in the soil, with nitrogen changing back and forth between the two forms. However, the balance can be shifted by soil conditions to cause a loss of ammonia. If soil dries, for instance, water is lost from the right side of the equation. As a result, the reaction shifts to the right and releases ammonia gas (see appendix 1 for an explanation). If the soil is made more alkaline (by liming, for instance), the reaction again shifts to the right because of an excess of hydroxyl ions. Thus, ammonia losses may occur in dry, alkaline, or recently limed soil.

In native habitats, including virgin forests or prairies, gains and losses in the cycle balance over time. However, farming changes the balance greatly in ways that increase nitrogen losses:

- Nitrogen is removed by crop harvest (figure 12-3).
- Cropped soil is more likely to erode, so nitrogen and other nutrients are carried off in running water.
- Irrigation increases percolation of water through the soil profile. Thus, losses of nitrate nitrogen by leaching increase on irrigated land.
- Liming may increase the loss of ammonia by volatilization.

To compensate for increased nitrogen losses and to meet the needs of modern high productivity, growers supply more nitrogen by manuring, growing legumes, or by fertilization. Figure 12-4 shows the nitrogen cycle as it operates on modern farms that raise both crops and animals. There is a strong trend away from a mixed farming operation toward one that specializes in either cash crops or raising animals. This trend improves economic efficiency, but exacts an obvious penalty in view of the nitrogen cycle. For the cash crop grower, more money must be spent on fertilizers. For the animal raiser, manure becomes a waste disposal problem (see chapter 15).

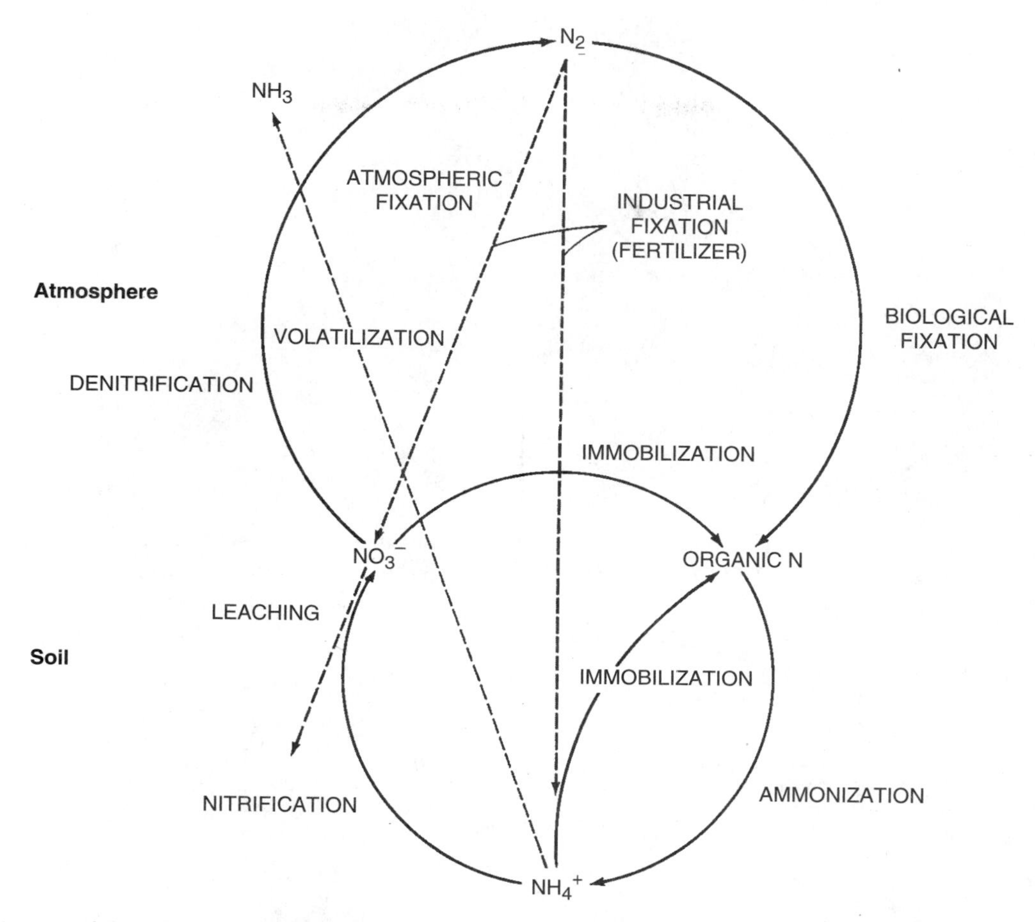

FIGURE 12-2 In this simplified version of the nitrogen cycle, dark lines indicate the biological processes described in chapter 7. The broken lines indicate nonbiological processes described in this chapter.

Forms of Nitrogen in the Soil. About 97 percent of soil nitrogen resides in organic matter, the soil's storehouse of nitrogen. At any time, only a small percentage of nitrogen is mineralized to usable forms. On the average, decay makes available about 90 pounds of mineral nitrogen per acre per year (see figure 12-5). However, it also follows that cropping systems that preserve soil organic matter, like no-till and some organic farming, also retain more soil nitrogen. The difference can be substantial. In one 1996 study, after twenty-three years of continuous corn, there were nine milligrams of mineralizable nitrogen per kilogram of soil (in the top 7.5 centimeters) under no-till, compared to 1.4 mg N under conventional tillage.[1]

Both mineral forms of nitrogen, ammonium and nitrates, are taken in by plants. In forest and woodland, ammonium is the most common form. Farm crops usually make more use of nitrate, either from nitrate fertilizer or

[1]Handayani, I. 1996. Soil carbon and nitrogen pools and transformations after twenty-three years of no-tillage and conventional tillage. Ph.D. Dissertation, University of Kentucky.

FIGURE 12-3 As a crop is harvested, nitrogen and other nutrients are being removed from the soil. The grower must replace these nutrients to raise next year's crop. *(Courtesy of USDA NRCS)*

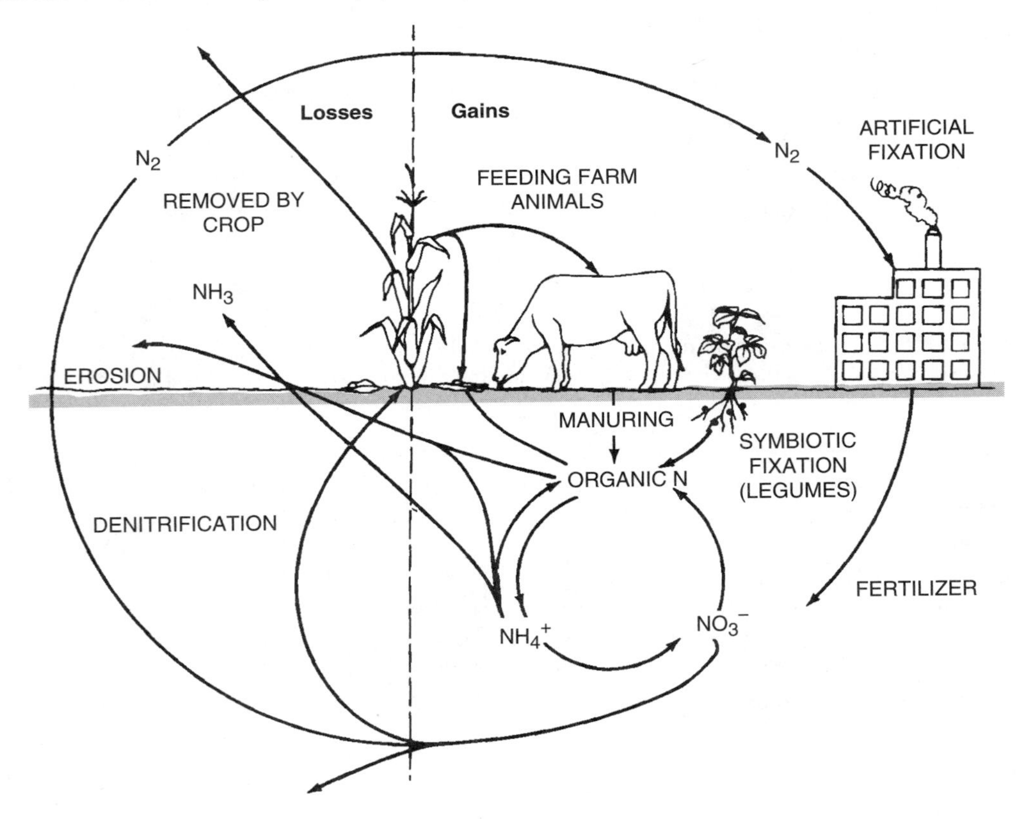

FIGURE 12-4 Agriculture breaks the natural nitrogen cycle of removing a crop, which removes nitrogen. Tillage also speeds nitrogen loss by leaching and erosion. Growers adjust nitrogen levels by fertilizing, manuring, and growing legumes.

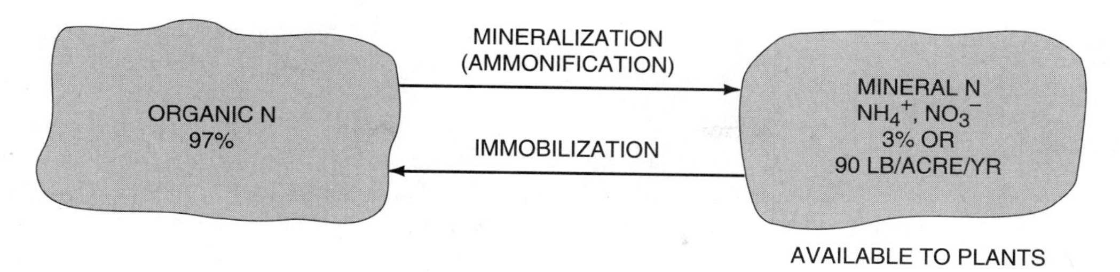

FIGURE 12-5 Organic matter serves as the storehouse of soil nitrogen; a small amount exists as usable nitrate and ammonium ions.

	Organic N	Ammonium N	Nitrate N
Storage	In humus	Adsorbed	Little storage
Losses	Mineralization, erosion	Volatilization, erosion	Leaching, denitrification
Plant Use	Not used	Usable	Usable
Changes	Mineralization	Immobilization, nitrification	Immobilization, denitrification

FIGURE 12-6 Characteristics of three types of soil nitrogen.

from nitrified ammonium. The two ions behave very differently in the soil (figure 12-6).

Ammonium nitrogen bears a positive charge. Negatively charged soil colloids attract the cation, protecting it from leaching. The nitrate ion, by contrast, moves freely in the soil because of its negative charge. Free movement allows nitrate to diffuse easily through soil to plant roots. However, nitrate losses from soil can be high. Nitrate ions leach out of the soil readily, and some may disappear as nitrogen gas in wet soil.

The amount of ammonium and nitrate nitrogen in the soil depends on the amount and type of nitrogen applied to the soil and the rates of nitrification and denitrification. Some nitrogen fertilizers contain nitrates. Most modern fertilizers, however, mainly provide ammonium nitrogen. Nitrifying soil bacteria change this to nitrates, the preferred form for crops. Nitrifying bacteria grow best in moist, loose, well-drained soil at a pH of 6.0 to 7.5. Nitrifying bacteria function poorly below 41°F and reach maximum activity between 85°F and 95°F. Thus, cold, wet, or acid soils slow the conversion of ammonium nitrogen to nitrate nitrogen.

Waterlogged soil prevents nitrifying bacteria from thriving. However, anaerobic denitrifying bacteria thrive in the same conditions. Denitrification causes the greatest loss of nitrogen when soil users apply nitrate fertilizers to wet soils. Similarly, overirrigation of turf wastes fertilizer by stimulating both denitrification and leaching.

Because of potential losses of nitrates, it is useful to control the ammonium nitrification rate. Several chemicals have been developed to inhibit (but not stop) nitrification. In practice, results have been variable, often of benefit but sometimes not.

Nitrogen Deficiency. In all plants, slow growth and stunting are the most obvious signs of nitrogen shortage. Because nitrogen is part of chlorophyll, nitrogen-deficient plants lack the dark green color of well-fed plants. This symptom is called **chlorosis.** Leaves turn light green, then yellow, starting with the lower leaves. In grasses, yellowing starts at the blade tips, progresses down the midvein, and finally the entire leaf yellows. In extreme cases, the leaf dries up, a symptom called firing. In broadleaf plants, leaves are small with overall yellowing.

Phosphorus

Phosphorus also spurs growth but to a lesser extent than nitrogen (figure 12-7). Phosphorus affects plant growth in a number of ways:

- Phosphorus is part of genetic material (chromosomes and genes) and so is involved in plant reproduction and cell division.
- Phosphorus is part of the chemical that stores and transfers energy in all living things. Without it, all biological reactions come to a halt.
- Phosphorus spurs early and rapid root growth and helps a young plant develop its roots.
- Phosphorus helps plants use water more efficiently by improving water uptake by roots.
- Phosphorus helps plants resist cold and disease, speeds crop maturity, aids blooming and fruiting, and improves the quality of grains and fruits.
- Phosphorus improves the efficiency of nitrogen uptake by plants, making better use of fertilizer nitrogen and reducing the risk of groundwater pollution due to nitrate leaching.
- Adequate soil phosphorus ensures that animal feeds will supply sufficient phosphorus.

In many ways, phosphorus acts to balance nitrogen. While nitrogen delays maturity, phosphorus hastens it. Nitrogen aids vegetative growth; phosphorus aids blooming and fruiting. As a rule of thumb, phosphorus is most important for crops from which we use the floral parts—that is, flowers, fruits, or seeds. One should not overapply this simple rule, however: nitrogen and phosphorus must both be sufficient for both vegetative and flower growth, and supplying more phosphorus than necessary does not stimulate more bloom.

Because phosphorus is needed for root growth, it is often a major element in starter fertilizers, those applied at planting. However, there is no evidence that amounts of phosphorus greater than adequate encourage heavier rooting. In fact, at low phosphorus levels, plants tend to favor roots over shoots to improve uptake, and in greenhouse production of bedding plants, the best root systems are achieved under low phosphorus rates.

In the natural world, phosphorus tends to be the primary nutrient limiting growth in tropical land ecosystems, on old soils, and in freshwater and some marine ecosystems.

FIGURE 12-7 The wheat plants on the right exhibit a phosphorus deficiency. *(Courtesy of Potash and Phosphate Institute)*

Forms of Phosphorus in the Soil. Soil phosphorus is provided by the weathering of minerals like the apatites, which are calcium phosphate minerals. As apatite weathers, it releases anions that can be used by plants. These anions are primary orthophosphate ($H_2PO_4^-$) and a secondary orthophosphate (HPO_4^{-2}). For simplicity, the text refers to them both as phosphates.

Many soils contain large amounts of phosphate, but much is unavailable to plants. Phosphate in insoluble forms that are not free for plant growth is said to be "fixed." The reactions that fix phosphate depend on soil pH. In strongly acid soil (pH 3.5–4.5), insoluble iron phosphate forms. Between pH 4.0 and 6.5, phosphorus reacts with aluminum. Calcium phosphates are important between pH 7.0 and 9.0. Maximum availability lies at pH 6.5 in mineral soils, but 6.0 to 7.0 is satisfactory for most crops.

Between 25 percent and 90 percent of all soil phosphorus resides in organic matter, an important storehouse of phosphorus. Figure 12-8 summarizes the forms of phosphate in the soil.

A typical acre of soil holds between 800 and 1,600 pounds of phosphorus in the plow layer. Of that total, only about four pounds is in solution at any time. As plants remove phosphate from solution, mineral and organic phosphate become soluble by mineralization and the activities of P-dissolving bacteria. At the height of the growing season, soluble phosphorus may be replaced from soil stores several times daily.

Because phosphorus availability is low in so many of the world's soils, plants have developed adaptations to improve access to it. These include mycorrhizal associations, specialized root systems, high root length densities, longer root hairs, and exudates given off by roots and mycorrhizae that free phosphorus. While these can be effective responses in natural ecosystems, crops often cannot tap soil stores fast enough to produce a full crop. For this reason, growers fertilize soil with phosphate to compensate for fixation.

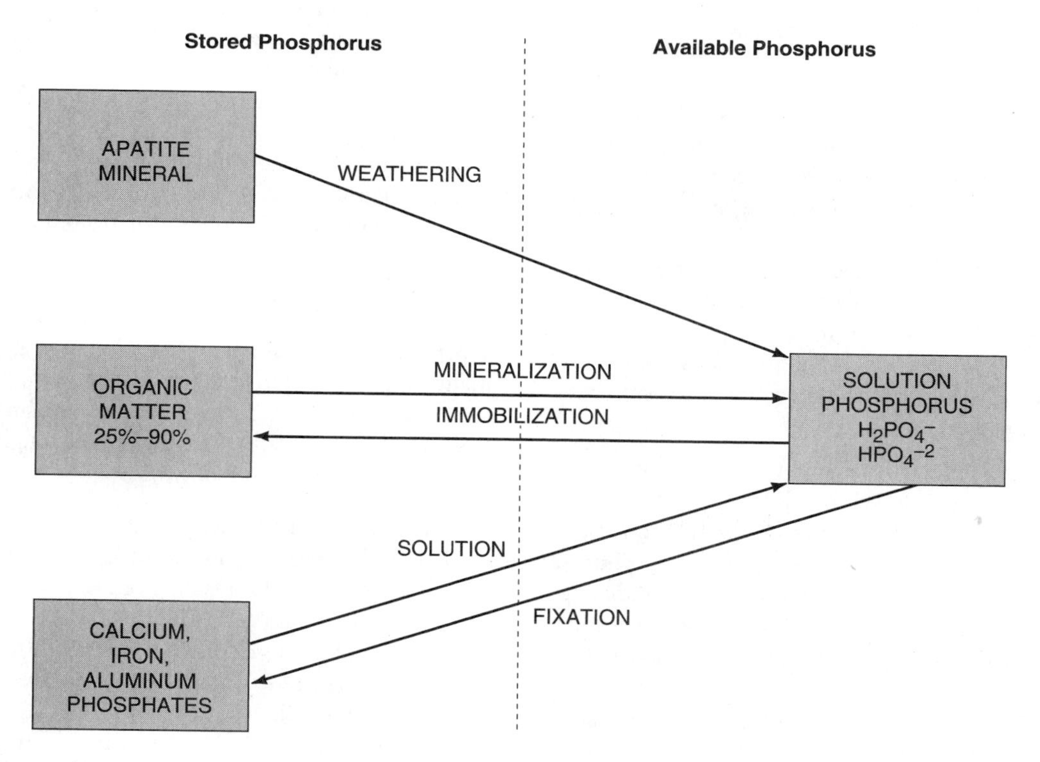

FIGURE 12-8 Phosphorus comes from weathering of minerals such as apatite. Most phosphorus is stored in organic and fixed forms. At any one time, only a small amount of phosphate ion is in solution and usable by plants.

Movement and Uptake in the Soil. Phosphorus moves very little in mineral soil, diffusing over a distance as small as one-quarter inch. This limited movement has important implications for soil management. It cannot leach downward in soil as do nitrates. Instead of leaching, phosphorus is more commonly lost by runoff, erosion, or blowing soil. It also increases the difficulty of plants in obtaining adequate phosphorus. Because of low mobility, it is critical that phosphate fertilizer be placed near seed when planted or mixed into soil near plant roots.

The uptake of phosphorus depends on a number of soil conditions:

- Soil pH largely sets the degree of fixation. Phosphorus is most free at a pH of 6.5 to 6.8.
- Dry soil stalls the diffusion of phosphorus to roots. Therefore, plants take up phosphate best in moist soils.
- Oxygen is needed for the breakdown of organic phosphates. Roots also need oxygen to take up nutrients. Thus, a loose, well-drained soil improves phosphorus uptake. Compacted or poorly drained soil reduces access.
- Cold soil slows the activity of microorganisms that place phosphorus in solution, slows diffusion to roots, and retards root growth. Root respiration also declines, depriving roots of the energy needed to absorb phosphorus. Phosphate shortages are commonly seen on cold, wet soils.
- The total nutrient balance is also important. Nitrogen, for instance, improves phosphorus uptake. Too much zinc seems to lower it.
- Mycorrhizae infection of plant roots helps the plant absorb phosphorus, especially in phosphorus-deficient soils.

A crop uses only 10 percent to 30 percent of the phosphate fertilizer applied to it. The rest goes into reserve and may be used by later crops. Many growers, in fact, have built up large reserves of soil phosphorus. With annual fertilizer applications, many turf and landscape areas are even better supplied. Only soil testing can tell soil users how much phosphorus crops need.

Deficiency. A shortage of phosphorus can cause stunting and fewer, smaller leaves. Plants remain dark green or may even become darker green than normal. Phosphorus-deficient plants often have a purple tint to leaves and stems, starting on lower, older leaves. A phosphorus shortage may delay maturity of several crops, including corn, cotton, soybeans, and others. Some crops, like carrots, develop poor root systems. On the other hand, excess soil phosphorus ties up several plant nutrients, such as iron.

Potassium

Potassium, often called potash, is a key plant nutrient. Plants consume more potassium than any other nutrient except nitrogen, and some plants, like Kentucky bluegrass, may use more. No organic compounds in a plant contain potassium, but many life processes need it. Potassium is dissolved in plant fluids, filling several regulatory functions. Potassium activates enzymes needed in formation of protein, starch, cellulose, and lignin. Thus, it is necessary for the development of thick cell walls and strong, rigid plant stems. Potassium regulates the opening and closing of leaf stoma (pores in the leaf that pass oxygen, carbon dioxide, and water vapor into and out of the leaf). Therefore, potassium is involved in the gas exchange needed for photosynthesis and in transpiration.

Potassium is instrumental in moving sugars produced by photosynthesis within the plant, so it is important in the development and ripening of fruits like apples or tomatoes. Similarly, potassium is needed for proper growth of root and tuber crops.

Potassium acts to balance the effects of nitrogen, and a particular nitrogen:potassium ratio is suggested for many crops. Nitrogen leads to soft growth, but potassium promotes a tougher growth. The toughness results from thicker cell walls. This increased toughness improves crops in a number of ways:

- Plants well stocked with potassium have strong stems that are less prone to lodging (figure 12-9). In corn, reduced lodging also results from the greater number of brace roots (figure 12-10).
- Well-fed plants fight disease. Potassium reduces diseases such as mildew in soybeans, wildfire in tobacco, and leaf and dollar spot in turfgrass.
- Potassium makes plants more winter-hardy and less likely to be injured by spring or fall frosts.

FIGURE 12-9 Lodging of corn from a potash deficiency.

- Potassium, by regulating the stoma, influences the transpiration rate. A plant well supplied with potassium transpires less and so makes better use of water supplies.

As an example, with adequate potassium, turf is less disease-prone, more winter hardy, and better resists wear and tear.

The more potassium in soil, the more plants take up. However, there is no evidence that supplying potassium beyond plant needs will additionally increase hardiness or toughness. In addition, excess potassium uptake may inhibit uptake of calcium or magnesium.

Forms of Potassium in the Soil. Weathering releases potassium ions into the soil solution from a number of common minerals such as feldspars and micas. This ion can be easily taken up by plant roots. Little potassium becomes part of soil organic matter, so most is stored in soil by adsorption and fixation.

FIGURE 12-10 Potassium promotes brace root development in corn. The K-deficient plant on the right has the smallest number of brace roots. The plant on the left has received all three nutrients and has the largest number of brace roots. *(Courtesy of Potash and Phosphate Institute)*

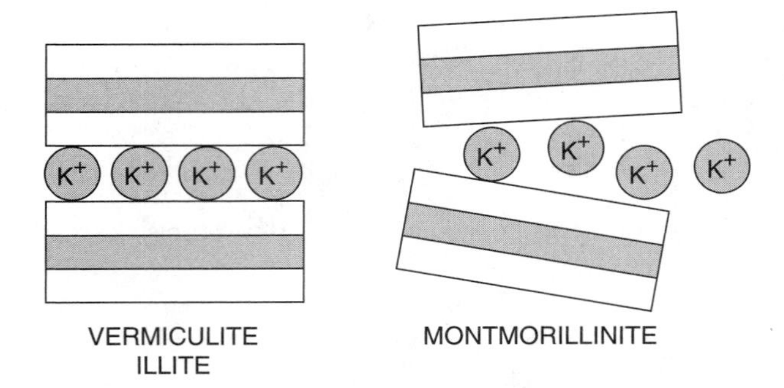

FIGURE 12-11 Potassium fixation occurs when potassium ions are trapped between the 2:1 layers of illite and vermiculite. The layers in montmorillinite open up so potassium does not get trapped.

Potassium ions bear a positive charge and so are adsorbed on soil colloids. In most mineral soils, a few pounds of potassium are dissolved in the solution of an acre of soil at any one time. In contrast, several hundred pounds per acre of exchangeable potassium occupy cation exchange sites.

Potassium can also be fixed by certain 2:1 clays, trapped between the 2:1 layers, as shown in figure 12-11. This potassium can be released slowly if the potassium level in the soil solution declines. Montmorillinite clay layers are so loose that potassium ions can enter and leave easily, allowing potassium to remain available. Figure 12-12 shows the forms of potassium.

Movement in the Soil. Potassium moves more readily in soil than does phosphorus, but less readily than nitrogen. Because potassium is held on clay or other colloids, it is least mobile in fine-textured soil and most readily leached from sandy soils.

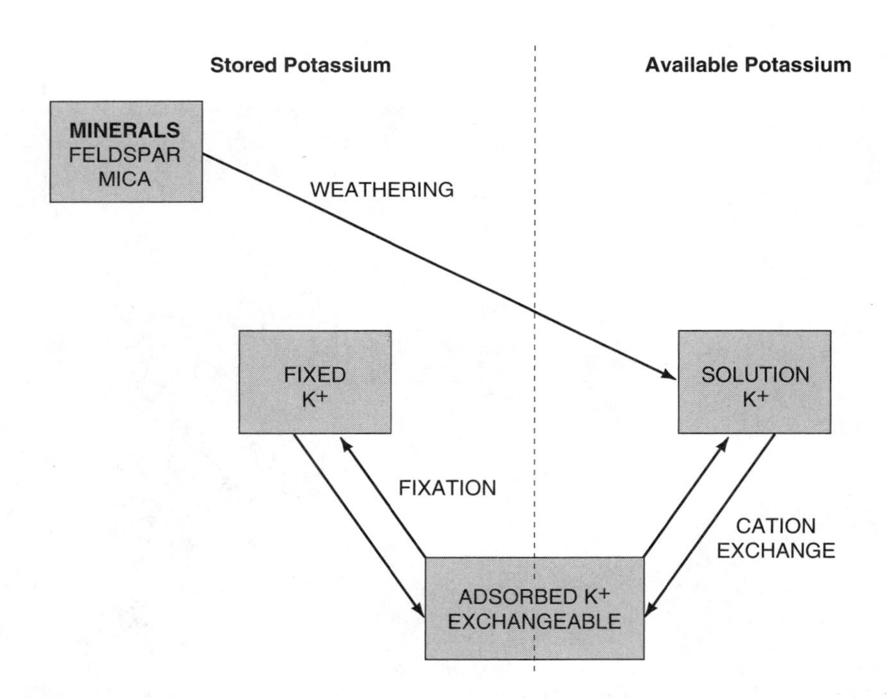

FIGURE 12-12 Potassium comes from weathering of minerals such as feldspar. Soil stores potassium by adsorbing some on soil colloids and fixing some in certain clay particles. Both exchangeable and solution potassium are available for plant use.

Most plant uptake of potassium occurs by diffusion. Because the element moves more readily than phosphorus, fertilizer placement is less crucial.

Deficiencies. Growers see potassium deficiencies less often than those of other primary nutrients. Shortages occur primarily in sandy, heavily leached soils, especially if irrigated, or in organic soils. Overfertilization with nitrogen can cause plant tissues to lack potassium. Dry, cold, or poorly aerated soil may also slow uptake. Potassium uptake is most rapid near neutral pH.

Plants show a lack of potassium by a "marginal scorch," or burnt look on the edges of the lower, older leaves. This symptom can be easily mistaken for moisture shortage during hot dry weather or for salt damage. In some cases, the margins merely turn yellow.

Secondary Nutrients

Calcium. Calcium, the nutrient used in third greatest amounts by most plants, is a critical component of both cell walls and membranes. In cell walls, much is found as calcium pectates, located especially in a layer in the outer part of the cell wall where it lends strength (figure 12-13). The crispness of apples, for instance, derives from a high calcium pectate content. Pectins are the same materials used to jell preserves, and one can make jellies using apple pectins. Calcium also stabilizes plant cell membranes to prevent leakiness.

Calcium also plays a role in protein formation and carbohydrate movement in plants, and plays a signaling, or regulatory, role in several plant functions—like directing roots to grow down rather than up. It also largely controls soil pH and helps soil aggregation.

Because of its role in cell walls and membranes, calcium shortages present the greatest problems where cells are actively dividing and enlarging, such as root and shoot tips (figure 12-14) and developing fruit. Calcium,

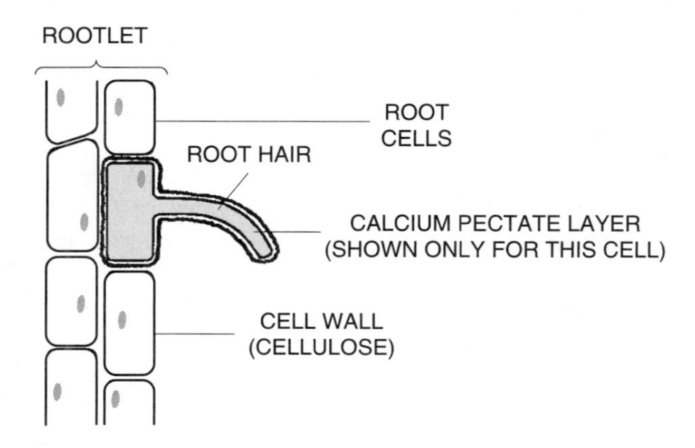

FIGURE 12-13 A wall of calcium pectate gives cells strength, as in this root hair.

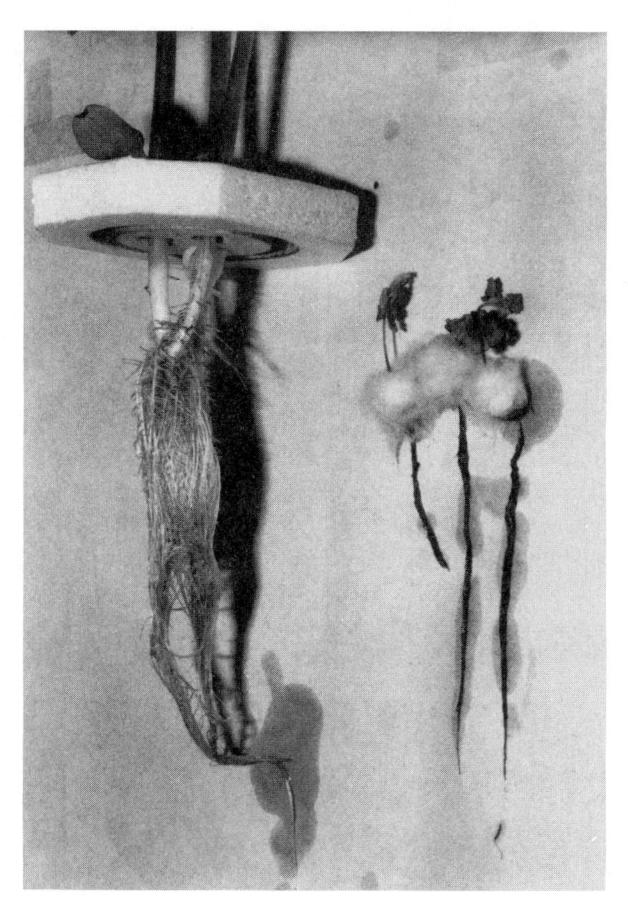

FIGURE 12-14 Effect of calcium deficiency on roots. Lacking calcium to strengthen cell walls, roots turn mushy and fail to develop, as shown by the plant on the right.

being relatively immobile in the plant, may not reach developing fruit or other fleshy plant parts rapidly enough to supply their needs. Water core of apples is a collapse of cells in rapidly growing fruit, leading to soft areas in the fruit. Similarly, blossom end rot in tomatoes is a weakening of tomato fruit at the end furthest from the sap stream, leading to cell collapse and attack by rot organisms. Similar conditions include bract burn in poinsettia and black-heart of celery. Growers often spray plants with calcium to reduce these problems. Adequate calcium, on the other hand, reduces insect and disease infestation and post-harvest decay of many fruits. While several horticultural crops often suffer these calcium shortages, calcium problems are more rare in other crops, especially in grasses, which have low calcium needs. Calcium shortages are most likely on acid, irrigated sands or where excessive potassium levels inhibit calcium uptake.

Calcium relations in plants also impact human diets. Much of plant calcium is tied up in forms unusable by humans. Those who cannot or do not consume dairy products must rely largely on vegetable sources, and care should be taken to find foods with more available calcium content, like green beans.

Calcium comes from weathering of common minerals including feldspars, limestone, or gypsum. These materials are so common that most soils contain enough calcium to supply most plant needs. Calcium is neither fixed in the soil nor held in organic matter. It is the main occupant of the cation exchange complex, and calcium storage depends on cation exchange capacity (figure 12-15).

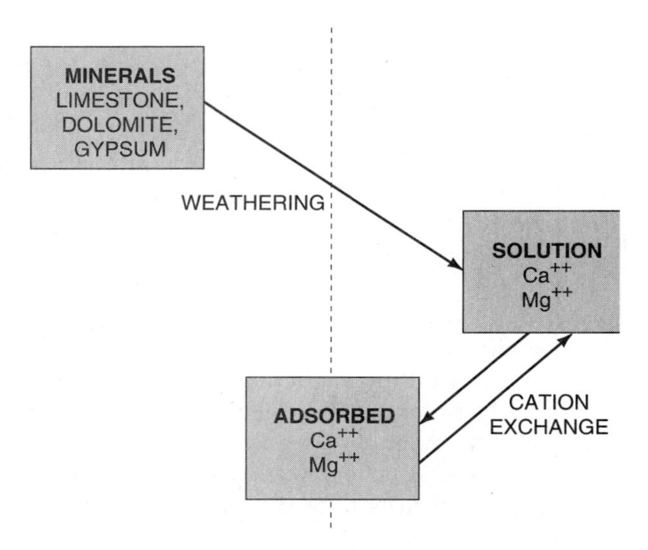

FIGURE 12-15 Weathering releases calcium and magnesium from soil minerals like dolomite. It can be stored on soil colloids.

Magnesium.

Magnesium resembles calcium chemically and in its action in soil. Its role in the plant differs, however. Magnesium is the essential ingredient in chlorophyll—each molecule has one magnesium atom at its center (see figure 12-17). Magnesium also aids the uptake of other elements, especially phosphorus. Like potassium, magnesium activates a number of important enzyme systems. Magnesium is involved in protein, carbohydrate, and fat synthesis, as well as a wide range of other compounds. Deficient plants offer less resistance to drought, cold, and disease.

Magnesium weathers from minerals as a cation (figure 12-15). However, clay holds magnesium less strongly than calcium, so it is more easily leached. As a result, low-magnesium soils are more common than low-calcium soils. Highly leached coarse soils are most likely to need fertilization with magnesium, especially if treated with low-magnesium lime. High levels of soil potassium may also induce a magnesium shortage in plants.

Hunger signs resulting from low levels of chlorophyll include chlorosis, a yellowing of the leaf, beginning with the older leaves. Forage low in magnesium is known to cause grass-tetany disease in cattle.

Sulfur.

Crops need less sulfur than the other macronutrients, but it is still a crucial nutrient. Several proteins include sulfur, and it is needed for making chlorophyll. It aids nodulation of legumes and seed production of all plants. Overall, sulfur improves protein and chlorophyll content, stress tolerance, animal nutrition, and the appearance of plant products. Alfalfa, members of the mustard family (including cabbage), and members of the onion family need much sulfur. The pungent flavors of those plants derive from sulfur compounds.

Most soil sulfur comes from the weathering of sulfate minerals such as gypsum. The sulfate anion is the form

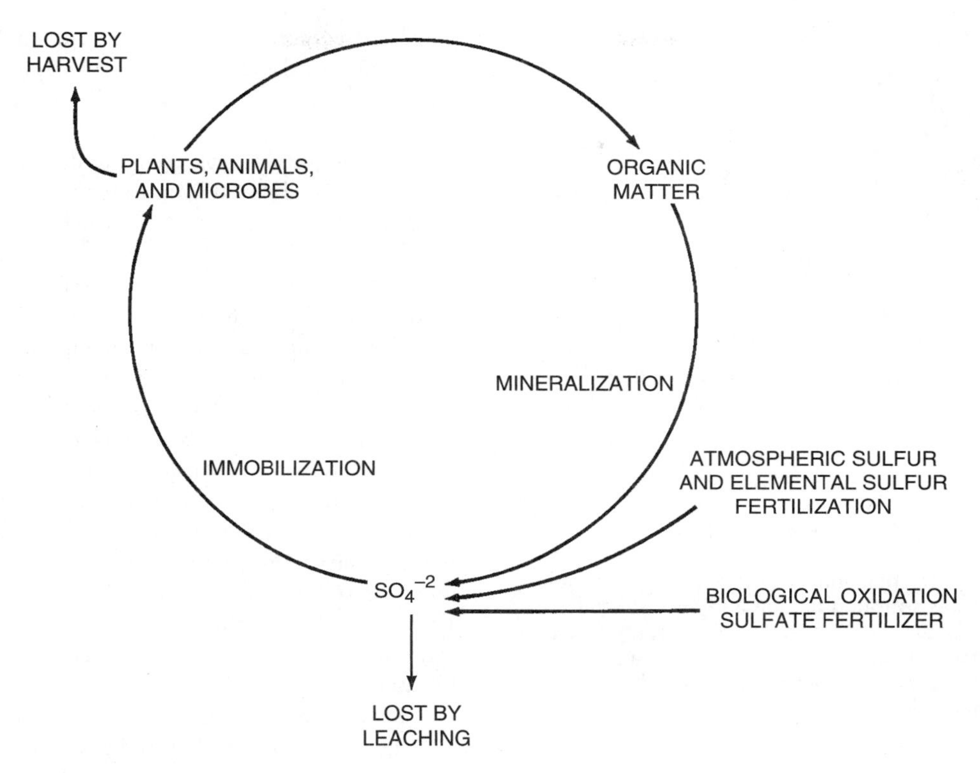

FIGURE 12-16 The sulfur cycle.

used by plants. Organic matter contains 70 percent to 90 percent of the soil sulfur; it is neither adsorbed nor fixed to any degree. Because it is readily leached, surface layers of soil are often low in sulfur. Figure 12-16 reviews the sulfur cycle.

Interestingly, acid rain supplies sulfur in many areas. In many parts of the country, sulfur from acid precipitation has been reduced by burning low-sulfur coal and by better clean-air controls on exhaust stacks. Older fertilizer types contained sulfur as a by-product of their manufacture. The fertilizers that are most popular now are much purer. Since pollution- and fertilizer-supplied sulfur have both been reduced, shortages are increasingly common. Use of sulfur fertilizer has increased rapidly, especially in the southeastern states. Leached and low-organic-matter soils are likely candidates for sulfur shortage. Soils high in organic matter or soils located near industrial centers are least likely to be short of sulfur.

Plants that are short of sulfur may be stunted and older leaves will be pale green, like those of nitrogen-deficient plants.

Trace Elements

Trace elements play many roles in plants, many difficult to understand without knowing plant chemistry. With the exception of boron and chlorine, trace elements are metals. These metals interact with special molecules, called **enzymes,** that control important biological reactions. Enzymes are "keys" that activate biological reactions in living systems. They are not consumed in the process. For instance, an iron enzyme controls one step in the formation of chlorophyll, but is not itself part of chlorophyll.

Very little of each enzyme is needed, because each is reused repeatedly. Therefore, very little of the trace elements that are part of enzymes is needed. Without

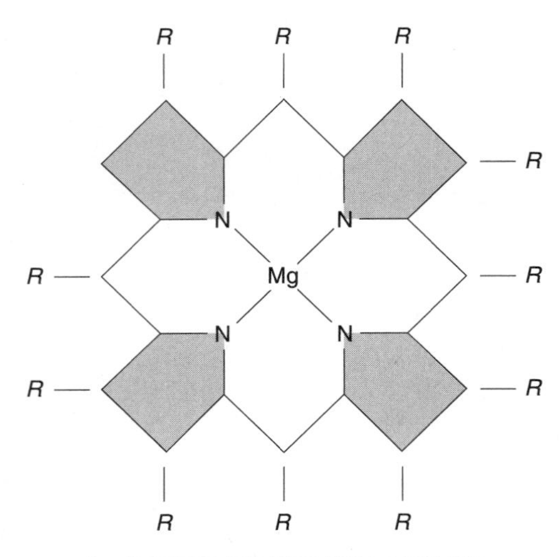

FIGURE 12-17 In chelation of a metal ion, very complex organic molecules surround the ion and form several bonds with it. Chelation protects the ion from other soil processes. Artificial trace element chelates are also common in fertilizers. The chelate pictured is the chlorophyll molecule.

this tiny amount, however, important processes suffer. On the other hand, an excess of a trace element can be toxic to plants or the animals feeding on them. The difference between enough and too much can be quite narrow, sometimes only a few pounds per acre. Growers should apply trace elements with caution, after proper soil and tissue testing.

Trace elements are stored in the soil in a somewhat different manner than macronutrients. Some trace elements are stored in slightly soluble compounds, or are involved to a small extent in cation exchange. Many trace elements combine with organic molecules in the soil to form very complex molecules called **chelates.** A chelate is a metal atom surrounded by a large organic molecule (figure 12-17). Chelates are an important form of storage for many trace elements.

Iron. Iron is part of many enzymes necessary in the formation of a number of chemicals, especially chlorophyll. Iron minerals are widespread in soil. Most soils have sufficient iron, but much is in the form of insoluble compounds, such as ferric hydroxide, Fe $(OH)_3$. Organic matter chelates some iron in the soil. Interestingly, some soil microbes living in the rhizosphere emit compounds that chelate iron, probably improving iron uptake by plants.

The solubility of iron compounds relates directly to pH, declining about 100 times for each rise of one pH point. Acid-loving plants suffer iron shortages when the pH rises above 5.0 or 6.0, while many plants become deficient at higher pH. Sorghum, soybeans, and flax are field crops sensitive to an iron shortage. Iron hunger is most likely in alkaline or calcareous soils, or with excesses of phosphate, zinc, copper, or manganese. Anything that inhibits nutrient uptake, like cold, wet soils, or drought, may induce iron deficiencies.

Iron chlorosis is the usual symptom of iron hunger. It is easy to see as an interveinal chlorosis on new, growing leaves (see figure 10-2). While leaf veins remain green, tissue between the veins becomes light green or yellow. In trees, branches begin to die back. Fruit and ornamental crops commonly show these symptoms. Examples include azaleas, pin oaks, and blueberries.

Various treatments are available to overcome a lack of iron: (1) soil pH can be lowered to free the iron; (2) soluble iron compounds such as iron sulfate may be mixed into the soil, sprayed on leaves, or even injected into the trunks of trees; (3) artificially prepared chelates may be used in the same way; and (4) animal manures can be mixed into the soil.

Manganese. Manganese resembles iron in that weathering releases a cation that is tied up in nonacid soil. Manganese acts with iron in the formation of chlorophyll. Manganese speeds seed germination and crop maturity and helps plants take up several other nutrients.

Manganese deficiencies are usually seen on calcareous soils or on soils that have been overlimed. The solubility of manganese decreases a hundredfold for each rise of one pH point. Soybeans grown on some slightly acid to alkaline soils of the Atlantic Coastal Plain are known to suffer manganese shortages. Dwarfing is a common symptom of manganese deficiency and is often seen in combination with chlorosis. Flecks of dead tissue, along with chlorosis, often appear on new leaves, as shown by some species of maple trees growing on

FIGURE 12-18 Manganese deficiency on a maple growing in alkaline soil.

alkaline soil (figure 12-18). When soil pH is below 5.0, so much manganese may be free that it reaches toxic levels.

Deficient soil can be treated by mixing manganese sulfate into the soil. Lowering pH by applying sulfur may also be helpful. Leaves may be sprayed with a solution of manganous sulfate or chelate. Oats, soybeans, sugar beets, and several vegetables are most likely to respond to manganese treatments. Liming cures manganese toxicity in acid soil.

Zinc. The zinc cation is weathered out of soil minerals, where it can be adsorbed, form a chelate, or form slightly soluble zinc compounds. Several biological reactions use zinc, including chlorophyll and protein production. Low zinc levels are widespread in many crops, including beans, corn, and rice. Some nutritionists have voiced fears that these shortages could be passed on to human consumers. The symptoms of zinc deficiency in plants are quite varied but include tight growth of small, closely spaced leaves, interveinal chlorosis, and dead spots on leaves.

Zinc is most available in acid soil, least available on alkaline or recently limed soils. Soils that have lost topsoil by leveling, terracing, or erosion may also be zinc poor. Low levels may also appear on very coarse soils, because the parent materials lacked zinc and the soils tend to be low in organic matter. Cold soils or excess levels of phosphate inhibit uptake. Like iron, a lack of zinc can be treated by fertilizing soil or spraying foliage with zinc compounds or chelates. Sewage sludge is an excellent source of zinc (see chapter 15). Corn, rice, and onions are most likely to respond to zinc treatment.

Copper. Copper is held by cation exchange and combines chemically with organic matter. Some organic-copper complexes are so stable that the copper is unavailable to plants. Copper is part of a number of important enzymes, especially for the formation of chlorophyll and lignin. Copper affects how well a plant resists disease and how well it controls moisture. Copper shortages also inhibit pollen formation, reducing fruit yields. Shortages are not common, but symptoms include poor fruiting, distorted new growth, stunting, and leaf bleaching. Shortages are most likely to be seen in either leached sands or peats and mucks. A few pounds per acre of copper sulfate mixed into the soil usually supplies all the copper that is needed. Carrots grown on organic soils may need extra copper, but small grains and other vegetables sometimes suffer as well.

Boron. Boron exists in the soil largely as boric acid, H_3BO_3, which is taken up by plants and gathers in organic matter near the soil surface. Fixation at high pH and leaching limit the amount of boron plants can use. Shortages sometimes appear if a soil is overlimed. Conditions that limit organic matter decay also limit the amount of free boron.

The functions of boron are not well understood. Unlike other micronutrients, it does not appear to be involved with any enzymes. It appears to be important for making cell membranes and walls, and for cell enlargement, an important growth process.

Boron deficiencies are fairly widespread in alkaline soils that limit its availability, in high rainfall areas where it leaches readily, and under drought conditions. A shortage of boron often appears as death of terminal buds, followed by tight, bushy growth, known as a "rosette." Thick, fleshy tissues like celery stems or sugar beets may get heart rot, and seed may fail to form in many plants. Boron toxicities are also common, mostly in arid regions.

A number of boron fertilizers may be applied to the soil or sprayed on plant leaves. The oldest form, the common laundry product borax, may be applied at the rate of a few pounds per acre. However, even slightly high boron levels hurt plants, so it should not be used without first testing the soil. Especially sensitive to excess boron are indoor foliage plants.

Molybdenum. Molybdenum, the nutrient with the smallest plant requirement except nickel, is necessary for proper nitrogen metabolism by plants and for nitrogen fixation by both symbiotic and free-living bacteria. Molybdate, MoO_4^{-2}, gathers in soil organic matter. Unlike other micronutrients, it is most available at a high soil pH. Shortages are most common on acid, leached, and low-organic-matter coarse soils, as well as acid peats.

Several crops, in addition to legumes, respond to treatment. Crops in the mustard family are especially sensitive. Whiptail of cauliflower, for instance, results from a lack of molybdenum. An ounce of a soluble molybdenum material, often mixed with phosphate fertilizer, will usually treat an acre of deficient soil. Frequently, liming releases enough of this trace element to cure shortages.

Chlorine, Nickel, and Others. The function of chlorine, a recently identified essential element, is not well understood. It is known to play a role in photosynthesis and may help regulate opening and closure of stomata. Chlorine is needed in very small amounts and is commonly found in the soil. Chlorine is thought to be never lacking in farm soils. However, it has been shown to increase grain yields in some soils of the Great Plains, most commonly where plant diseases have been a problem. Toxicity is observed in such sensitive species as beans and cotton, particularly in drier regions of the world.

Nickel is needed by plants and microorganisms for the proper metabolism of the simple nitrogen compound urea and possibly for other uses. It may also help plants resist disease. Nickel is the element most recently shown to be an essential element, making it the seventeenth essential element.

A number of other elements contribute to nutrition of certain plants, though they are not currently considered to be universal essential elements. Legumes need cobalt for nitrogen-fixing. Some grasses and horsetail need silicon; it also is needed for best yields in rice and sugarcane. Research also shows that high silicon content strengthens cell walls and reduces disease infections and insect attack. Sodium appears to be required for many plants native to sodium-rich soils. Plants that need sodium also include many species that have special types of photosynthesis adapted to hot, sunny climates. These include cacti, succulents, and many warm-season grasses. In some areas of the country, selenium and cobalt may be needed in animal forage.

SUMMARY

The fourteen mineral nutrients perform many important tasks in the plant. Of the major elements, nitrogen promotes rapid succulent growth. Phosphorus gives early root growth, blooming, and resistance to pest and weather damage. Potassium lends toughness, strength, and pest resistance. Plants need a balance of these three nutrients for strong, vigorous, and healthy growth.

Anyone who grows crops should know how nutrients behave in the soil. An important consideration, for instance, is how a nutrient is stored in the soil. Some nutrients, such as nitrogen and boron, are stored predominantly in organic matter. Some nutrients, such as calcium and magnesium, are adsorbed primarily on soil colloids. Many nutrients are part of slightly soluble compounds, including phosphorus and iron. Many trace elements, like copper, react with organic matter in the soil to form chelates. Most nutrients are found in several of these forms.

Other important traits of nutrients include their solubility and mobility. The solubility of most nutrients depends on pH. For example, phosphorus compounds are most soluble between pH 6.5 to 6.8. Highly mobile nutrients, like nitrate nitrogen, can leach easily from the soil. Elements that move only a short distance, like phosphates, must be placed where roots or seeds will use them.

Plants grow best when each nutrient is present in the right amount. A lack of any one nutrient causes poor or abnormal growth. In addition, plants need a balance of nutrients. To achieve this balance, soil testing should be completed before fertilization is started.

REVIEW

1. Prepare a table with four columns for the four major pools of nutrients in the soil. Place each of the macronutrients in the correct columns. In the soil solution column, write the chemical form the nutrient is in.
2. Soil professionals whose background is agronomic often consider calcium deficiencies an uncommon problem, but horticulturists deal with them often. Explain the difference.
3. Compare and contrast the roles of the primary macronutrients in top growth, root growth, flowering, hardiness, toughness, and pest resistance.
4. How is nitrogen lost from the soil? How can these losses be reduced?
5. Soils in natural ecosystems tend to contain very low amounts of dissolved nitrogen at any given moment compared to agricultural soils. Why might this be true? What could be ecological consequences of the higher amounts in agricultural soils?
6. You want to fertilize a large tree with N, P, and K. Discuss the importance of fertilizer placement for each of these elements for successful fertilization.
7. Some elements move readily in the plant, and when deficiencies occur, the plant moves those elements out of the older leaves into new growth. Deficiency symptoms thus tend to occur on old leaves first. Other elements do not move so readily, and symptoms occur on new leaves first. Go through this chapter and categorize the nutrients on this basis.
8. Excess amounts of many nutrients can also have negative consequences on plant growth. Describe several examples.
9. Compare the kinds of natural ecosystems where nitrogen tends to be growth-limiting to those where phosphorus tends to be. These are broad-scale generalizations.
10. Both Chapter 10 and this one describe reasons why plants growing under low light should be fertilized less than those in full sun (especially with nitrogen). What are those reasons?

ENRICHMENT ACTIVITIES

1. For color images of nutrient deficiencies, try this Web site and answer these questions: <http://www.back-to-basics.net/nds>

    a. Describe deficiency symptoms of iron on apples.

    b. Describe deficiency symptoms of phosphorus on corn.

    c. Describe deficiency symptoms of calcium on tomatoes.

    Note that symptoms are often different on different plants. If you click back to the site's home page, you will find other interesting articles about soil fertility.

2. The nitrogen, phosphorus, and potassium content of the tissues in many crops can be found at: <http://www.nrcs.gov/technical/land/pubs/nlapp1a.html>.

    Compare the NPK content of alfalfa hay to small grain hay. Using information on the site, calculate how much NPK is removed from a field in each ton of corn silage.

3. For more information on the fate of nutrient ions in the soil, try this Web site: <http://www.uog.edu/cals/site/users/soil/soil/fertft2a.html>.

4. For more information on the nitrogen cycle, especially N fixation, see: <http://helios.bto.ed.ac.uk/bto/microbes/nitrogen.htm>.

CHAPTER 13

Soil Sampling and Testing

OBJECTIVES

After completing this chapter, you should be able to:

- explain why soils are tested
- sample soils correctly
- describe soil testing
- interpret soil-test reports
- explain how plant tissue tests are used

TERMS TO KNOW

composite sample
green tissue test
precision farming
site-specific management
soil sampling
soil testing
tissue testing

Why Test Soils?

Fertilizing can increase yields, and increased crop yields add to a grower's income. However, because fertilizers cost money, a grower must use the amount that is most profitable. Further, improper fertilization contributes to environmental problems. How does a grower know how much fertilizer to apply for the best return at the least environmental risk?

Figure 13-1 shows a stylized relationship between nutrient levels in the plant tissue and productivity, divided into four levels:

- **Level I: Deficient.** The nutrient is clearly deficient; growth and productivity are affected. After the missing mineral is applied, growth response is strong and profitable.
- **Level II: Sufficient.** A critical level is reached that satisfies plant needs. More fertilizer may increase yields slightly, but not enough to pay for fertilizer.
- **Level III: High.** Nutrient levels are high, yields are maximum. Additional nutrients would be stored in the plant (luxury consumption). Fertilization could also shift the plant to Level IV or contribute to water pollution.
- **Level IV: Toxic.** Nutrient levels in plant tissue are so high as to be toxic. Yields decline.

Growers can use three methods to find nutrient shortages in plants:

- Visual inspection of crops for deficiency signs may uncover clear shortages. Unfortunately, this method notes only critical shortages after yield damage has occurred. In addition, visible symptoms may be unreliable. Chlorosis, for example, may result from low nitrogen, nematode feeding, dry or salty soil, diseases, or other problems unrelated to soil nutrient levels.
- Soil tests measure nutrient levels in soil as well as other soil features. Growers depend on these tests to determine the lime and fertilizer needs for crops. Soil tests have limits, however. Conditions that affect nutrient uptake, such as wet soils, cannot be detected in the laboratory.

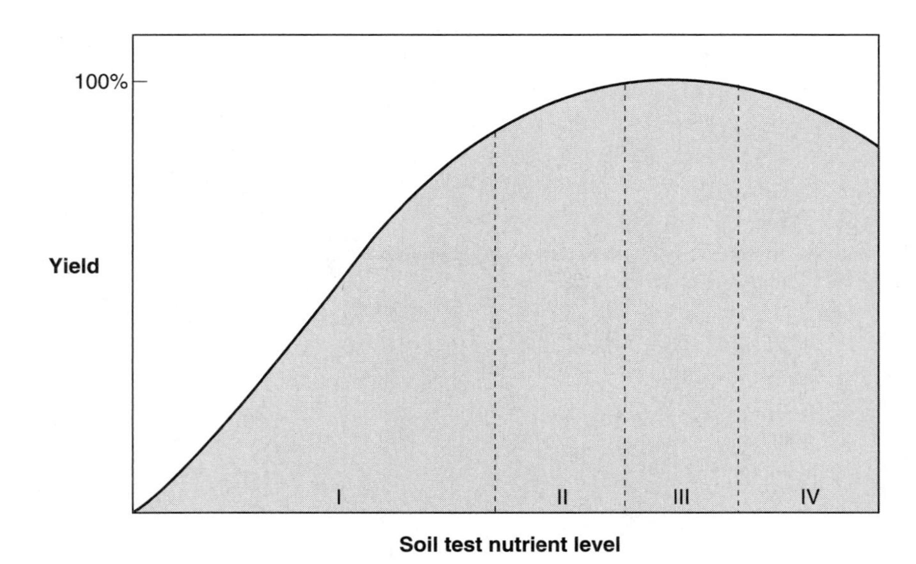

FIGURE 13-1 The yield of a crop is directly related to the nutrient level of the soil. Fertilization is most profitable for crops in Level I.

- Tissue testing measures nutrient levels in plant tissue itself. This type of testing may uncover problems soil testing misses.

Of the three methods described, soil testing is most important for a majority of crops, especially annual farm crops. A soil test can be performed early in the season to allow the grower time to supply the necessary nutrients before the crop is planted.

There are three separate activities involved in soil testing. (1) **Soil sampling:** the grower samples the soil and sends the sample to a testing center. (2) **Soil testing:** the soil laboratory tests the sample and makes a recommendation to the grower. (3) The grower acts on the recommendation from the testing center. First let's look at soil sampling.

Soil Sampling

Soil laboratories use the most modern testing methods and tools. However, the material to be tested is the sample provided by the grower. This means that test results are no better than the sample itself. Sampling methods are described here, but local recommendations may vary. Ask the local testing center or Cooperative Extension agent to recommend the best sampling method for your area.

Testing Frequency. Frequency of soil testing depends on the crop and how it is grown. For most annual farm crops, sampling every two or three years should be adequate. Intensive crops like fruits or vegetables benefit from annual sampling, and greenhouse crops are tested even more often. Soils should be tested before any crop is planted that occupies the soil for longer than one season, such as turf, trees, or perennial forages. This practice allows the grower to mix potash and phosphate into the soil, or to adjust pH, before planting.

Any change in cropping practices should be preceded by thorough soil testing. For example, if a grower intends to shift from regular to conservation tillage, the soil should be tested before the first year. A grower changing crops, such as a corn grower trying out sunflowers, should also test the soil before planting the new crop.

Growers may sample soil whenever it is not frozen. Fall is probably the best time, because the grower can then use the winter months to respond to the results and order fertilizer.

Selecting the Sampling Area. Land should be divided into sampling areas of uniform texture, topography, and cropping history. For example, a coarse-textured field should be sampled separately from a nearby fine-textured field. If half a field received manure last year and half did not, each section should be tested separately. Sampling areas vary in size from a homeowner's garden to a maximum of twenty acres on farm fields. Some labs now recommend using a soil map to determine sampling areas. Each soil type in this system becomes a sampling area. One would need to continue to consider cropping history.

A sampling area will be fertilized as a unit, using the same rate throughout the area. Even in this area, all the soil is not the same. Differences are averaged by carefully taking many samples from random spots in the sampling area and combining them for a **composite sample.**

Depth of Testing. For field crops grown by conventional tillage, the top six to nine inches of soil should be sampled. For no-till systems, a special pH sample should be taken from the top two inches to test an acid layer that forms at the soil surface. For the ridge system of reduced tillage, samples should be removed from the sides of the ridges. Chapter 16 describes these tillage systems.

Sod or pasture need only be sampled to a two- to three-inch depth. Tree crops, on the other hand, may need to be sampled as deep as eighteen to twenty-four inches. Testing laboratories suggest that nitrogen tests be made on samples taken as deep as three feet. The deeper samples measure how much nitrate is in the subsoil.

Sampling Procedure. For each sampling area, follow these procedures:

1. As shown in figure 13-2, gather many topsoil subsamples from random spots in the field. Take samples no closer than 100 to 300 feet from such odd areas as dirt roads, barns, or fencerows. Also avoid dead furrows, fertilizer spills, and other spots with unusual conditions. Large areas need fifteen or more subsamples.
2. Scrape away surface litter at each testing spot and remove a sample of the soil. Augers or soil sampling tubes (figure 13-3) are convenient sampling tools. A spade can also be used. Dig a V-shaped hole, remove a half-inch slice from the side of the hole, and shave away most of the sample on the blade as

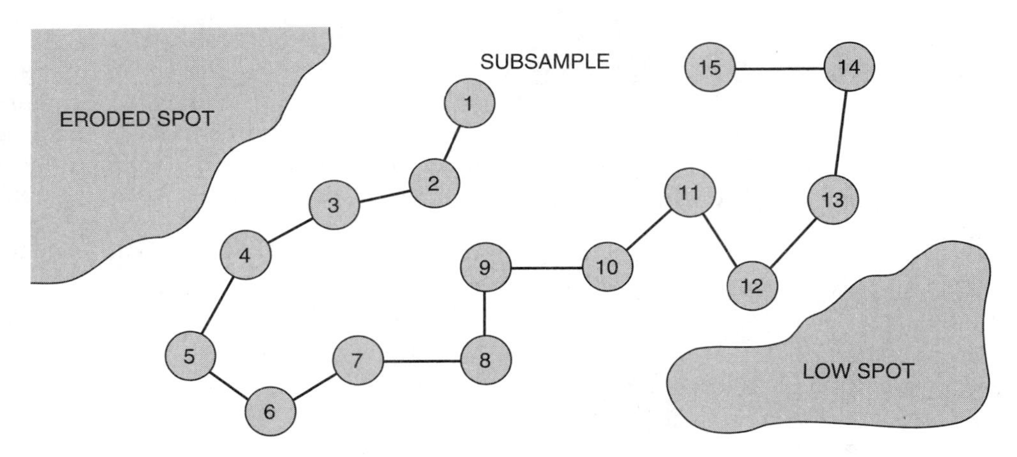

FIGURE 13-2 When taking soil samples, up to fifteen subsamples should be taken at random locations. Samples are not taken from low spots or other unusual areas. Mix these subsamples to form a composite sample. *(Source: University of Minnesota, Agricultural Extension Service)*

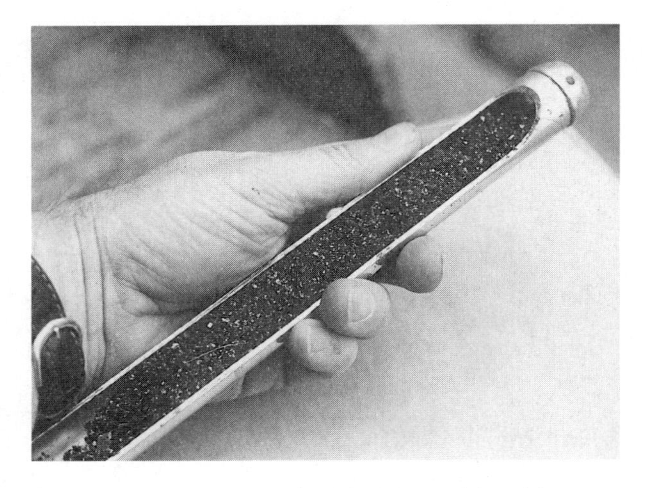

FIGURE 13-3 A soil-testing tube is a useful tool for sampling soil.

shown in figure 13-4. Each soil sample should include soil from the entire testing depth. Drop each soil sample in a clean plastic bucket as it is collected.

3. Mix all subsamples from one sampling area, and remove about one cup of soil. This composite sample represents the average soil in the field. Label the composite sample and let it dry in the air. Do not oven-dry the sample.
4. Fill a mailing container with the dried composite sample. Mark the container according to the instructions provided by the testing center. Complete the sample sheet (figure 13-5), including the intended crop, production goals, cropping history, and other necessary information.
5. Mail the samples to the laboratory, or deliver them to the lab or county extension agent. The sample containers and sheets can be obtained from the soil laboratory or extension agent.

Sampling Greenhouse and Container Plants.

The fertility program for greenhouse and container plants is very different from the plan for field crops. Field growers depend mainly on nutrient reserves in the soil, like organic nitrogen or exchangeable potassium, and fertilize to add the extra nutrients needed for best growth. Container growers, in contrast, use potting mixes (see chapter 17) that contain essentially no nutrient reserves. Thus, all nutrients the plants need must be supplied by fertilization. This need is increased by the small soil volume, high watering rates, and soluble salt problems of greenhouse plants.

Greenhouse or container soils should be sampled before being planted to the crop. They should then be tested periodically during the growing period. The soil should also be tested at any sign of a growth problem. In sampling potted plants, the top half-inch medium, which is likely to be high in salts from capillary rise, is first scraped away. A core of soil is then removed. The core must include soil

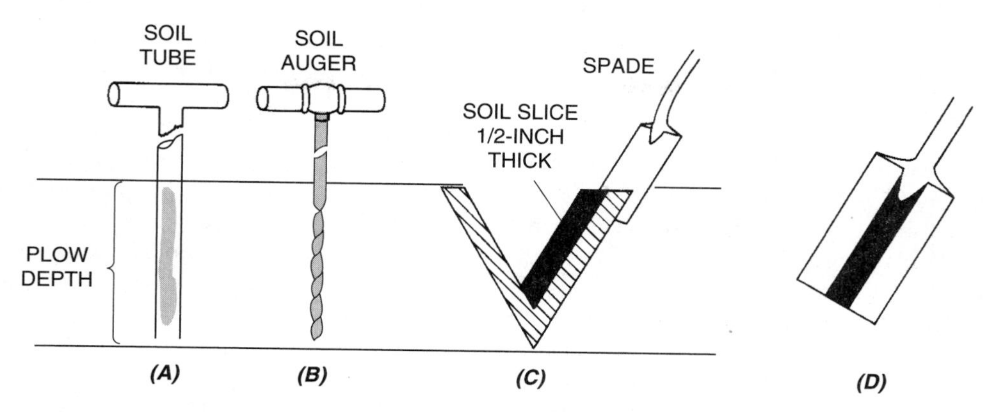

FIGURE 13-4 Soil samples can be taken to the plow depth using *(A)* a soil tube, *(B)* a soil auger, or *(C)* a common spade. In using a spade, a half-inch slice of soil is taken from the side of a triangular hole. As shown in *(D)*, most of the soil in this sample is removed to leave a one-inch strip on the blade.

Read instructions on back

SOIL SAMPLE INFORMATION SHEET
Soil Testing Laboratory, University of Minnesota
St. Paul, MN 55108

Leave this space blank

Name Edward Johns
(PLEASE PRINT ALL INFORMATION)

Address RR2, Wyoming, MN 55407
ZIP CODE

Date 10/18/84

Location: County Chisago
Township Wyoming
Fee payment enclosed $15.00
Charge fee to account

Firm submitting samples:
Name
Address
ZIP CODE

Sample Identification					Crop Information						Check (✓) Tests Requested									
	Sample designation				Last Crop		Next Crop		Crop Following Year		Testing fees* for each sample ($):									
Laboratory number (do not write below)	Field number or letter	Sample number	Depth of Sampling	(✓) Check if irrigated	Crop 19 84	Yield	Crop 19 85	Expected Yield	Crop 19 86	Expected Yield	4.00 Regular series	2.00 Nitrate	3.00 Zinc	3.00 Sulfur	3.00 Soluble salts	1.00 pH and lime only	2.00 Organic Matter (%)	2.00 Magnesium	3.00 Copper	5.00 Boron
	A	1	9"		Soybeans	25	Corn	130	Corn	130	✓		✓							
	B	2	9"		Oats	70	Soybeans	30	Corn	130	✓									
	C	3	9"		Corn	120	Oats	70	Soybeans	30	✓									

Sample Designation	Soil Series (Type) or Mapping Unit	Section-Quarter	Remarks
A	Zimmerman loamy sand	15-SW 1/4	
B	Braham fine sandy loam	15-SW 1/4	
C	Hayden loam	15-SE 1/4	

*Fees subject to change

Note: Regular series includes: pH, lime requirement, P, K, O.M. (relative level), texture.
Copper test is reliable for organic soils only.
Be sure to give the sampling depth for the nitrate test.

(P-992A)

FIGURE 13-5 Completed sample forms accompany submitted soil samples. *(Source: University of Minnesota, Soil Testing Laboratory)*

from the top to the bottom of the pot or flat. Several containers should be tested, perhaps six to a bench, and composite samples prepared. Beds or benches should be tested to eight inches or to the depth of the bench.

Sampling in Site-Specific Management.

Site-specific management and **precision farming** are terms naming a nutrient management system that relies heavily on exhaustive soil sampling. The system recognizes that nutrient requirements can vary greatly over small areas of the field—a problem that has been partially solved by composite samples that average soils over the sampling area. In site-specific farming, those small areas are instead sampled, tested, and fertilized separately. This system is made possible by computers and other new technology.

Precision farming begins by dividing fields into a grid of about 2.5-acre (one-hectare) cells. As many as ten samples are pulled from each cell and mixed to create a composite sample that represents that cell. The testing results are entered into computer mapping software that creates a field map of fertilizer needs that is very site specific. Subsequent fertilization then varies throughout the field based on the field map.

The grid may be established by various methods, but most advanced is the Global Positioning Satellite (GPS) system. Radio receivers mounted on farm equipment, receiving signals from a system of U.S. navigation satellites in orbit, can locate tractor position within a few feet. GPS is both a tool to construct the grid and to continuously locate farm equipment on that grid as it traverses a field. Thus, an on-board computer with GPS, a field map derived from soil sampling, and a computer-operated fertilizer spreader enable very precise fertilization application.

Soil Testing

There are two basic ways to test soil samples. The oldest method uses chemical reactions that produce color changes. The exact color depends on the amount of available minerals in the soil. In the case of the pH test, color depends on soil pH. With test kits used by home gardeners, a certain amount of soil is placed in a test tube. A measured amount of chemical, or reagent, is dripped into the tube (figure 13-6). The color of the resulting mixture is then compared with

FIGURE 13-6 To use a test kit, add a few drops of reagent to a bit of soil in a test tube, then compare the resulting color to a standard color chart.

known standards, and the amount of nutrient is read from the standard.

Simple chemical tests are easy to use, but not reliable. One problem is the human element, because the results are based on the technique of the tester and what he or she sees. In student labs, for instance, the results of several tests on the same soil sample may be surprisingly different. However, such tests can be useful in the field, where one cannot drag equipment about or find a source of electricity.

Modern laboratory testing follows a basic three-step process of extraction, measurement, and interpretation. First, nutrients and acidity are extracted by filtering water or other solutions through a soil sample. These extractions remove most of the free nutrients in the soil as well as some adsorbed on soil colloids, like exchangeable potassium. There are various chemicals used for this purpose, and different labs may use different ones for the same nutrient.

Next, labs measure the concentration of nutrients and pH in the extracted solution. Such modern tools as the pH meter (figure 13-7) and spectrophotometer rapidly and accurately measure the extracted samples. A spectrophotometer, for example, passes a beam of light through a test solution and measures the amount of light absorbed. The more light absorbed, the stronger the solution.

Lastly, the lab interprets the results with charts. One chart, for instance, might tell how much of the nutrient the soil has, in pounds per acre, for some tested level in the solution. Other charts recommend how much fertilizer or lime is needed to bring the soil to a proper level.

Laboratory results, however, are only reliable if they have been validated on soils similar to the one being sampled. That is, the tests must be based on research about fertilization, nutrient levels, and yields or crop performance on soils like the sample soil. There are also differences in climates and crops. Usually this means that a grower should use a local laboratory whose recommendations are based on local conditions. Such labs may be associated with a university or a state agricultural experiment station. A number of private firms also do soil and plant tissue testing. Different labs may get different results due to testing procedures; this should not be a concern unless one switches labs. Some testing labs are geared to production agriculture, and their results may be less valid for horticulturists or other soil users.

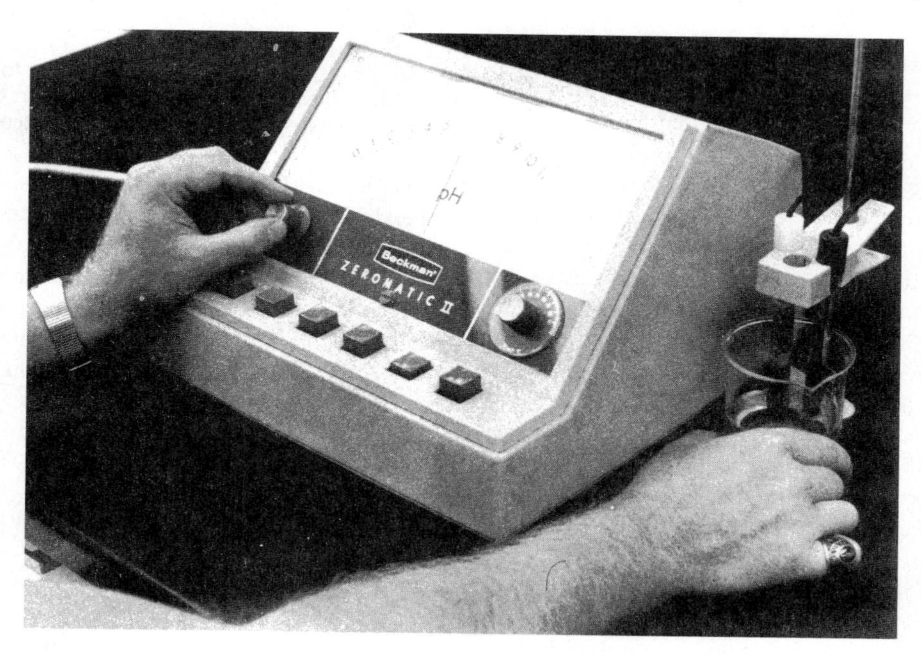

FIGURE 13-7 The pH meter provides the most accurate measure of a soil's acidity or alkalinity.

Few testing labs routinely test for all fourteen mineral nutrients. Rather, a lab offers tests useful for the area it serves. Generally, a standard series includes the following tests:

- Texture is determined by feeling the soil sample or by mechanical analysis.
- Organic matter is measured by comparing the soil color with known standards or other methods.
- The pH and the buffer pH are measured with a pH meter (figure 13-7). From the pH and buffer pH, the technician can suggest lime application rates.
- Phosphorus is washed from the sample with an acid solution. The solution then is put in a spectrophotometer. This method measures soluble phosphorus only; insoluble forms are not measured. The resulting value, called available phosphorus, indicates how much phosphorus is free for plant growth. Several different tests may be used, and they do not always get the same results. If the tests are validated for your area, all are satisfactory.
- Potassium is washed from the sample with a solution that replaces potash on the cation exchange sites. The resulting value indicates the exchangeable potassium, the amount that is readily available for plant growth.

Nitrogen is difficult to test accurately, so there may not be a standard test for it in your area. Labs base nitrogen needs on the amount of nitrogen stored in the soil organic matter, the effect of last year's crop, other cropping practices, and the grower's production goals. However, in areas of low rainfall where leaching and denitrification are low, nitrate nitrogen can carry over from the previous year. Here, a nitrate-nitrogen test is valuable. Also, as concern for nitrate leaching into groundwater has grown in recent years, nitrogen recommendations based on root zone nitrates have increased in popularity. The top two or three feet of soil should be sampled for nitrates.

Other tests may be standard or optional, depending on the laboratory. These include testing for soluble salts, cation exchange capacity, calcium, sulfur, magnesium, and trace elements known to be a problem in the area. Growers of containerized plants and nurseryworkers should ask for a soluble salt test and complete nutrient analysis.

After testing the sample, the lab issues a computer-generated report. The report includes test results, interpretation of the results, and fertilizer and lime recommendations. Figure 13-8 shows a sample report for strawberries in the home garden.

SOIL TEST REPORT
UNIVERSITY OF MINNESOTA
SOIL TESTING LABORATORY

DAKOTA COUNTY AVTI
LANDSCAPE DEPT

Soil Science Department
Horticultural Science Department
Agricultural Extension Service
Agricultural Experiment Station
Cooperating

Report NO: 07648
Laboratory NO: 21945
Date Received: 11/22/82
Date Reported: 11/23/82

Soil Test Results

Sample Designation	Soil Texture	Soil pH	Buffer Index	Organic Matter	P Phosphorus (lb/A)	K Potassium (lb/A)	Soluble Salts (mmhos)
A	CLAY LOAM	5.6	6.2	HIGH	44	274	

INTERPRETATION OF SOIL TESTS

Soil Nutrient Levels: Low Medium High

pH
Phosphorus (P)
Potassium (K)
Soluble Salts

RECOMMENDATIONS FOR: SMALL FRUITS

LIME RECOMMENDATION: NONE RECOMMENDED

THE AMOUNT OF EACH PLANT NUTRIENT TO APPLY:

NITROGEN (N)	+	PHOSPHATE (P2O5)	+	POTASH (K2O)
0.20 LBS./100 SQ.FT.	+	0.05 LBS./100 SQ.FT.	+	0.05 LBS./100 SQ.FT.

APPLY A FERTILIZER THAT SUPPLIES THE RECOMMENDED AMOUNTS OF PLANT NUTRIENTS.

FOR NEW FRUIT PLANTINGS, INCORPORATE THE FERTILIZER INTO THE SOIL PRIOR TO PLANTING.

FOR STRAWBERRIES, SIDEDRESS THE FERTILIZER AT RENOVATION, IMMEDIATELY AFTER HARVEST.

FOR RASPBERRIES, APPLY THE FERTILIZER IN EARLY SPRING, BEFORE GROWTH STARTS.

County: DAKOTA
For additional information contact your County Extension Agent:

Receipt: This is to acknowledge receipt of payment in the amount of $ 4.00

Customer copy

FIGURE 13-8 A sample soil-test report for a homeowner growing strawberries. Note that the report consists of three parts: results, interpretation, and recommendations.

Part of the interpretation is to translate the raw numbers as low, medium, or high. The numbers attached to these categories vary from lab to lab. A "low" category implies that a yield response to fertilization is highly likely, whereas a yield response to a "high" category is unlikely.

Grower Testing. A number of relatively inexpensive, simple electronic devices like the digital pocket pH meter shown in figure 13-9 are now available to growers. These include pH meters, conductivity meters (soil salinity), and even devices for measuring nutrients. Greenhouse growers and container nurseries make extensive use of such devices, allowing rapid and timely

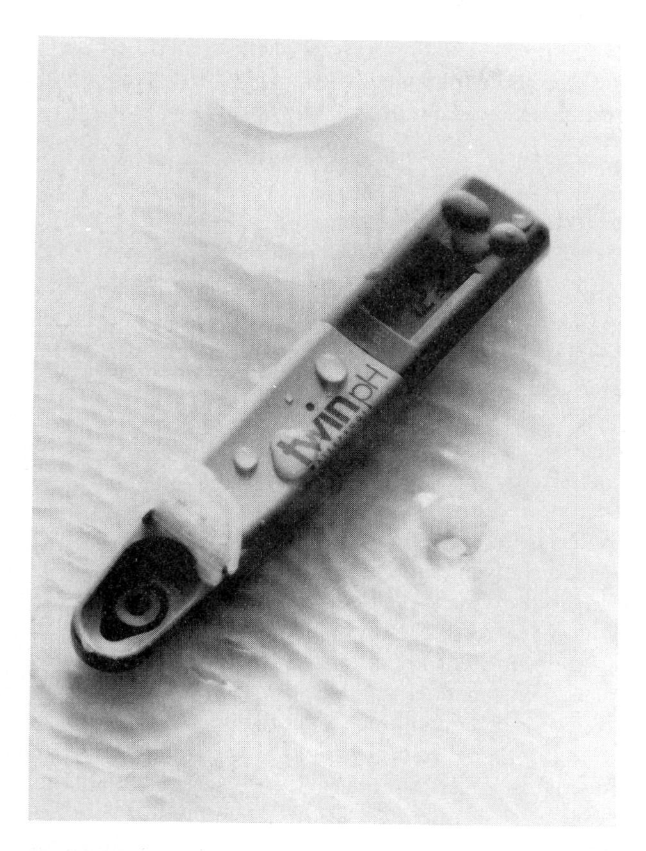

FIGURE 13-9 This pocket pH meter is one convenient testing device available to greenhouse growers and other soil users. *(Courtesy of Spectrum Technologies, Inc.)*

tests on crops requiring constant soil monitoring. Other soil users may find the devices helpful as well.

Most often, one part of the soil sample is mixed with two parts water and allowed to sit for about a half hour. The mixture is filtered to remove solid material, and the remaining liquid is tested with one of the devices. Tables are available to help interpret the results. Keep in mind that these tests use different procedures than those of commercial laboratories, so values cannot be directly compared.

Tissue Testing

Plant tissue tests in combination with soil tests give the most complete picture of nutrient status in the plant. In **tissue testing,** nutrients in the plant itself, rather than nutrients in the soil, are tested. These tests are useful for "smoking out" trace element problems and may be more reliable than soil tests. Tissue tests may distinguish symptoms of nutrient problems from disease or other problems. Tissue tests may also indicate if some soil condition is hindering nutrient uptake. Some growers employ tissue testing to check the effectiveness of their fertilizer programs.

Plant tissue tests are also very useful for tree and vine crops in nurseries, vineyards, or orchards. Their root systems are much more extensive than are those of annual crops. Thus, it is often difficult to determine exactly where feeding roots are and at what depth a soil sample should be taken. For these reasons, tissue tests are useful in monitoring the nutritional status of the crop.

Nutrient levels vary sharply in different plant tissues of different ages. Before sending samples to a laboratory, be sure to determine the plant part used and the growth stage required. For example, soybean samples are collected from fully open leaves at first flowering. Apple samples are taken from leaves in the middle of shoots eight to twelve weeks after full bloom. A grower should follow the instructions of the actual laboratory being used. A general method for sampling plants follows:

1. Sample about ten to fifteen plants using the recommended plant parts. The parts should be clean of soil or dust. Sample only the intended species. Do not, for example, mix both clover and grass samples from a pasture. Do not include dead materials, and avoid damaged parts unless they are the intended sample.
2. If necessary, use water to clean dust or soil from the leaves.
3. Air-dry the samples before shipment.
4. Fill out the information sheet completely. Include any recent soil-test results, if suggested by the laboratory.
5. Ship the samples to the laboratory in a heavy paper sample bag. Leaves may mold if shipped in a plastic bag.

Green Tissue Tests. A simpler form of tissue test is the **green tissue test** or plant sap test. In this test, plant sap in the leaf petioles or young stems is tested for nutrient levels. Portable test kits are available that can be used in the field. Some kits use test papers treated with testing reagents. To use these, squeeze plant sap on the test spots. Other test kits use glass vials and spot plates (figure 13-10). These tests require bits of leaf petioles to be mixed with a liquid reagent in a vial and the color is noted on a spot plate.

(*A*)

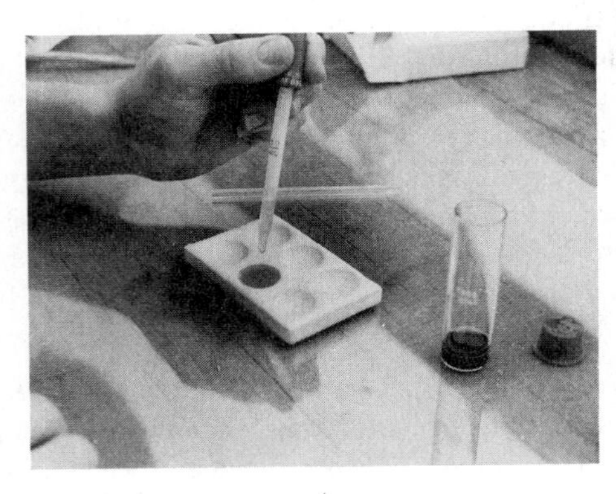

(*B*)

FIGURE 13-10 A portable plant sap test kit can be used to measure the concentrations of nutrients in the sap of crop plants. Here, geranium petioles are cut up *(A)* and treated with a reagent *(B)*.

As in tissue testing, several plants should be tested. Best results are obtained by comparing test results from both deficient plants and nearby healthy plants.

SUMMARY

Growers need to make effective use of fertilizers. Soil testing is the best tool growers have to decide how much fertilizer is needed to avoid both underfertilization and overfertilization.

The first step is to sample the soil. Each area to be sampled should be uniform. Many subsamples are collected and mixed to form a composite sample from which a small amount is removed and sent to a private testing laboratory or university soil-testing center. A sampling information sheet accompanies the sample to provide data the laboratory needs to make useful recommendations.

Soil-testing laboratories use modern equipment to measure nutrient levels quickly and precisely. A computer then generates test results, interprets the data, and makes fertilizer and lime recommendations.

Plant tissue tests are used less often, but when added to soil tests, a more complete picture of plant nutrient levels is obtained. Growers who wish to try tissue tests should consult with the testing laboratory for instructions.

Commercial laboratory testing provides more reliable results, but portable test kits and devices are available to growers for both soil and plant sap tests.

REVIEW

1. Describe the three main steps of soil testing in a commercial laboratory.
2. Why should a grower not rely on visual symptoms as a way to detect nutrient shortages?
3. When is fertilization most cost-effective?
4. Why do container plants need more frequent testing?
5. How does soil sampling in precision farming differ from that of standard practices?
6. Why might a grower best patronize a local testing laboratory?
7. How do we obtain a valid composite sample—that is, one that accurately reflects soil of the field being sampled?

8. What might be soil or growing conditions that might limit plant growth not detected by a soil test?
9. Why should soil be tested before a new crop—or even a new garden in someone's yard—is planted?
10. A case study: A greenhouse grower conducted both a soil test and a tissue test on a potted geranium showing some unknown damage. Because the symptoms did not resemble any known pest problems, a nutrient problem was suspected. The grower sampled both healthy plants and affected plants and compared the results. Most nutrients were within acceptable range in both soil and tissue, but iron was very low in the soil and extremely high in the plant tissue, especially on infected plants. Speculate on an explanation for the difference between the two tests. Why test healthy plants? What might have happened had the grower done only the soil test?

ENRICHMENT ACTIVITIES

1. Sample a test area and send the sample to a local testing lab.
2. Use a soil-test kit to test the same sample in your school. Compare the results of your test with the lab results.
3. Tour a soil testing facility.
4. A good explanation for how soil tests are developed is available on the Internet at: <http://www.ext.colostate.edu/pubs/crops/00501.html>.
5. Check for various commercially available soil-testing equipment on the Internet by searching for terms such as "soil-testing equipment" or "pH meter."

CHAPTER 14

Fertilizers

OBJECTIVES

After completing this chapter, you should be able to:

- distinguish forms of fertilizers
- describe fertilizer sources for each nutrient
- perform important fertilizer calculations
- describe how to use fertilizers
- list two effects fertilizers have on the soil

TERMS TO KNOW

banding
broadcasting
calcium carbonate equivalent
chelates
complete fertilizers
fertigating
fertilizer
fertilizer analysis
fertilizer burn
fertilizer filler
fertilizer grade
fertilizer ratio
fluid fertilizers
foliar feeding
fritted trace elements
granules
mixed fertilizers
nutrient carriers
pop-up fertilizers
pressurized liquid
prills
pulverized fertilizers
sidedressing
split application
starter fertilizer
topdressing

Before plant nutrients were scientifically identified, growers knew some materials helped plants prosper. Early farmers probably noticed the lush growth of grass around animal droppings and began using them to raise better crops. Lime, ashes, dead fish, and ground bones have all been used. Even human waste ("night soil") has been an important fertilizer in some cultures. Most American growers now use chemical fertilizers to supply the elements their crops need.

Fertilizer is material applied to soil or plants to supply essential elements. Some states legally define minimum requirements for a material to be sold as fertilizer. Fertilizers can be grouped into four categories—mineral, organic, synthetic organic, or inorganic (figure 14-1):

- Mineral fertilizers are ground rocks containing nutrients. Dolomitic lime, for example, is a fine source of calcium and magnesium. Most minerals have low nutrient content and dissolve very slowly, so their usefulness as fertilizers is limited.
- Organic fertilizers are organic materials, such as animal manure, that contain nutrients. Many can be considered "slow-release" fertilizers because nutrients are released slowly over the growing season as the organic matter decays. Many organic materials contain low amounts of nutrients and may therefore be expensive nutrient sources. In some states, "dilute" organic materials may not be legally labeled as fertilizers because their nutrient content is too low.

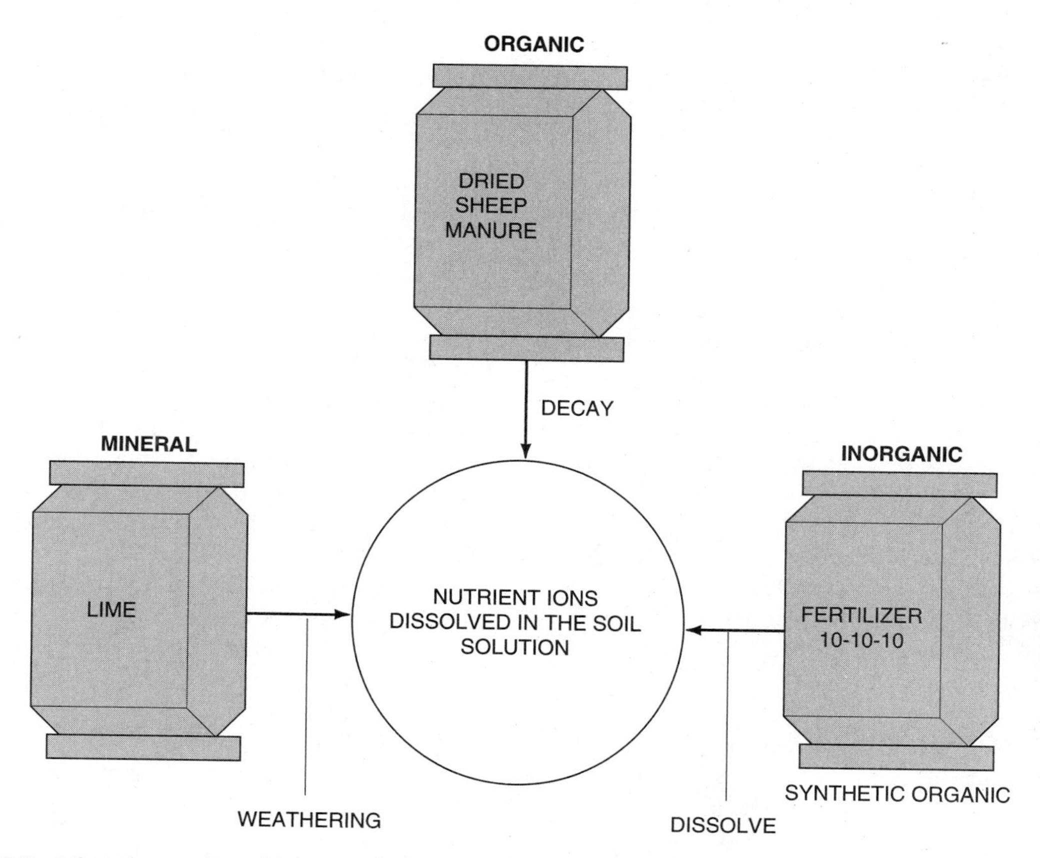

FIGURE 14-1 Mineral, organic, and inorganic fertilizers weather, decay, or dissolve to release nutrient ions taken up by plants.

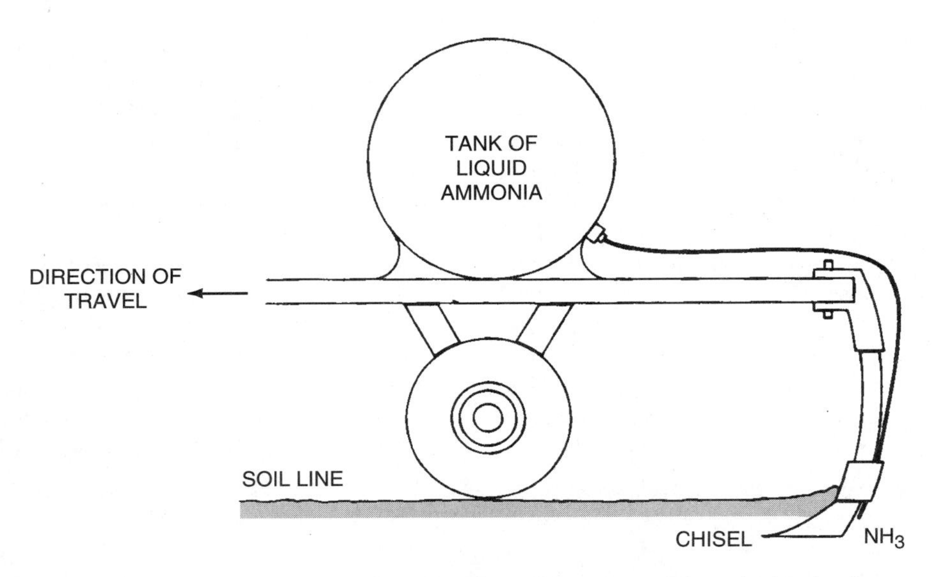

FIGURE 14-2 Anhydrous ammonia, a pressurized liquid, is injected into the soil through chisels. The ammonia dissolves in soil water to become ammonium ions.

- Synthetic organic fertilizers are manufactured by industry but are chemically organic (contain carbon and hydrogen). Urea is readily available to plants, but others, mostly urea derivatives, are made to be slow-release. Nutrient content is generally high compared to mineral or natural organic fertilizers.
- Inorganic fertilizers are mined or manufactured and are chemically inorganic. Most dissolve quickly in the soil for rapid growth response and have a high nutrient content.

Fertilizers are provided in a number of forms, giving growers several choices of application methods. The forms can be divided into four main groups: pressurized liquids, fluids, dry fertilizers, and slow-release fertilizers.

Pressurized Liquids. Anhydrous ammonia is the primary **pressurized liquid.** Ammonia is a gas at normal temperatures and pressure, but changes to a liquid when cooled to −28°F. It can then be stored in large, high-pressure or refrigerated tanks. Smaller, wheeled tanks are filled from the storage tanks and are driven to the field to be fertilized.

The liquid is applied by injecting it into the soil (figure 14-2). Pressure in the tank forces the liquid through tubes to special chisels that are pulled underground. When it reaches the soil, the liquid evaporates rapidly.

Fluid Fertilizers. **Fluid fertilizers** are also liquid, but they are not under pressure. In the case of nitrogen, they are often called nonpressure solutions. The most common fluid fertilizers are solutions (figure 14-3). In a true solution, fertilizer dissolves in water to form a clear liquid and will not settle out. Solutions can only be made of chemicals that are soluble in water.

Less water-soluble chemicals can be made into fluids called suspensions. A finely ground fertilizer is coated with a special clay and mixed with water. The treatment helps the fertilizer grains suspend in water to form a cloudy liquid. If the liquid is not kept stirred, however, the grains will settle out.

Fluid fertilizers are popular because they can be applied in many ways: sprayed on a field or on turf, injected into soil, mixed into irrigation water (figure 14-3), or sprayed on crop foliage. They can also be mixed with fluid lime, herbicides, and other crop chemicals, with care to ensure that they mix without problems.

FIGURE 14-3 Liquid feed being used in a flood irrigation system in Arizona. Fluid fertilizer is being injected into the irrigation lateral. *(Courtesy of USDA)*

Organic Material	Percentage, Dry Weight Basis N	P_2O_5	K_2O
Bat guano	10.0	4.0	2.0
Blood meal	12.0	2.0	1.0
Fish meal	10.0	6.0	—
Cotton seed meal	6.0	3.0	1.5
Soybean meal	7.0	1.2	1.5
Bone meal, raw	3.0	22.0	—
Bone meal, steamed	1.0	15.0	—
Wood ashes	—	1.0	4.0

FIGURE 14-4 Organic fertilizers do not give a rapid response, but mineralized nutrients become available over time.

Dry Fertilizers. Dry fertilizers are applied to soil, where they dissolve quickly in soil water to release nutrients. Dry fertilizers are available in three types:

- **Pulverized fertilizers** are made by crushing fertilizers into a powder. They are dusty, making them unpleasant to handle or spread evenly. Some pulverized fertilizers absorb moisture from the air, causing them to cake during storage.
- **Granules** are much easier to use. The manufacturer treats the material so it has large, more evenly sized grains. Granules spread evenly and easily, with much less dust. However, some dust, or "fines," still causes problems. Granules are coated to reduce moisture absorption.
- **Prills** are smooth, round, and dust-free. They are made by a different process than granules. Prills have superior flowing and spreading qualities and are free of fines. Growers find them easy to use. Prills are also coated to prevent caking during storage.

Slow-Release Fertilizers. Slow-release fertilizers are also dry. The nutrients they contain dissolve into the soil solution slowly over a period of several weeks up to a few months. Slow-release fertilizers are too costly for common agricultural use, but they are widely used in horticulture; for example, as turf fertilizers. Slow-release fertilizers benefit growers by releasing nutrients only as fast as crops can use them, so little is lost to leaching and adverse impacts on the environment are limited.

Fertilizer Materials

Few fertilizers are pure elements. Rather, most consist of compounds that release nutrients in forms useful to plants. These compounds are called **nutrient carriers.**

Nitrogen Carriers. Organic nitrogen exists in several forms (figure 14-4). Decay changes the organic nitrogen to ammonia, which, in turn, changes to nitrates. Many organic nitrogen sources are too expensive for farmers and are used primarily by home gardeners. However, some sources, such as manure, continue to be used by many farmers.

The first commercial nitrogen fertilizer was actually organic—guano, or bat and bird droppings. Later, deposits of saltpeter (sodium nitrate) were mined in Chili. Today fertilizer companies manufacture most nitrogen carriers by the Haber process.

The Haber process uses nitrogen from the air. Natural gas is used as a source of hydrogen, which is combined

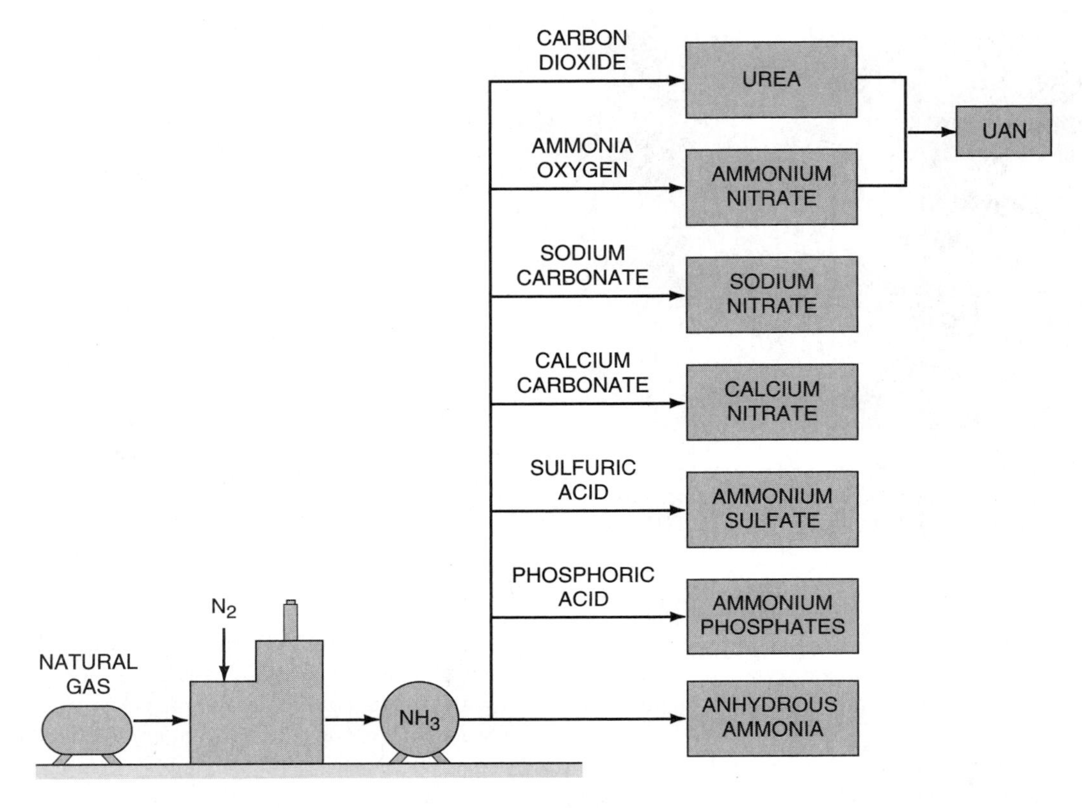

FIGURE 14-5 The Haber process is the source of most chemical nitrogen fertilizers.

with gaseous nitrogen. This reaction, in the presence of heat, pressure, and an iron catalyst, produces ammonia:

$$3H_2 + N_2 \rightarrow 2NH_3$$

Ammonia can be used as a fertilizer or changed to other forms, as shown in figure 14-5.

- Anhydrous ammonia, 82 percent nitrogen, results directly from the Haber process. The word "anhydrous" means "without water." It is the main pressured liquid and is applied as shown in figure 14-2. Anhydrous ammonia is the cheapest, strongest form of nitrogen. In the soil, ammonia reacts with water to release ammonium ions. If not injected deep enough, especially in sandy soil, it may evaporate and be lost from the soil. Ammonia destroys lung tissue if inhaled, so care should be taken during transport and application to ensure that none escapes.
- Aqua ammonia, consisting of 20 percent nitrogen, comes from dissolving ammonia in water to form a low-pressure solution. The use of aqua ammonia has declined as growers switch to other fluid fertilizers.
- Ammonium nitrate, containing 33 percent nitrogen, is half ammonium nitrogen and half nitrate. This is a good general-purpose dry material that is easy to handle and apply. It absorbs moisture from the air, causing it to cake. To prevent ammonium nitrate from hardening, it should not be left in open bags or piles.
- Ammonium sulfate, 21 percent nitrogen, contains sulfur as well as nitrogen. It is a dry fertilizer. Ammonium sulfate is very acid-forming and is ideal for acid-loving plants. It is not suitable as a starter fertilizer for plants that grow best at a neutral pH. Any ammonium fertil-

Nitrogen Carrier	% Nitrogen	% Ammonium	% Nitrate	Effect on pH
Anhydrous ammonia	82.0	82.0	0	Very acid
Ammonium nitrate	33.0	16.7	16.6	Acid
Ammonium sulfate	21.0	21.0	0	Very acid
Aqua ammonia	24.0	24.0	0	Acid
Nitrate of soda	16.0	0	16.0	Basic
Calcium nitrate	15.5	0	15.5	Basic
Urea	46.0	46.0*	0	Acid
UAN (nitrogen solution)	32.0	22	10.0	Acid
Urea formaldehyde†	37.0	37.0	0	Acid
Sulfur-coated urea†	39.0	39.0*	0	Acid
IBDU†	30.0	—	—	—

*Urea becomes ammonia in the soil. †Slow-release fertilizers.

FIGURE 14-6 Common nitrogen carriers. Other sources are listed under potash and phosphate sources.

izer, including ammonium nitrate or sulfate, can lose nitrogen by volatilization when spread on recently limed or calcareous soils. To prevent the loss of nitrogen, the material is best mixed into the soil.

- Nitrate of soda (sodium nitrate, 16 percent nitrogen) is commonly used on tobacco. Unlike most other nitrogen sources, it raises soil pH. Calcium nitrate is a similar but less salty material. Both materials are dry but water soluble.
- Urea, 46 percent nitrogen, is a synthetic organic material ($CO(NH_2)_2$). In soil, urea rapidly breaks down to ammonia. If urea is left on the top of the soil, some ammonia may escape into the air. Urea is rapidly becoming the most popular dry fertilizer, because it can be produced more cheaply than ammonium nitrate. It is also used to make fluid fertilizer.
- Urea-ammonium nitrate (UAN) is a nitrogen solution. UAN is made by mixing liquid urea and ammonium nitrate to make either a 28 percent N solution or a 32 percent N solution.
- Urea-formaldehyde (UF), IBDU, and SCU (sulfur-coated urea) are slow-release synthetic organic materials. These are used primarily to fertilize turfgrasses and potted plants. They are too costly for general agriculture use.

Figure 14-6 summarizes the characteristics of these and other nitrogen forms.

FIGURE 14-7 Phosphate fertilizers come from mined phosphate rock. *(Courtesy of Potash and Phosphate Institute)*

Phosphorus Carriers. Phosphorus fertilizers are obtained from the mining of rock phosphate in Florida and other states (figure 14-7). Phosphate rock contains the mineral apatite, or calcium phosphate. The ground

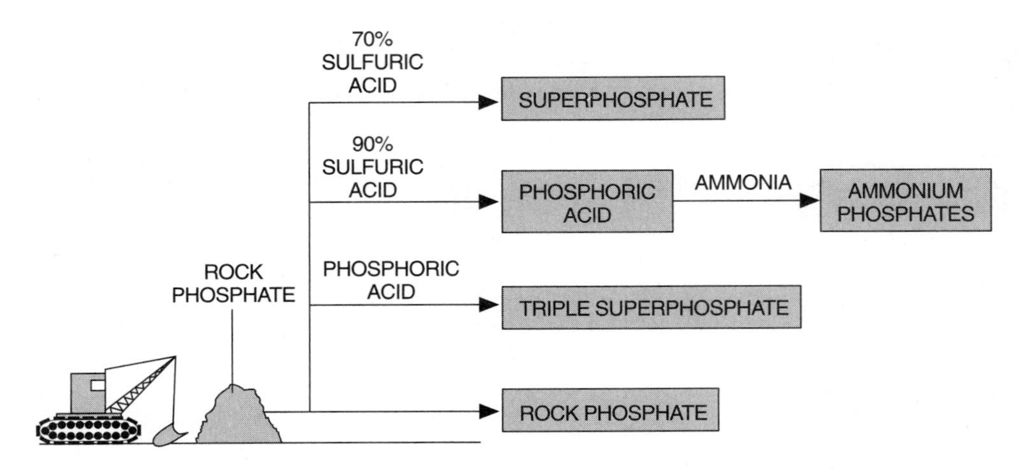

FIGURE 14-8 Phosphate rock is treated with acids to produce common fertilizers.

rock can be applied directly to the soil, but is usually treated with acid to break down the apatite into simpler compounds (figure 14-8). Figure 14-4 lists some organic sources of phosphorus and figure 14-9 summarizes the following carriers:

- Rock phosphate, containing between 25 percent and 35 percent phosphate, and calcium, may be spread on the soil but will dissolve slowly. It is an example of a mineral fertilizer. Rock phosphate works best when finely ground and used on acid soils (pH below 6.0) or on soils high in organic matter. Acid soils break down the apatite.
- Superphosphate, containing 20 percent phosphate, results from the reaction of rock phosphate with sulfuric acid (figure 14-8). This material is half gypsum (calcium sulfate) and half calcium phosphate. Being lower in phosphate than other carriers, it is no longer used by most growers. Growers find it most useful when sulfur or calcium is lacking in the soil.
- Triple superphosphate, with 46 percent phosphate, is also treated rock phosphate. It is higher in phosphorus than regular superphosphate and contains no sulfur or calcium. Triple superphosphate is a popular fertilizer. Both superphosphates contain small amounts of flourine, making them unsuitable for the many potted foliage plants that are highly sensitive to flourine.
- Ammonium phosphates are made by mixing phosphoric acid with ammonia. Two similar compounds result from the reaction, monoammonium phosphate (53 percent phosphorus), and diammonium phosphate (46 percent phosphorus). These are often referred to as MAP and DAP. Both compounds are water soluble. They are used either as dry fertilizers or as fluids. Ammonium polyphosphates (35 percent to 62 percent phosphorus) are similar materials and are used to make fluid fertilizers. It has been found that the presence of ammonium ions improves the uptake of phosphate by plants, making these chemicals very effective phosphorus sources. Ammonium phosphates also supply nitrogen and are rapidly

	Percentage Available		
Phosphate Carrier	P_2O_5	P	N
Rock phosphate	25–35	11–15	—
Superphosphate	20	8.7	—
Triple superphosphate	46	20	—
Monoammonium phosphate	48–55	21–24	11–13
Diammonium phosphate	46–53	20–23	18–21
Ammonium polyphosphate	36–62	16–27	10–15
Phosphoric acid	53	23	—

FIGURE 14-9 Common phosphate fertilizers.

(A)

(B)

FIGURE 14-10 *(A)* Potash ore is processed in a plant such as this. *(B)* In the process, KCl is separated by flotation from other salts. *(Courtesy of Potash and Phosphate Institute)*

becoming the most popular phosphate fertilizers. They are widely used in bulk blends.

- Bone meal and manure are both organic sources of phosphate. Bone meal is made by grinding bones that are a by-product of the meat-packing industry. Homeowners use bone meal as a phosphate and calcium fertilizer.

Potassium Carriers. Potash mines in New Mexico, Utah, California, and Canada produce most of the potash used by American growers. The deposits are mixtures of several potassium, sodium, and magnesium salts. Producers process the salts to separate and purify the potassium (figure 14-10). Figure 14-4 lists some organic potassium sources, and figure 14-11 summarizes the primary carriers.

Potash Carrier	Percentage Available K_2O	Nitrogen
Muriate of potash	60	—
Sulfate of potash	49	—
Potassium nitrate	44	13
Sulfate potash magnesia	22	—

FIGURE 14-11 Common potash carriers.

- Muriate of potash (potassium chloride), 60 percent potash, accounts for 97 percent of all potassium fertilization. It costs less than other carriers and dissolves readily in water. Because it contains chlorine, other sources may be better for crops sensitive to chlorine. It is also highly saline.
- Sulfate of potash (potassium sulfate) contains 49 percent potash and is used to a minor extent in dry fertilizers. It can be used to make a suspension. It is preferred for tobacco because of the sensitivity of this crop to chlorine. It also adds sulfur to the soil.
- Nitrate of potash (potassium nitrate) contains 13 percent nitrogen and 44 percent potash. It is a common fertilizer for container plants. Although it is mostly used in a dry form, it can also be used in solution.
- Sulfate of potash-magnesia is also useful for chlorine-sensitive crops and where soil salinity is a problem. It consists of 22 percent potash, 11 percent magnesium, and 22 percent sulfur. It may be used for soils that lack sufficient amounts of magnesium and sulfur.
- Wood ashes and manure are also good potash sources.
- Granite meal is a mineral fertilizer used by some growers who prefer not to use chemical fertilizers. This material is a finely ground, gritty waste product of the monument and building-stone industry. Granite dust is too insoluble to be of any immediate use to plants but adds to the "bank" of soil potash. Figure 2-5 lists nutrients in a typical granite.

Secondary Elements. Mineral fertilizers supply most of the calcium, magnesium, and sulfur required by plants. The most important fertilizers include lime,

Material	Percentage Available Ca	Mg	S	Effect on pH
Calcitic lime	31.7	—	—	Basic
Dolomitic lime	21.5	11.4	—	Basic
Gypsum	22.5	—	12.0	Neutral
Hydrated lime	46.1	—	—	Basic
Burned lime	60.3	—	—	Basic
Magnesia	—	55.0	—	Basic
Magnesium sulfate (Epsom salts)	—	11.0	14.5	Neutral
Potassium magnesium sulfate	—	11.0	22.0	Neutral
Flowers of sulfur	—	—	30–100	Acidic

FIGURE 14-12 Common sources of secondary elements.

Trace Element	Sulfate FTE	Salts	Chelates	Others	Treatment
Boron	X			Borax	BC borax
Copper	X	X	X	Oxide	BC or B sulfate
Iron	X	X	X		F chelate, acidify soil
Manganese	X	X	X	Oxide	BC or B sulfate
Zinc	X	X	X		B chelate
Molybdenum	X			Sodium molybate, molybdic acid	Mix with NPK, liming soil

Key: FTE = Fritted trace elements BC = Broadcast B = Banding F = Foliar feeding

FIGURE 14-13 The most commonly used sources of trace elements are marked with an "X." The treatment is the most common effective type.

gypsum, and sulfur. The finer the grind, the more quickly these materials act. Many of the minerals affect soil pH. Figure 14-12 lists several sources of secondary elements.

Trace Elements. Each trace element is available in a number of chemical forms. Most trace elements, however, are commonly used in the following forms, which are summarized in figure 14-13:

- Sulfate salts are inexpensive and dissolve easily in water. They can be used as dry or fluid fertilizers. Trace elements are also available in other types of salts and oxides.
- **Fritted trace elements** (FTE) are a safe way to apply trace elements. FTE are made by adding salts to molten glass, which is poured into cold running water. The glass cools and shatters; the pieces are then ground into fine powder. Frits are dry fertilizers that dissolve slowly in the soil.
- **Chelates** are common, useful, water-soluble trace element fertilizers, generally used in fluid form.

Mixed Fertilizers

Growers could apply a fertilizer that contains a single nutrient, which would mean a fertilization operation for

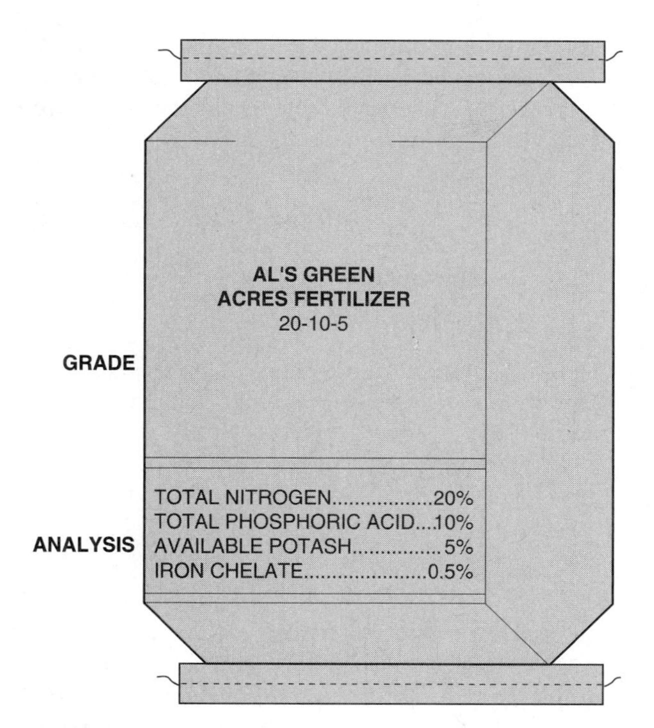

FIGURE 14-14 Grade is expressed by three numbers: nitrogen, phosphate, potash. All three are present here, so it is a complete fertilizer. Also shown is analysis, a listing of the percent of all nutrients present.

each nutrient. It is often more convenient to use fertilizers that contain several nutrients. Such fertilizers are made by mixing several of the carriers just described.

Fertilizer Analysis and Grade. The contents of a bag of fertilizer may be listed in two ways. Some bags list **fertilizer analysis,** which lists the fertilizer elements in the bag and their percent content (figure 14-14). Such a list could include any of the fourteen mineral elements. All bags of fertilizer should show the **fertilizer grade,** which indicates the primary nutrient content. Grade lists the content as a sequence of three numbers that tell, in order, the percentage of nitrogen (N), phosphoric acid (P_2O_5), and potash (K_2O). Grade is often referred to as "N-P-K," which stands for nitrogen, phosphorus, and potassium, in that order.

For example, a fertilizer with the grade 0-0-60 is 60 percent potash with no nitrogen or phosphate. To decide how much potash is in the bag or ton of the fertilizer, multiply the weight times the percentage. Thus, one ton of muriate of potash contains the following amount of potash:

$$\text{Potash} = \frac{2{,}000 \text{ pounds} \times 60 \text{ percent}}{100} = 1{,}200 \text{ pounds}$$

Fertilizer containing only one element is called a single-grade fertilizer. Many fertilizers contain two or three nutrients and are called **mixed fertilizers. Complete fertilizers** have all three of the primary elements. Note that "complete fertilizer" does not mean that all fourteen mineral nutrients are included.

To determine the amount of each nutrient in a complete fertilizer, the percentage of the nutrient is multiplied by the weight of fertilizer. For example, in a fifty-pound bag of 20-10-20:

$$\text{Nitrogen} = \frac{50 \text{ pounds} \times 20 \text{ percent}}{100} = 10 \text{ pounds}$$

$$\text{Phosphate} = \frac{50 \text{ pounds} \times 10 \text{ percent}}{100} = 5 \text{ pounds}$$

$$\text{Potash} = \frac{50 \text{ pounds} \times 20 \text{ percent}}{100} = 10 \text{ pounds}$$

Additional information may also be found in the analysis, like the percentage nitrogen that is ammoniacal and the percentage that is nitrate. Some fertilizers, especially those blended for turf, contain nitrogen sources that dissolve slowly. These will be identified as water-insoluble nitrogen (WIN) or slow-release nitrogen (SRN).

Grade may also identify a secondary nutrient as a fourth number in the traditional NPK. For example, calcium nitrate may carry the grade 15-0-0-39Ca, meaning the material is 39 percent calcium. Similarly, sulfur (S) or magnesium (Mg) is found as a fourth number.

Contents of Fertilizers. Fertilizer grades never total 100 percent. For example, 10-10-10 fertilizer is 30 percent nutrient and 70 percent other ingredients. What are those other ingredients? Primarily, the remainder of the fertilizer is the weight of the other elements that are part of the carrier, such as hydrogen and oxygen.

A small percentage is **fertilizer filler** and conditioner. Fillers may be sand, clay granules, ground limestone, or ground corncobs. They are used to bring a load of bulk fertilizer to a weight of one ton. Conditioners improve the quality of the fertilizer and make it easier to use.

Fertilizer Ratio. **Fertilizer ratio** states the relative amounts of nitrogen, phosphate, and potash in fertilizers. Ratios are useful when comparing two fertilizers, as shown in the following examples:

	Grade	Ratio
(a)	10-10-10	1-1-1
(b)	20-20-20	1-1-1
(c)	6-12-12	1-2-2
(d)	5-15-30	1-3-6

Note that *(a)* and *(b)* have the same ratio. This means that one fertilizer can be used in place of the other. Applying one ton of 10-10-10 is the same as applying one-half ton of 20-20-20. A grower may select fertilizer with the ratio suggested by soil-test reports. For instance, if the report suggested 100 pounds of nitrogen, 50 pounds of phosphate, and 75 pounds of potash per acre, a single fertilizer with the ratio of 4-2-3 would be ideal.

Elements and Oxides. The way fertilizer grade is listed leads to some confusion. Most people think of fertilizer grade as "NPK." This is read as nitrogen, phosphorus, and potassium. Actually, nitrogen is listed as the element, but the other two nutrients are listed in their oxide forms. The true grade should be listed as N-P_2O_5-K_2O, which is read as nitrogen, phosphoric acid, and potash.

As an example of the confusion, consider the fertilizer 20-10-10. The numbers lead one to expect 200 pounds of phosphorus in a ton of this fertilizer. Actually, one ton contains only 88 pounds of phosphorus. The amounts of nutrients in a ton of 20-10-10 can be listed in the elemental and oxide forms:

	Oxide	Element
N	400	400
P	200	88
K	200	166

When reading soil-test reports or other recommendations, always check which form is being used. To convert between phosphorus/phosphoric acid and potassium/potash, the following formulas are used:

(a) $P \times 2.29 = P_2O_5$

(b) $P_2O_5 \times 0.44 = P$

(c) $K \times 1.2 = K_2O$

(d) $K2O \times 0.83 = K$

As an example of the use of one of the formulas, determine how much actual potassium is contained in one ton of 0-0-60:

$$K = \frac{2{,}000 \times 60 \text{ percent}}{100} \times 0.83 = 996 \text{ lb}$$

Calculations for Blending. Growers can buy premixed fertilizer, but a limited number of ratios are available. Fertilizer can be custom blended to obtain the analysis and ratio that best suits the needs of the grower. To blend fertilizers, it is necessary to determine how much of each carrier is needed to produce the final mixed fertilizer. The following formula can be applied to each carrier:

$$Z = \frac{A \times B}{C}$$

where

Z = pounds of carrier for each element

A = pounds of mixed fertilizer needed

B = percentage of the element needed

C = percentage of the element in the carrier

As an example, let's determine how a ton of 10-10-20 can be blended from the following carriers:

Ammonium nitrate	33-0-0
Triple superphosphate	0-46-0
Muriate of potash	0-0-60

Calculations: $Z = \frac{A \times B}{C}$

$$\text{Ammonium nitrate} = \frac{2{,}000 \times 10}{33} = 606 \text{ lb}$$

$$\text{Triple superphosphate} = \frac{2{,}000 \times 10}{46} = 435 \text{ lb}$$

$$\text{Muriate of potash} = \frac{2{,}000 \times 20}{60} = 667 \text{ lb}$$

Total carriers = 1,708 lb

A total of 1,708 pounds of carrier will be blended. To bring the total to 2,000 pounds, 292 pounds of filler will be added to the mix.

Selecting Fertilizer

Growers can choose from an array of fertilizers. Factors influencing the selection include the crop, time of year, application method, and cost.

For most crops, the form of the fertilizer is not critical. One choice is between nitrate and ammonium nitrogen. Plants absorb both ions, but the preference for most agricultural crops is for the nitrate form. However, under warm, moist conditions, ammonium nitrogen nitrifies to the nitrate nitrogen in four to six weeks. For that reason, ammonium and nitrate usually have similar effects on crop growth. On the other hand, nitrates are lost more easily from the soil.

In a few cases, either ammonia or nitrates work best. There are a few simple rules to guide the grower in selecting the best form:

- Nitrates are preferred for early-spring planting of cool-season crops.
- Ammonia is better for fall fertilization, because less nitrogen will leach out of the soil before spring.
- Fertilizers for container plants should favor nitrates over ammonium nitrogen. Roots growing in a pot are easily damaged by excess ammonia. Ammonia favors soft, rapid growth in container plants, while nitrates promote a tougher, more restrained development.
- Many acid-loving plants prefer ammonium nitrogen.

Some crops, including tobacco, are sensitive to chlorine. For these crops, low-chlorine fertilizers should be chosen. In some applications, growers need to be concerned about a fertilizer's affect on soil pH or salinity, as discussed later in this chapter.

Growers also base their fertilizer selections on the means used to apply fertilizer. For example, fertilizers applied through the irrigation system must be water soluble. Several other recommendations are noted later in the chapter.

Fertilizer selection commonly depends upon price—the least costly fertilizer per pound of plant food is commonly selected. Cost can be calculated as follows, using nitrogen as an example:

$$\text{Price/lb N} = \frac{\text{Price per ton}}{2{,}000 \times \%\text{N}} \times 100$$

For instance, the price of nitrogen in a ton of ammonium nitrate (33-0-0) that costs \$200 would be:

$$\text{Price/lb N} = \frac{200}{2{,}000 \times 33} \times 100 = \$.30 \text{ per lb nitrogen}$$

The same calculation can be made for a single bag of fertilizer; just substitute the weight of the bag and the price per bag. Similarly, the cost of potash and phosphate may be computed by substituting their values for nitrogen in the formula. These figures allow a grower to compare the cost of different fertilizer elements.

Applying Fertilizer

Fertilizers can be applied before a crop is planted, during planting, after it is growing, or in some combination of the three methods. Preplant feeding brings the field to a good nutrient level before a crop is planted. On a fine soil with little leaching and a high cation exchange capacity, this one feeding may supply all the nutrients needed for the season. While this is the main fertilization for most crops, there are three drawbacks:

- Phosphate is not concentrated near young seedlings. Phosphates do not move much in the soil, and young seedlings are limited in their ability to forage for nutrients. The same is true, to a lesser extent, of potash.
- In coarse, low CEC soils, much of the nitrogen applied in a preplant feeding will leach away before plants can use it.
- Applying fertilizer before planting does not match the needs of the crop. Generally, the crop needs most fertilizer when growing rapidly later in the season.

Crop Stage	Fertilization
(1) Preplant	1/6
(2) 8 leaves	1/6
(3) 12–15 leaves	1/2
(4) Early tassel	1/6

FIGURE 14-15 A model nitrogen schedule for fertigated corn on sandy soils. The recommended nitrogen is divided into four applications at the fractions indicated. This reduces leaching losses and supplies nitrogen according to crop needs.

FIGURE 14-16 Preplant application of anhydrous ammonia by soil injection on an Iowa no-till cornfield. *(Courtesy of USDA)*

Timing nitrogen application closest to maximum crop use reduces nitrogen losses and chances of groundwater contamination.

Fertilizer applied while planting, called **starter fertilizer,** allows phosphate to be placed near the seed. Fertilization after planting solves the other two problems. This is usually done by dividing the year's fertilizer into two or more parts, one applied before planting and the rest used later in the season in one or more applications. For example, corn may be fertilized before planting, then again thirty days after planting. This **split application** reduces the loss of nitrogen by leaching and allows a grower to apply fertilizer when the crop has the greatest need for it. Fertilization with irrigation allows the grower to make several applications to closely match growth needs. Figure 14-15 presents a model schedule for **fertigating** (fertilizing with irrigation) corn. See extension agents for local recommendations.

For perennial crops like hay or fruit, later fertilizations must follow a preplant application. A single preplant feeding will not meet the nutrient needs of the crop in later years. Thus, fertilizer is added yearly.

Now let's look at the methods used to apply fertilizers at these different times.

Fertilizing Before Planting

BROADCASTING. The simplest way to fertilize before planting is by **broadcasting.** Machinery, and sometimes aircraft, is used to spread dry fertilizers evenly on the soil surface. Fluid fertilizers also can be sprayed on the soil. For phosphate and potash, the material should then be "plowed down," or mixed into the soil, before the crop is planted. This step is important because these nutrients do not leach very far into the soil, and, if left on the surface, will not reach the root zone and may run off into surface water. Broadcasting is quite popular because bulk blends can be applied rapidly.

SOIL INJECTION. This is also known as knifing or chiseling, and can be used before crops are planted. Most commonly, anhydrous ammonia is chiseled into the whole field (figure 14-16). Fluid fertilizers may also be chiseled into the soil.

Fertilizing at Planting

BANDING. The most common method of fertilizing at planting is called **banding.** A seed planter places a band of dry fertilizer two inches below and one inch to the side of the seeds (figures 14-17 and 14-18). Banding is the most efficient way to apply phosphate, and sometimes potash. The placement is not close enough to hurt the seed but is close enough that young roots quickly find the band of fertilizer. Because the phosphate fertilizer is packed in a narrow band, there is less soil/fertilizer

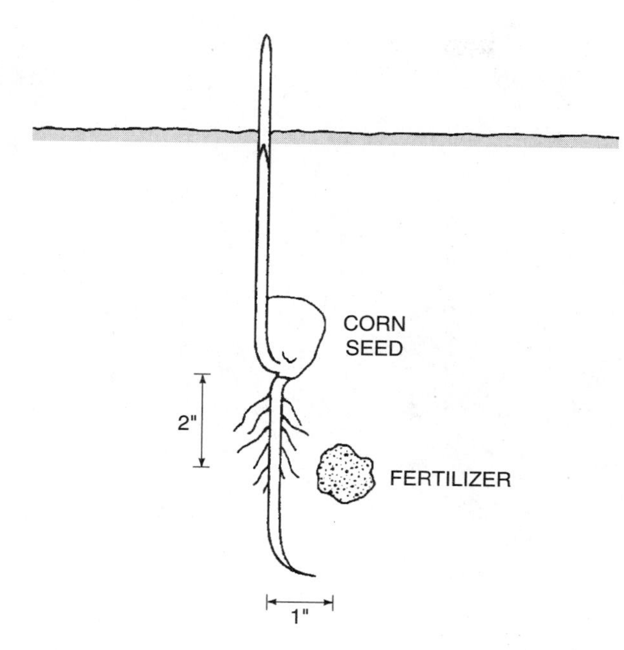

FIGURE 14-17 In banding, the seed planter places a band of fertilizer below and to the side of seeds. This placement puts concentrated fertilizer where it will be immediately available to the young seedling without damaging roots.

contact, reducing phosphate fixation. Fertilizers used for banding usually have an N:P ratio of 1:2 or 1:4—nitrogen aids in the uptake of the phosphorus. Banding is an excellent way to apply phosphate, but it slows down the planting operation.

Banding is recommended when certain soil conditions restrict phosphate nutrient uptake, including cold, wet, or compacted soil; acid or alkaline soils; low phosphate soils; and in conservation tillage.

POP-UP FERTILIZERS. **Pop-up fertilizers** are placed in the row with the seeds, rather than beside the seed as in banding. These fertilizers are quite effective in cold soils. Generally, only small amounts are applied to prevent damage to the seed. Fertilizers for pop-up use should

- be water-soluble complete fertilizers high in phosphate.
- have a low salt index (salt can injure the seedling).
- not produce any free ammonia, because ammonia will also injure the seedling; this excludes urea, UAN, and diammonium phosphate.
- be either fluid (figure 14-19) or dry fertilizers.

TRANSPLANT SOLUTIONS. Dilute fertilizer solutions are often used to water-in newly planted transplants like tomatoes or bedding plants. Vegetable transplanting machines are designed to dispense a cup solution to each plant as it is placed in the ground. In smaller transplanting operations, each plant may be similarly treated by hand. Fertilizer ratios of 1-2-1 are typical.

Fertilizing after Planting

TOPDRESSING. **Topdressing** is similar to broadcasting, except that fertilizer is spread over a growing crop and is not mixed into the soil. Either dry or fluid fertilizers can be used. Farmers often topdress perennial crops like hay to replace lost nutrients. This method is also used to feed grains, range, and lawns. Topdressing works best for nitrogen, much less well for phosphorus.

Turf managers use these terms differently. Spreading fertilizer on established turf is called broadcasting, while topdressing is spreading a thin layer of soil, compost, or other amendment over existing turf.

SIDEDRESSING. **Sidedressing** is a way of making post-plant applications to row crops (figure 14-20). Sidedressing is done by fertilizing along the crop row. Commonly, this is done by knifing ammonia into the soil. Sidedressing is the most popular way to make split applications.

FERTIGATION. A third way to fertilize a growing crop is to inject fertilizer into irrigation water (figure 14-21). This method is only as efficient and uniform as the irrigation system. Fertigation works best in sprinkler or drip irrigation but can be used with surface irrigation. The best fertilizers for fertigation are clear solutions, commonly the nitrogen source UAN. Ammonium phosphates and potassium nitrate can be used to supply phosphate and potash. Chelated trace elements also can be used. Fertigation is especially important to the greenhouse industry.

FIGURE 14-18 Banding fertilizer. *(Courtesy of American Society of Agronomy)*

FIGURE 14-19 Starter fertilizers are applied next to seeds. *(Courtesy of National Fertilizer Solutions Association)*

FIGURE 14-20 Sidedressing no-till corn with a fluid fertilizer on an Iowa farm. Such split applications improve fertilization efficiency and reduce leaching losses. *(Courtesy of USDA)*

Foliar Feeding. Growers sometimes fertilize by spraying solutions directly on the leaves of the crop. The nutrients are absorbed through the stomata (openings in leaves that allow gases to move in and out). The quickest response of any method of fertilization is obtained with **foliar feeding** and so this method may be used as a quick cure for a deficiency. The results are usually short-lived; thus, it may be necessary to repeat feedings several times. Usually it accompanies a complete program of soil fertility.

Plants use large amounts of the major elements, so it is hard to supply enough by foliar feeding. Sprays strong enough to supply much nutrient value can burn the leaves and damage application equipment.

The most practical use of foliar sprays is to solve trace element shortages. Plants need trace elements in small quantities, but problems often arise because trace elements are tied up in the soil. Spraying the leaves bypasses soil problems. The most common example is spraying iron chelates to relieve iron shortages.

FIGURE 14-21 During fertigation, fertilizer solution is pumped out of the tank and injected into the irrigation system. *(Courtesy of National Fertilizer Solutions Association)*

Horticultural Fertilization. Horticulture practitioners use the preceding fertilization techniques as well. Soil injection and banding have limited use; the first as a method of sidedressing field-grown nursery stock and the second when planting vegetable seeds. The others are widely used. Horticulturists also employ several specialty techniques. Examples include injecting fertilizer into holes under trees, placing fertilizer into tree trunks, mixing fertilizer into potting soils, and others. Chapter 17 details horticultural practices.

Fertilizer Effects on Soils

Effects of Fertilizer on Soil pH. Fertilizers may change soil pH. For instance, some acid-forming fertilizers shorten the time between lime applications. One may even pick fertilizers to help change pH in a desired way. For instance, acid-loving plants may be fertilized with acid fertilizers. Acidification by fertilizers in other crops may be counteracted by liming. The effects of fertilizers on pH are summarized as follows:

- Potassium fertilizers do not cause a lasting pH change.
- Superphosphates also do not cause a lasting effect on pH. Phosphoric acid is highly acidic and has been used to counter a rising pH. Ammonium phosphates are also acidic.
- Ammonium fertilizers and urea tend to lower pH because of the liberation of hydrogen ions during nitrification—reaction (*k*) chapter 11. Ammonium sulfate, which contains both sulfur and ammonia, is highly acidic.
- Nitrate fertilizers, like calcium and sodium nitrates, have a basic reaction in the soil.

A measure of potential acidity may be printed on fertilizer labels as the **calcium carbonate equivalent,** the weight of calcium carbonate that will neutralize potential acidity of a given weight of fertilizer.

Effect of Fertilizers on Soluble Salts. Most fertilizers are salts. A high concentration of fertilizer salts prevents seed sprouting, slows water uptake by plants, and injures plants. This is why the banding method places fertilizers no closer than two inches to the seeds.

FIGURE 14-22 Fertilizer burn. Excess fertilizer burned the margins of this maple leaf.

Misapplying or spilling fertilizer may injure a plant (figure 14-22). This causes death or browning of some leaves, known as **fertilizer burn.** Homeowners who fill their fertilizer spreaders on the lawn risk patches of dead grass. Starter fertilizers place salts very close to sprouting seeds, which are sensitive to damage from fertilizers. These fertilizers, such as the pop-up type, should have low salinity.

Fertilizer salinity is of greatest concern to container plant growers. The constant high rate of feeding easily results in a buildup of fertilizer salts. For this reason, feeding with a product low in salts is best for greenhouse growers. Farmers with saline soils may also choose such fertilizers.

SUMMARY

A fertilizer is a substance used to supply essential elements. Fertilizers may be finely ground minerals, natural or synthetic organic materials, or inorganic chemicals made by industry.

The Haber process fixes nitrogen from the air to make ammonia. Ammonia, in turn, is the base for most other nitrogen carriers. Phosphate and potash result from mining. Factories treat rock phosphate to produce superphosphate and purify potash deposits to make potassium fertilizers.

Ground-up minerals supply most of the secondary elements. They may also be obtained from other fertilizers.

For instance, superphosphate also contains calcium and sulfur. A wide variety of compounds deliver trace elements, including sulfates, oxides, fritted trace elements, and chelates.

Fertilizers come in a number of physical forms that allow many methods of use. Dry blends come in large grains that can be scattered on the soil, as in broadcasting or topdressing. They can also be banded next to seeds or used as pop-up fertilizers.

Fluid fertilizers, applied as liquids, can be sprayed on the ground for broadcasting or topdressing, injected into the soil, added to irrigation water, or sprayed on plant leaves. The high-pressure liquid, anhydrous ammonia, is used to prepare a field for planting or to sidedress a row crop.

Slow-release fertilizers, unlike other forms, release nutrients slowly. They find their greatest use in turf and in growing potted plants.

Mixed fertilizers contain two or three primary elements. The fertilizer grade lists the percentage of each primary element in the form of nitrogen, phosphate, and potash. These numbers can be used to determine how much of each nutrient is contained in a fertilizer, how much they cost, and fertilizer ratios. Growers can also determine how much of each carrier must be blended to make a mixed fertilizer.

Growers have many fertilizers from which to choose. Obviously, the fertilizer should fit the needs of the crop and the method of use. Cost is often the most important factor.

REVIEW

1. Why does the price of natural gas affect the price of fertilizers?
2. Explain how large amounts of nitrogen fertilizer can be wasted by volatilization. How can it be prevented?
3. Assume you use a complete fertilizer containing urea, superphosphate, and potassium chloride. Describe what might happen to all the nutrients added. How might they be used, stored, or lost?
4. Figure the cost per pound of nitrogen for each of the following:
   a. ammonium nitrate at $150 per ton
   b. anhydrous ammonia at $210 per ton
   c. urea at $5.00 per fifty-pound bag
5. How much of each of the following must be mixed to make one ton of a 5-10-20: urea, superphosphate, muriate of potash, and filler?
6. Name and describe an organic, a mineral, and an inorganic source of phosphorus.
7. Identify whether each of the following has an acidic, basic, or neutral effect on the soil. Give examples of how this information might influence fertilizer selection.
   a. sodium nitrate
   b. ammonium phosphate
   c. gypsum
   d. dolomitic lime
   e. superphosphate
8. What value is there in split applications of fertilizer?
9. Under what situation is banding fertilizer called for?
10. Research fertilizer use trends over the past fifty years and describe them in a short essay. Typing "fertilizer use trends" into your favorite Web browser will supply information.

ENRICHMENT ACTIVITIES

1. Use a soil test of your school grounds and local recommendations to develop a fertilizer program for an important crop in your area.
2. Safety practices are an important issue when working with fertilizers, especially anhydrous ammonia. At this Internet site, click on Pesticide and Agricultural Hazards: <http://safety.coafes.umn.edu/>.
3. Check nitrogen price trends over recent years at this site for international fertilizer prices. <http://fadinap.org/int_prices>. Why do you think there was a price spike in 2001? If you don't know, use your browser to find out why.
4. Before the Haber process was developed, guano and sodium nitrate (saltpeter) deposits were so rare and valuable that they helped spark the War of the Pacific between Chile and Bolivia in 1879. There were also sea battles between Germany and Britain early in World War I over shipments from the same fields. Why would armies fight over fertilizers? Check out the story by typing "War of the Pacific and Chile" or something similar into a Web browser.
5. Use the Internet to learn more about potash or phosphate mining and issues surrounding them.
6. This on-line "Efficient Fertilizer use Manual" provides detail about the subject of this chapter: <www.back-to-basics.net/efu/efu.html>.

CHAPTER 15

Organic Amendments

OBJECTIVES

After completing this chapter, you should be able to:

- explain the benefits of organic amendments
- describe how to use animal manure
- describe how to use biosolids
- explain large-scale composting
- list environmental side effects of fertilizers and amendments

TERMS TO KNOW

best management practices
biosolids
eutrophication
hypoxia
mesophilic
organic amendment
sewage sludge
thermophilic

An **organic amendment,** as defined here, contains both plant nutrients and large amounts of organic matter. Organic amendments are used to both fertilize and amend the soil. In this chapter, we stress their use as fertilizers.

While inorganic fertilizers effectively raise soil nutrient levels, these organic amendments also improve soil health and quality in ways that benefit the soil user:

- Organic amendments contain a combination of nutrients, including secondary and trace elements. Part of these nutrients are readily used by plants and part are slowly released by decay.
- The organic matter acts as the main soil storehouse of many nutrients, including nitrates, phosphates, sulfates, and others.
- Organic amendments contain large amounts of organic matter to improve the physical condition of the soil and increase its cation exchange capacity.
- Organic amendments support the growth of beneficial living organisms in the soil.
- Organic amendments often produce a greater yield than a complete fertilizer applied at the same NPK rate, especially on sandy soils. Further, improved yields may continue long after organic amendments are applied, unlike the shorter-term benefits of inorganic fertilizers.

Society also benefits from a grower's use of these materials because many are waste products that can be best and most safely utilized as soil amendments. Other options, such as landfilling, incineration, and ocean dumping, carry greater risks to the environment and waste their nutrient and organic matter content.

Of the many organic materials available, three account for the greatest use: animal manures, biosolids, and compost.

Animal Manure

It is ironic that for many farms today manure has become a waste disposal problem. This is an abrupt change, for throughout history people have long relied on animals as a source of soil nutrients. However, many farms do still use this resource. Properly used, manure offers many benefits; improperly used, it poses many problems.

Benefits of Manure. When handled and applied correctly, manure benefits growers several ways:

- Manure is a fertilizer with good amounts of nitrogen and potash. Phosphorus and calcium are present as are lesser amounts of sulfur and magnesium. Most manures also have traces of several micronutrients. Figure 15-1 provides examples of the nutrient content of several manures.
- Manure adds organic matter to the soil. Organic solids make up 20 percent to 40 percent of manure. This matter decays readily because of its high nitrogen content. Nitrogen tie-up occurs only if the manure includes a lot of straw or wood shavings used as animal bedding.
- Manure has longer-lasting effects than an equivalent amount of chemical fertilizer. Improved yields may continue years after manure stops being added to the soil.

	Pounds/Ton					
Animal	N	P_2O_5	K_2O	S	Ca	Mg
Dairy cattle	10	4	8	1	6	2
Beef cattle	11	8	10	1	3	2
Poultry	23	11	10	3	36	6
Swine	10	3	8	3	11	2
Sheep	28	4	20	2	11	4
Horse	13	5	13	—	—	—

FIGURE 15-1 Nutrient content of several animal manures, in pounds of nutrients per ton of manure.

Problems of Manure. Manure can also pose problems for the environment. The growth in recent years of large feeding operations, generating large amounts of manure in a small area, raises the potential for environmental side effects. These include:

- excessive application of phosphorus to land. While manure is not high in phosphorus, it builds up over repeat applications. With high soil-phosphate levels, runoff elevates phosphate levels in surface waters, with results detailed later in the chapter.
- excessive rates of manure application to land. This occurs most commonly where more animals are being raised than there is land available to safely receive their manure.
- leaching of nitrates under animal confinement areas.
- runoff of nitrates and organic materials into lakes and streams.
- large spills from manure lagoons.
- generation of gaseous air pollutants such as hydrogen sulfide (H_2S), which has human health effects, methane (CH_4), a greenhouse gas, and ammonia (NH_3), which can dissolve in local surface waters.

Under the Federal Clean Water Act of 1972, large feedlots are considered to be point sources of water pollution (see chapter 8), so their operations can be regulated. For growers today, the goal is to make the most efficient use of manures for profit while minimizing environmental problems.

Content of Manure. Manure includes both solids and liquids, which, for the most part, are the feces and urine of the animal. The solid part may also include bedding. As figure 15-2 shows, the solid part of the manure contains most of the phosphate. Most of the potash is in the liquid part. Urine holds about half the nitrogen in manure, primarily in the form of urea and similar compounds. The rest of the nitrogen is contained in the animal feces.

Several factors determine the amount of nutrient in manure, including the type of animal. In general, sheep and poultry manure has a high nitrogen content; the manure of cattle, pigs, and horses has a lower nitrogen content. The amount and type of bedding also influences nutrient content since it thins out the manure. If manure contains a large amount of high C:N ratio bedding, nitrogen tie-up can even occur in the soil for a time. The amount and type of rations and the age and health of the animal are also important factors.

Figure 15-1 lists average values of nutrient content for several animal manures. To change these values to the standard percentages used in commercial fertilizers, divide by 20. This operation changes pounds per ton to percent. For example, poultry manure contains 25, 11, and 10 pounds per ton of nitrogen, phosphate, and potash respectively. Dividing these values by 20, its NPK becomes 1.2-0.5-0.5. This is much weaker than commercial fertilizer, mainly because manure is largely water and organic carbon. Manure must be applied in quantities of tons per acre rather than pounds per acre.

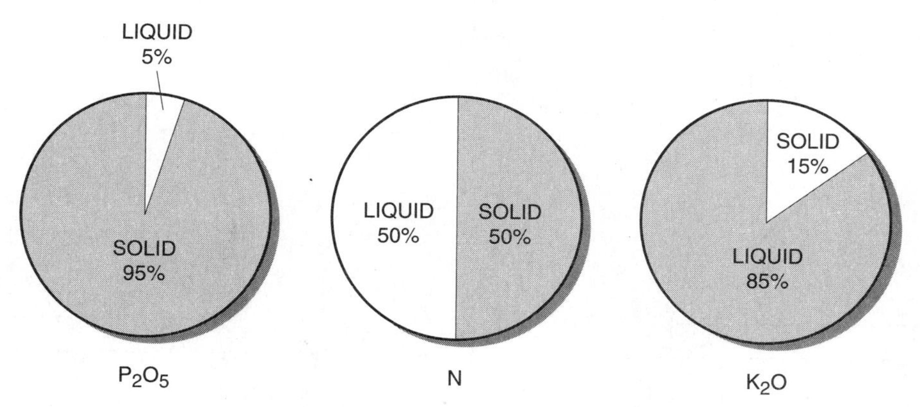

FIGURE 15-2 Most of the potash in manure is contained in the urine, and the phosphate is contained primarily in the feces. Nitrogen is distributed equally between the two parts.

Part of the secret to using manure is to keep its nutrient value intact and prevent large losses.

Nutrient Losses from Manure. Urine contains about 50 percent of the nutrient value of manure. If this part of the manure is lost, most of the potassium and much of the nitrogen will also be lost. Urine is lost when it seeps into the ground through barn floors or in feedlots. A great deal of urine simply drains away from manure heaps.

Sharp nitrogen losses occur if the manure begins to decay before it is spread. As much as 90 percent of the nitrogen can be lost within three weeks if manure is poorly handled. The losses occur when urea changes to ammonia gas during decay (see the discussion of volatilization in chapter 12). The loss is most rapid when it is warm and the concentration of urea is highest. Water in the manure dilutes the urea and slows the change. The following storage conditions promote nitrogen loss:

- High air temperatures speed decay, with resulting nitrogen loss.
- Heat in a manure pile during decay also speeds up losses.
- As manure dries out, urea becomes more concentrated. Therefore, as manure dries out during storage, ammonia enters the air rapidly.
- Freezing also speeds up losses. As the water in the pile begins to freeze, urea is concentrated in the remaining unfrozen water. The higher concentration speeds up losses.

Nutrient losses continue after manure is spread in the field. Ammonia continues to escape unless the manure is mixed quickly into the soil. Runoff and leaching increase the loss. Spreading manure on frozen, sloping land increases the chances of manure being lost to runoff.

Decay organisms respiring in a manure pile change organic carbon to carbon dioxide gas. Organic matter decay explains why manure piles shrink over time. Because the organic matter would be a desirable addition to the soil, it can be considered a loss as well.

Figure 15-3 summarizes the ways in which nutrients are lost from manure.

Handling Manure. The best way to handle manure is to spread it immediately on unfrozen ground and then plow it into the soil. In this way, the grower prevents the loss of ammonia gas that occurs during storage and in the field. However, in some regions, this technique is not practical in every season.

If manure cannot be mixed into the soil immediately, it should be stored properly and then applied when it can be plowed into the soil. The actual loss of nitrogen varies with handling and storage systems. As noted in figure 15-4, piles in an open lot, exposed to sun, rain, and air movement, will lose about half their nitrogen.

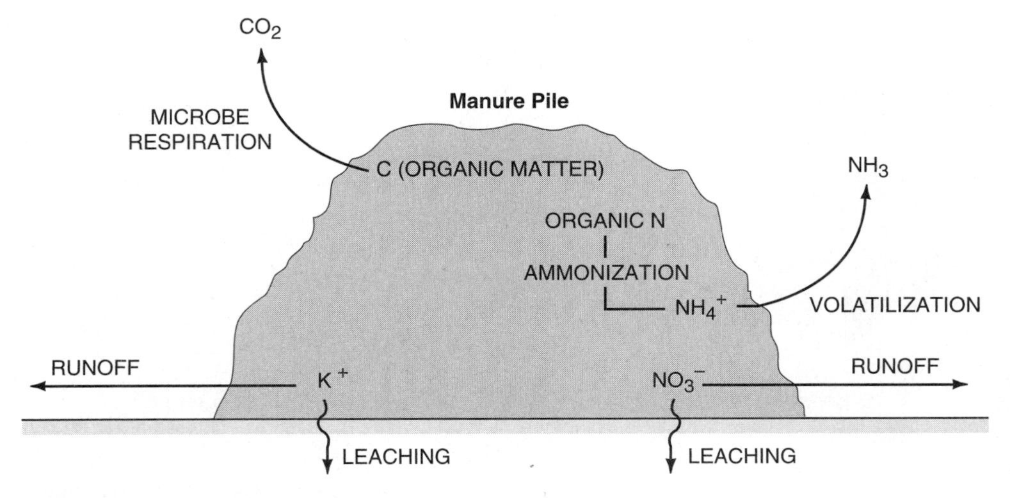

FIGURE 15-3 Much of the nitrogen and potash can be easily lost during manure storage.

Method	Nitrogen Loss
Solid Systems	percent
Daily scrape & haul	15–35
Manure pack	20–40
Open lot	40–60
Deep pit (poultry)	15–35
Liquid Systems	
Anaerobic deep pit	15–30
Aboveground storage	15–30
Earthen storage pit	20–40
Lagoon	70–80

FIGURE 15-4 Nitrogen losses from animal manures as affected by method of handling and storage. *(From Utilization of Animal Manures as Fertilizer; Sutton, Nelson, and Jones, Purdue University)*

Long-term storage in lagoons is even worse. Better is short-term storage of solid or liquid manures in proper storage structures. Good storage facilities have concrete floors and walls and a roof to stop drainage losses and slow down the drying of the manure.

Liquid manure handling systems, except for large lagoons, are the best way of saving nutrients. In these systems, growers store liquid animal wastes in concrete pits (figure 15-5) or tanks. The manure is about 90 percent or more liquid and can be handled by pumps. The liquid manure can be spread on fields by machinery (figure 15-6) or even by gun irrigation.

Freshly spread liquid manure should be plowed into the soil immediately. In warm weather, 20 percent of the nitrogen volatilizes within six hours. The best system uses a tank that knifes the liquid into the ground. Knifing stops the loss of gaseous ammonia and reduces odors (figure 15-6).

FIGURE 15-5 A livestock waste-holding tank under construction. *(Courtesy of USDA NRCS)*

FIGURE 15-6 Injecting liquid manure into a Nebraska cornfield. Injection stops the loss of ammonia gas, reduces odors, and makes pollution from runoff less likely. *(Courtesy of Ag-Chem Equipment Co., Inc.)*

Method	Types of Manure	Nitrogen Loss
Broadcast without	Solid	15–30%
incorporation	Liquid	10–25%
Broadcast with	Solid	1–5%
incorporation	Liquid	1–5%
Injection	Liquid	0–2%
Irrigation	Liquid	30–40%

FIGURE 15-7 Nitrogen losses from manure to the air as affected by method of application. *(From Utilization of Animal Manure as Fertilizer; Sutton, Nelson, and Jones; Purdue University)*

Figure 15-7 summarizes nitrogen losses from various methods of application.

Best Management Practices. As noted earlier, it is a grower's task to make the most efficient use of animal manures without inflicting damage on the environment. To this end, a number of **Best Management Practices** have been proposed:

- Test manure and soil for nutrient levels. Charts like figure 15-1 merely suggest possible levels; actual tests are needed to measure exact nutrient content. With tests, the grower can reduce fertilization by an amount equivalent to the nutrients in the manure.
- Base manure application rates on phosphorus needs rather than nitrogen. The latter is more commonly done, but in many soils it leads to heavy phosphorus soil loading. Basing application rates on phosphorus will often result in manure being spread more thinly over more acres of land, which could be difficult for operations lacking the needed acres. In such situations, reducing phosphorus in animal rations reduces its level in manure, so it can be spread more thickly.[1]
- Incorporate all manures into the soil as soon as possible.
- Where there is not enough land for safe spreading of all the manure, a grower might compost the excess and sell it.

Biosolids

The spreading of human waste on soil has a long history in many societies. In the United States, **sewage sludge** began to be spread on land in the early 1970s in response to federal clean water and air laws. The Environmental Protection Agency (EPA) estimates that about half the sewage sludge produced in the United States is spread on land.

Four major options exist for handling sewage sludge: ocean dumping (now prohibited), incineration, landfilling, and land application. The last method may be safest because the material is diluted by application over large areas, filtered by the soil matrix, and decomposed by soil organisms. Further, it contains nutrients and organic matter useful as a soil amendment but wasted in the other treatments.

[1]Powell, J., Wu, Z., & Satter, L. 2001. Dairy diet effects on phosphorus cycles of cropland. J. Soil Water Cons. 56(1): 22-26.

Land application is a form of recycling by closing a nutrient loop. Farm fertilizers are absorbed by crops, the crops are eaten by people, and some of the nutrients are flushed into the sewage system. If the sewage is then spread on farm fields, nutrients return to the land and the nutrient loop is closed. In recognition of the value of this material, it is generally now called **biosolids:** the primarily organic solid product yielded by municipal wastewater treatment that can be beneficially recycled.

Biosolids Problems. Biosolids may be a valuable amendment, but they carry potential risks needing regulation. The Environmental Protection Agency (EPA) issued a set of rules in 1993, since amended, governing proper sludge treatment and application. These rules especially concern three hazards that may be present in biosolids: pollutants, human pathogens, and human pest vectors.

POLLUTANTS. Biosolids contain small amounts of such pollutants as heavy metals like cadmium or lead, and organic pollutants such as pesticides. The EPA currently regulates nine heavy metals that can be toxic to animals—as in lead poisoning—or damaging to plants. Those sludges with the lowest heavy metal levels may be spread on land with the least regulatory control; those with the highest cannot be used at all. The intermediate group can be spread on land with certain restrictions and monitoring requirements.

HUMAN PATHOGENS. Biosolids may contain human disease organisms and parasites, and while these eventually die in soil, they may present risks in some situations. The presence of these pathogens gives rise to four possible hazards:

- Pathogens may contaminate crops grown for human food on biosolids-treated field.
- Pathogens may infect animals grazing on or eating hay from treated land.
- Pathogens could be carried into surface waters by runoff.
- Direct contact with nearby residents.

Biosolids are also classified according to their pathogen content. Class A biosolids are essentially free of pathogens and may be safely applied anywhere, including home gardens, with little regulation. Class B biosolids, less treated, may be applied to other lands under certain restrictions.

HUMAN VECTORS. Vectors are insects or rodents that may carry human diseases and that may be attracted to fields treated with biosolids. EPA regulations require sludge treatments that make biosolids less attractive to vectors when spread on land.

OTHER PROBLEMS. Biosolids may also present problems of nutrient pollution (see later in chapter), soluble salts, or odor. Application guidelines are designed to reduce these problems as well as the major issues described above.

Application Guidelines. Federal rules set standards for biosolids application to protect public health and the environment; states may set stricter standards. These rules apply to common sites of land application, including:

- agricultural land, which receives nearly 80 percent of land-applied sludge.
- public contact land, such as golf courses, ball fields, or nurseries.
- reclamation sites, such as strip mines, where biosolids can greatly improve the site's ability to support vegetation.
- forests, where biosolids have improved the growth of trees.
- lawns and gardens, which have the strictest requirements for being pollutant and pathogen free.

The EPA classifies biosolids that can be spread on land as Exceptional Quality (EQ) or High Quality (HQ). EQ biosolids meet all the standards for all three problem areas, while HQ conforms to slightly lower standards. The EPA accepts EQ biosolids for home and garden use, and they can be used like other fertilizers with little restriction. HQ biosolids can be applied to other land categories, with certain restrictions. These include, for example:

- avoiding land application where it might affect a threatened or endangered species.
- no application to flooded, frozen, or snow-covered sites to avoid entry into surface waters or wetlands in runoff.
- no application within ten meters (about thirty feet) of wetlands or surface waters.

Federal guidelines specify that biosolids be applied at rates based on nitrogen needs of the crop. This rate reduces chances of nutrient pollution and is appropriate as a fertilizer rate.

Biosolids sold in bulk may be applied to land at the recommended rate until an upper limit for heavy-metal loading in the soil is reached (e.g. 2,000 kg zinc per hectare, or about 800 pounds per acre); then application must be discontinued. Biosolids sold in containers can be applied at less than a specified annual loading for pollutants (e.g. 140 kg/hectare/year zinc, or about 56 pounds/acre). Obviously, these rates require record keeping and monitoring on the part of biosolid producers and users.

Biosolids can be handled much like manure: injected into the soil (figure 15-8), irrigated, or spread on the ground (figure 15-9). Because injection reduces odors, discourages vermin, and lessens runoff losses, the EPA prefers such application. Surface-applied biosolids should be tilled into the soil within seventy-two hours.

Criticism of Biosolids. Some disagree with current regulation of land-spreading of biosolids. It has been pointed out that the EPA allows much higher levels of pollutants than many European countries. Critics also question many EPA assumptions and tests. For instance, critics note that projected rates of percolation into soil do not account for the presence of biopores or other preferential flow paths.[2] Critics have also challenged assumptions about the amount of heavy metals entering the human diet.

Compost

Composting is defined in chapter 8 as a method of causing the decay of organic matter in a pile above the ground. That chapter described a typical home compost pile. Composting also presents larger-scale commercial uses as well. Examples of important composting operations include composting of yard wastes by cities and landscape companies (figure 15-10), other municipal solid wastes, food processing wastes, or sewage sludge. Nurseries and

FIGURE 15-8 The truck is loaded with biosolids at the treatment plant and is driven to the farm, where the sludge is injected about ten inches into the soil. *(Courtesy of Ag-Chem Equipment Co., Inc.)*

[2]Harrison, E., McBride, M. & Bouldin, D. 1999. The Case for Caution: Recommendations for Land Application of Sewage Sludge and an Appraisal of the US EPA Part 503 Sludge Rules. Cornell Waste Management Institute, Cornell University, Ithaca, New York.

FIGURE 15-9 After being spread on the soil, biosolids are tilled into the soil as soon as possible to reduce odors, ammonia loss, and the chance of pollution. *(Courtesy of Ag-Chem Equipment Co., Inc.)*

FIGURE 15-10 A windrow of yard wastes being composted at a resort in Hawaii. Note the irrigation pipe used to moisten the pile. *(Courtesy of Katy Deshotels-Moore, Hualalai Resort, Hawaii)*

FIGURE 15-11 Finished compost at this Hawaiian resort is used as a soil amendment and topdressing on the grounds. *(Courtesy of Katy Deshotels-Moore, Hualalai Resort, Hawaii)*

greenhouses compost organic materials like wood chips for use in potting soils, and farmers may compost manure and bedding as well as animal carcasses. The compost that results from these operations may be used as a soil amendment by the operator (figure 15-11), given away to homeowners for their gardens, or sold as a commercial product. In the process, a variety of wastes are put to good use.

Composting offers three benefits over simply spreading uncomposted organic matter on the soil. First, composting reduces the weight and volume of organic material, making it easier to handle and ship. Second, it reduces the carbon:nitrogen ratio of materials like woodchips or leaves, eliminating the problem of nitrogen tie-up when compost is added to soil. Third, the heat generated in a compost pile kills most plant or human pathogens and weed seeds. Of course, the pile must be mixed a few times to ensure that the outer edges of the pile are pulled into the hot interior.

A properly prepared compost pile mixes carbonaceous and nitrogenous materials to achieve a C:N ratio of about 30:1. At this ratio, nitrogen is preserved because it remains immobilized; at a lower ratio, much will be free to leach or volatilize. At much higher ratios, the process is very slow. The pile is also kept moist but not wet, at a moisture level of about 50 percent. The pile must be moist to permit rapid decay, but excess moisture creates anaerobic conditions. Anaerobic piles decay very slowly, generate unpleasant odors, and produce "sour compost" containing organic acids and other chemicals that damage plants.

In commercial composting, wastes may be shredded before it is piled in long windrows (figure 15-10) and moistened. The composting process now follows three stages. In the first, **mesophilic,** stage, organisms that prefer moderate temperatures begin the decay process, and temperature begins to rise. In the second, **thermophilic,** stage, heat-loving microbes replace mesophilic ones, and temperature rises to around 150°F. During this stage, temperatures are monitored and whenever it begins to drop, the pile is turned to bring in fresh oxygen and organic matter, and temperatures rise again. The bulk of decay occurs during the thermophilic stage. When temperatures drop for good, a second mesophilic stage takes a month or two, during which stabilization of the compost occurs.

Well-prepared compost is well decayed and stabilized. It is low in heavy metals and soluble salts with a pH between 5.0 and 8.0. Particle size should be around a millimeter in size, with little or no foreign matter, such as pop-can tabs. The C:N ratio should be between 15 to 25:1, and it should be free of the toxic residues of anaerobic decay. It should also be mostly organic matter, with little soil mixed in. Figure 15-12 shows nutrient analysis of a turkey manure/bedding compost from one composting operation.

Component	Pounds/Ton
Total nitrogen	44.00
Phosphate (P_2O_5)	68.00
Potash (K_2O)	38.00
Calcium carbonate	160.00
Magnesium	8.00
Sulfur	12.50
Sodium	5.56
Iron	6.80
Aluminum	4.24
Manganese	0.73
Copper	0.59
Zinc	0.50
Organic matter	1000.00

FIGURE 15-12 Average analysis of one wood shaving/turkey manure compost product. *(Courtesy of Agri-Brand Compost, Holden Farms, Inc.)*

Fertilizer and the Environment

Fertilizers aid the productivity of American farms, but they have not been without their problems. Fertilizers, manures, and sludges all can cause pollution and human health problems. The grower who is environmentally aware acts to keep these problems to a minimum by practicing proper nutrient management.

Animal and Human Health. The main health problem is the effect of nitrates on animal and human infants, and ruminant animals like cattle. Small amounts of nitrates in drinking water cause an infant anemia called "blue-baby disease" (methemoglobinemia), in which the ability of the blood to carry oxygen is reduced. In some rural parts of the United States, water is no longer fully safe for human or animal infants.

A certain amount of natural nitrate leaches into groundwater from normal mineralization of organic matter and other natural nitrogen inputs. Grower additions such as fertilization increase the nitrate load. Thus, groundwater may be contaminated by nitrates from water percolating through fertilized soil. This has occurred most often on irrigated outwash sands with water tables near the surface. These soils, naturally droughty and infertile, tend to be heavily irrigated and fertilized. Natural rainfall or irrigation can then leach the nitrates into the shallow groundwater.

Interestingly, some plants tend to store nitrates in their leaves. These "nitrate accumulators" include spinach, corn, and sorghum. This may result in high-nitrate forages that can harm livestock. It is uncertain if any human health problems have also occurred.

Eutrophication and Hypoxia. Nutrient losses from farms and landscapes cause **eutrophication,** an increase of algae growth in bodies of water. In many ecosystems this is a slow, long-term but natural process.

However, fertilizer inputs dramatically speed up the process.

In freshwater systems, phosphorus most often limits algal growth so phosphate is a major cause of eutrophication in lakes and streams. In coastal marine ecosystems, nitrogen tends to be more limiting. It leads to loss of water clarity, less dissolved oxygen in the water, damaged aquatic ecosystems, and reduced value for swimming or fishing. Runoff and erosion from farm fields and turf areas can all contribute to the process. When algae begin to die, their decay robs the water of oxygen. Severe oxygen loss, called **hypoxia,** results in major losses of aquatic life, including fish.

The largest areas of low-oxygen conditions occur in hypoxic zones of the coastal oceans of United States, Europe, and Japan. These areas of severe hypoxia have grown in recent decades where river systems enter the ocean. Here nutrient inputs, particularly nitrates, cause an offshore growth of algae. These settle to the bottom, and their decay reduces oxygen levels in subsurface water. The effects of this hypoxia can seriously damage commercial and recreational fisheries.

The Gulf of Mexico hypoxia zone ranks as the largest in the United States. This zone along the coast of Louisiana and Texas can reach an area the size of Connecticut during the summer. It is fed by an average annual input from the Mississippi River of some 1.5 million metric tons of nitrates, mostly from the Corn Belt but also from other croplands of the river basin. The Council for Agricultural Science and Technology (CAST) reports that an estimated 55 percent of this input comes from fertilizers and another 25 percent from legume crops.[3]

Energy Costs. Fertilizers have a high energy cost, particularly nitrogen fertilizers. Each ton of industrially fixed nitrogen consumes 1.5 tons of natural gas, and several percent of the nation's yearly fuel bill goes to making fertilizer. This high energy bill raises the cost for farming and even threatens farms if supplies become limited. It also means much of our agriculture suffers a negative energy balance; that is, fewer calories of food are produced than calories of energy consumed to grow that food. Organic farmers cite this factor as one reason for avoiding chemical fertilizers.

Human Changes in Global Nitrogen Cycles. Ecologist Peter Vitousek and others have argued that humans have greatly altered the global nitrogen cycle.[4] Most of the world's nitrogen pool (N_2 gas) is useless to most organisms. Humans now create as much active nitrogen like nitrates or ammonia as all the natural processes combined, produced by burning of fossil fuels, fertilization, and cultivation of legumes. For instance, fertilizer factories fix about 80 million metric tons of nitrogen annually. Much of this nitrogen eventually moves elsewhere in water or in the atmosphere. The Gulf of Mexico hypoxic zone is one consequence.

Some of the results are well known, such as those described above. Others remain unclear. Nitrogen oxides, released from fertilized soil by denitrification, act as strong greenhouse gases, add to smog, and acidify soils and waters. Nitrogen enrichment is also changing the nature and health of natural ecosystems. While details are beyond the scope of this text, it can be an interesting topic for further study in soil classes.

Best Management Practices. Growing the crops needed by humanity requires plant nutrients in some form, but we can use cropping systems that make the most efficient use of them. Best Management Practices, or BMPs, are practical cropping systems that reduce environmental impacts of practices like fertilization. BMPs can greatly reduce the total amount of nutrients needed and lost.

It is estimated, for example, that only half the nitrogen applied to crops actually ends up in those crops—the rest is lost to water or the atmosphere. This not only creates nutrient problems, but also costs growers money: the CAST report suggests that $410 million worth of nitrogen fertilizers is annually lost down the Mississippi River.

---

[3]CAST. 1999. Gulf of Mexico Hypoxia: Land and Sea Interactions.

[4]Vitousek, P., et al. 1997. Human Alteration of the Global Nitrogen Cycle: Sources and Consequences. *Ecological Applications* 7(3): 737-750.

Many of the practices described elsewhere in this text are BMPs that improve nutrient use. Examples include:

- practices like conservation tillage that reduce erosion and runoff.
- efficient irrigation practices that reduce nitrate leaching and denitrification on irrigated lands.
- proper manure handling to increase nutrient retention and decrease such losses as leaching and runoff from feedlots and fields.
- fertilization practices that deliver fertilizers at times and rates best suited to plant growth, such as split applications.
- precise fertilizer applications resulting from good soil testing with credits taken for manure and legumes. Precision agriculture is a good tool.
- careful use of fertilizers on lawns and golf courses, such as sweeping up granules and clippings from sidewalks.
- use of slow-release fertilizers and recycling of all irrigation water in container-growing nurseries and greenhouses.
- many of the practices of organic and sustainable agriculture.

SUMMARY

Manure, biosolids, and compost provide a double benefit to growers—they contain nutrients to promote crop growth and organic matter to improve the soil. Applying these materials to land is a good alternative to other disposal methods.

Manure is highest in carbon, nitrogen, and potash. It also contains phosphates and secondary and trace elements. Proper handling of manure reduces nutrient losses and lowers the chance of polluting surface or groundwater. Manure is best spread on unfrozen land as soon as possible after collection and then tilled into the soil. If this is not practical, the grower should store the manure in sealed, covered pits. Some liquid systems also work well to preserve nutrients.

Biosolids are recycled "human manure" and can be handled much like animal manure. However, possible health problems from heavy metals and human pathogens mean that it must be used according to state and federal EPA guidelines. Biosolids are applied according to the nitrogen rate to a variety of crops and sites.

Composting is a way to reduce the C:N ratio of organic materials and to kill harmful organisms. For example, composting sludge with wood chips stops nitrogen tie-up from the chips while killing pathogens in the sludge.

Any nutrient source—fertilizer, biosolids, or manure—can harm the environment if used improperly. Nutrients and pathogens can wash into surface waters or leach into groundwater, causing pollution and human and animal health problems. Using inorganic and organic fertilizers in the suggested ways and rates and avoiding erosion are important ways to reduce these problems.

Farm fertilizers increase national energy use and force agriculture to rely heavily on fossil fuels. For most growers, making the most efficient use of fertilizer, biosolids, manure, and legumes is the best answer.

REVIEW

1. Grasses respond much more in growth to nitrogen than do legumes. What would be the consequences of spreading manure on mixed grass/legume pastures?
2. Nutrient pollution is particularly a hazard where large animal operations own too little land to receive the amount of manure generated. Suggest some answers.
3. Discuss factors that will determine how much of the nutrient value of a manure as it is excreted by an animal will end up being used by crops in a field on which it has been spread.
4. What must be true of any biosolid applied to home lawns and gardens?

5. Occasionally large manure lagoons spill their contents into local surface waters. Based on this chapter, what effects would this have on those waters?
6. What happens if a compost pile becomes too dry? Too wet?
7. Some cities burn sludge instead of spreading it. Think about benefits and problems of spreading and burning, then decide which you think is the best policy. Defend your choice.
8. Why can heavy phosphorus loading occur on lands receiving manures?
9. Describe practices that would reduce nutrient pollution from land.
10. Describe policies and regulations about manure or biosolid application to land in your state. Information can be obtained by entering "manures and (name of state)" into your favorite web browser.

ENRICHMENT ACTIVITIES

1. Visit a facility that works with one of the amendments described here, like a composting facility, sewage treatment plant, or manure storage facility.
2. Build and manage a compost pile.
3. For further details on manure handling, read the Council for Agricultural Science and Technology 1996 report "Integrated Animal Waste Management."
4. Find Internet sites on manure handling, like the one at:

   <http://www.extension.umn.edu/distribution/cropsystems/DC7401.html>.
5. Proper manure management involves land application at the correct rate, which means calibrating the spreader. Here is an Internet site with instructions: <http://www.ext.colostate.edu/pubs/crops/00561.html>.
6. Read "Agricultural Nutrient Management and Environmental Quality: Position of the Soil Science Society of America" at <http://www.soils.org/agmgt/2000.html>.
7. Research the Gulf of Mexico hypoxia zone on the Internet.
8. The Ecological Society of America offers an article by Vitousek and others on human alteration of the global nitrogen cycle at <http://esa.sdsc.edu/tilman.htm>.

   Other Internet resources on the topic can be found by use of a search engine.

CHAPTER 16

Tillage and Cropping Systems

OBJECTIVES

After completing this chapter, you should be able to:

- explain the reasons for and effects of tillage
- describe conventional and conservation tillage
- list several cropping systems
- briefly describe organic and sustainable agriculture

TERMS TO KNOW

allelopathy
conservation tillage
conventional tillage
cover cropping
crop rotation
disc plow
double cropping
dryland farming
fallow
finishing harrow
lister plow
moldboard plow
organic farming
primary tillage
rangeland
row crops
saline seep
secondary tillage
small grains
sustainable agriculture
tillage

To produce crops, a grower places seeds in contact with the soil, provides nutrients, controls pests, and manages soil water. These activities usually involve some form of tillage. There are many ways to work the soil and different situations require different methods. Each method has an effect on the crops and the soil. This chapter looks at some standard tillage and cropping systems.

Uses of Tillage

Tillage is working the soil to provide a favorable environment for seed placement and germination and crop growth. In the United States, mechanization and research have led to a variety of tillage systems. Regardless of the method of tillage used, a grower has three basic goals: (1) weed control, (2) alteration of physical soil conditions, and (3) management of crop residues.

Weed Control. Tillage for weed control can be divided into two time periods: before crop planting and after crop planting. Before planting, tillage prepares a weed-free seedbed that greatly simplifies weed control during the growing season. Tillage destroys young seedlings, and repeated tillage operations may also weaken perennial weeds. After planting, cultivation continues to destroy or bury emerging seedlings. However, deep cultivation or cultivation late in the season may sever crop roots and reduce yields.

The importance of tillage for weed control has declined with increases in both herbicide use and tillage systems designed around herbicide use. Some herbicides require incorporation into the soil by shallow tillage. However, with increased interest in organic agriculture, renewed emphasis on mechanical tillage may be expected.

Physical Soil Conditions. Tillage alters physical soil properties, such as structure, moisture, and temperature. Tillage during seedbed preparation stirs and loosens soil, improves aeration, and creates a suitable medium for growth. Deep tillage and subsoiling may temporarily break up subsoil compaction.

However, tillage causes a long-term decline in physical structure. The decline is partly due to losses of soil organic matter that result from tillage. Repeated tillage operations crush some soil aggregates. Wheel traffic compacts the soil, especially wet soils, and tillage pans may form. Soil aggregates on the surface of bare soil shatter from raindrop impact, causing crusts that hinder seed germination and shed water. The bare soil resulting from many forms of tillage erodes easily. Recent changes in tillage aim to reduce these adverse side effects.

Tillage also affects the moisture level and temperature of soil. Tilled soil usually warms up earlier in the spring, allowing earlier seeding and better germination. In areas where soil tends to be wet or cold in the spring, crops may be planted on ridges created by tillage. The ridges warm and dry faster than the rest of the soil.

Shallow cultivation of crust-forming soils may improve crop yield even where herbicides are used to control weeds. By breaking up crusts, cultivation improves water infiltration and reduces runoff. Such cultivation should be just deep enough to break the crust.

Crop Residue Management. After most crops are harvested, residues like stalks or leaves remain in the field. The amount of residue depends on the type of crop, how well it grew, and how it is harvested. For example, corn leaves about 8,500 pounds of residue per acre for a 150-bushel corn crop, and about 5,600 pounds of residue for a 100-bushel crop. If the corn is harvested for silage rather than grain, little residue is left in the field. Figure 16-1 lists residues for several crops.

There are several ways growers manage crop residues, depending on objectives. Moldboard plowing buries crop residues, resulting in a clean field that is easy to plant and cultivate. In semiarid grain-growing areas, special tillage tools, including rodweeders and sweeps, till under the surface to kill weeds but leave residues on the surface to protect against wind erosion. Conservation tillage in more humid climates leaves residues on the surface to protect against water erosion. Figure 16-2 lists the amounts of residue left on the soil surface from various tillage tools.

In addition to crop residues, tillage may also incorporate phosphates, potash, and lime into the root zone. Tillage also incorporates sewage sludge, manures, and nitrogen sources like urea that volatilize if left on the soil surface.

Seedbed Preparation. The three reasons for tillage come together in preparing a seedbed to ensure that the soil meets the needs of germinating seeds. Seeds need a moist soil at the right temperature with sufficient air for

Crop	Approximate Residue per Bushel Grain (lb/acre)		Sample Yield (bu/acre)		Sample Residue (lb/acre)
Barley	80	×	50	=	4,000
Corn	56		125		7,000
Flax	80		15		1,200
Oats	60		32		4,300
Rye	100		30		3,000
Sorghum	60		50		3,000
Soybeans	50		40		2,000
Wheat	100		40		4,000

FIGURE 16-1 Crop residues in pounds per acre for several crops. To obtain these values, the number of pounds residue per bushel of grain is multiplied by the per-acre yield. Sample yields may not represent yields in your area.

Implement	Estimated Percentage of Residue Remaining After Each Operation
Inverting tools	
Moldboard plow	5
Lister plow	20
Mixing tools	
Field cultivator	80
Chisel plow, spear point	80
Chisel plow, twisted point	50
Rototill to 6 inches	25
Rototill to 3 inches	50
Tandem disc to 6 inches	25
Tandem disc to 3 inches	50
Spring-tooth harrow	60
Spike-tooth harrow	70
Subsurface tools	
Blades or sweeps	90
Rodweeders	90

FIGURE 16-2 Percentage residue remaining after one pass over the field for various tillage tools. If two or more tillage operations are practiced, each operation after the first uncovers about half the amount it covers.

seed respiration. The soil should be loose enough for good aeration, but compact enough around the seed for good soil/seed contact. It should be free of clods that prevent proper seed/soil contact and seedling emergence (figure 16-3).

Seedbed smoothness and the amount of allowable crop residues depend upon seed size and type of planter. Large seeds, like corn and soybeans, germinate in a fairly rough, cloddy seedbed. Very small seeds, like alfalfa seed, germinate best in a very fine, firm seedbed. A seedbed free of crop residue is easiest to plant in, but conservation tillage demands that crop residues be left on the surface to control erosion. Most older seed planters only operate on a fairly smooth, clean seedbed. Many modern planters can plant through crop residues and clods, preparing correct soil conditions near the seed.

Conventional Tillage

Conventional tillage, the primary form of tillage since invention of the moldboard plow, involves two stages. First, **primary tillage** breaks up the soil and buries crop residues. Primary tillage is often accomplished with an inverting implement, like the plow or lister plow, that inverts or tips over the top few inches of soil. **Secondary tillage** produces a fine seedbed by a series of operations that break up the soil into smaller and smaller chunks. Secondary tillage involves mixing implements like harrows. The following discussion describes these operations in more detail.

FIGURE 16-3 A conventional seedbed is smooth and free of crop residues. It is excellent for seed germination but is prone to water and wind erosion. *(Courtesy of John Deere Company)*

Plowing. The traditional primary plowing tool is the **moldboard plow** (figure 16-4). The moldboard shears off a section of soil, tips it upside down, and fractures it along several planes. In the process, any organic material on the soil surface is buried. The moldboard plow leaves the surface very rough with a series of ridges and furrows.

Moldboard plows work best in moist soil; in wet or dry soil the operation uses more power and the results are poor. For wet or dry soils, a **disc plow** works better. A series of three to ten large (two to two and one-half feet) discs are mounted on a frame at an angle to the direction of travel. The discs cut into the soil as they rotate and roll the soil over.

Subsoilers like the one shown in figure 16-5 are used to shatter tillage pans or natural soil pans. Subsoiling should be done when the soil is dry, because if pans are moist, they do not shatter. Deep plowing can temporarily help water infiltration and root penetration into the subsoil. Usually, however, compacted layers reform as soil is exposed to further wheel traffic and tillage.

Harrowing. Harrowing is usually a two-step process. In the first stage, ridges left from plowing are smoothed

FIGURE 16-4 This ten-bottom moldboard plow flips over the top seven or eight inches of soil and buries crop residues. *(Courtesy of John Deere Company)*

FIGURE 16-5 A subsoiling chisel plow is used to break up a tillage pan or other compacted layer. *(Courtesy of Year-A-Round Cab Company)*

out and large clods broken up. Then smaller lumps are pulverized and a fine seedbed is produced.

Growers commonly begin the operation with a disc harrow (figure 16-6). The typical tandem disc has four gangs of discs set like the four arms of an X. The front two gangs turn the soil inward, and the back two turn it back out. A disc tends to compound compaction problems because it shatters soil aggregates but does not dig deep enough to loosen compaction. Spring-tooth harrows and field cultivators (figure 16-7) may be used rather than the disc. A long, springy C-shaped tooth and a spear point or broad shovel digs into the soil, dragging clods to the surface and breaking them up.

A **finishing harrow,** or drag, completes the job of pulverizing the soil. Figure 16-8 shows a drag being pulled behind a spring-tooth harrow.

The steps just described are often modified. If the soil has good tilth, deep tillage by plowing may not always be needed. In such cases, the tandem disc shown in figure 16-6 is heavy enough to be used alone. Growers often combine operations, hitching several tillage tools behind the tractor. Any time a grower can eliminate a pass through the field, compaction is reduced, and time and fuel are saved.

FIGURE 16-6 This tandem disc with 24-inch blades is heavy enough for both primary and secondary tillage. *(Courtesy of John Deere Company)*

FIGURE 16-7 A field cultivator can be used for weed control on fallow ground, for secondary tillage, and for primary tillage where low levels of surface residue are present. *(Courtesy of John Deere Company)*

FIGURE 16-8 The last step in preparing a fine seedbed is the finishing harrow. Note that the harrow is attached to a field cultivator-one trip through the field does two jobs. Combining operations saves time and fuel and reduces compaction. *(Courtesy of John Deere Company)*

Lister Plowing. **Lister plows** are equipped with two moldboards mounted back to back, resulting in a pattern of ten-inch-high ridges and furrows across the field (figure 16-9). In humid regions, crops may be planted on the warmer, drier ridges. In arid areas, they may be planted in the moist soil of the lister furrow. Listing can also help protect the soil from wind erosion. Listing on the contour captures water to improve water use and reduce water erosion.

PREPARATION OF FURROW-IRRIGATED FIELDS. Additional steps are needed to prepare furrow-irrigated fields. After the standard primary and secondary tillage, the grower carefully levels the field with a blade to ensure the proper grade for flow of surface-applied water. Then the field is listed with a special tool to create ridges and furrows.

Timing and Depth of Plowing. Farmers in the eastern United States can plow in either fall or spring. Fall plowing gives the farmer a head start on spring planting by warming and drying the soil. Freezing and thawing on fine-textured soil breaks up large lumps, making it easier to develop a good seedbed. The benefits of fall plowing are especially important with fine-textured soils with somewhat poor drainage.

Spring plowing leaves crop stubble in the field over winter to capture snow and reduce erosion. To conserve soil, one plows in the spring unless there are overriding reasons for fall plowing.

FIGURE 16-9 A listed field in the Rio Grande Valley of Texas.

In the western United States, where moisture preservation is critical, plowing immediately after harvest gives more time for the soil to store moisture if weeds are controlled. However, leaving soil bare increases the risk of erosion.

The standard plow depth of seven to eight inches gives the best results. Shallow plowing results in a poor seedbed, and deeper plowing takes more power without noticeably improving yields.

Conservation Tillage

Conservation tillage is a program of crop residue management aimed at reducing erosion (figure 16-10). Rather than plowing under crop residues, some or all of the residue is left on the soil surface. The definition of

FIGURE 16-10 Runoff from conventional- and conservation-tilled fields. The runoff sample from a conservation-tilled field (bottom sample) contains far less soil than the sample off a field prepared by conventional tillage. *(Courtesy of USDA ARS)*

conservation tillage has required that, at planting, 30 percent or more of the soil surface be covered with crop residues. USDA 1994 conservation standards specify instead that enough residue remain on the surface to reduce soil losses below a tolerable level as calculated by current erosion prediction methods. Erosion prediction and soil loss tolerance are described in chapter 18.

Conservation tillage reduces water and wind erosion by at least 40 percent to 50 percent. In areas where moisture can be limiting, conservation tillage increases soil moisture by improving infiltration and reducing runoff, reducing evaporation, and trapping snow. Because of reduced runoff, fewer pesticides and nutrients leave the field. Conservation tillage also improves organic matter content near the soil surface, as described in chapter 6. Conservation tillage, therefore, is one of the most important Best Management Practices for soil and water conservation.

Conservation-tilled soil tends to be cooler than clean-tilled soil because of light reflection off the mulch and increased soil moisture. In warm climates, cooler soil benefits production, but may hinder growth in northern states. Diffusion and mass flow of nutrients improves in the moister soil, increasing nutrient uptake.

Other benefits of conservation tillage are obtained from fewer trips across the field. These benefits include less time in field work and lower fuel costs. At times, compaction is reduced because of less wheel traffic. Conservation tillage may also require fewer implements, thus reducing equipment costs per acre. Conservation tillage provides better habitat for pheasants and other wildlife. For instance, no-till fields have been shown to provide improved habitat for nesting ducks in North Dakota and nesting bobwhite quail in Tennessee.

Because of soil conservation and economic benefits of conservation tillage, its use has spread rapidly in recent decades. The USDA reports that in 1995 about 35 percent of United States agricultural land was under conservation tillage, another 25 percent under other reduced tillage methods, and 40 percent under conventional tillage. As technology improves, conservation-tillage use is expected to grow.

Conservation tillage covers several different tillage methods:

Mulch-Till or Chisel-Plow. A chisel plow (figure 16-11), which loosens the soil but does not invert it, is used for primary tillage. Chisel plowing to eight inches

FIGURE 16-11 The chisel plow is the primary tillage tool in the mulch-till system. *(Courtesy of USDA NRCS)*

FIGURE 16-12 Residues and ridges left by chisel plow on soybean grounds. *(Courtesy of USDA NRCS)*

leaves the soil rough with about 50 percent to 80 percent residue cover (figure 16-12). Light discing reduces residues to between 30 percent and 50 percent. Seeds are then planted through the remaining residues. After planting, cultivation and herbicides control weeds.

Strip-Till. With no primary tillage, a specialized implement tills a band of soil and plants seeds into the band. Another type of implement sweeps residues off a strip into the middle of the rows. The planting operation bares about one-fourth of the soil surface, leaving about 50 percent of crop residues.

Ridge-Till. The ridge-till system excels in cool, moist conditions. Seed is planted on six-inch ridges (figure 16-13) with crop residues swept into the shallow furrows. About two-thirds of crop residues remain after planting. Cultivation with special tools minimizes residue burial and rebuilds ridges for the coming year.

The ridges in this system warm up and dry more quickly than soil in other tillage systems. In addition, if oriented across the slope, they further reduce runoff and erosion. If oriented perpendicular to prevailing winds, they further reduce wind erosion and help to trap snow. The roots of plants on the ridges also grow separate from the compacted zone between the ridges where wheel-traffic occurs.

FIGURE 16-13 Soybeans planted ridge-till in an Iowa farm. Note the corn residues swept into the row middles. Planting is done on the ridge where soil warms and dries more quickly than in other conservation-tillage systems. *(Courtesy of USDA)*

No-Till. In this method soil is barely disturbed (figure 16-14). Specialized planters cut a slot through the residues, insert the seed and fertilizer, and close the slot. About 90 percent of the soil surface is untouched. Contact, systemic, and preemergent herbicides are used to control weeds with no cultivation.

Because no-till involves the least soil disturbance, it maximizes the benefits of conservation tillage. By not

FIGURE 16-14 Soybeans planted no-till into wheat residues on an Arkansas farm, reducing erosion and retaining moisture. *(Courtesy of USDA)*

disturbing the soil surface, it preserves the tops of earthworm and other channels at the surface, greatly improving water infiltration. No-till best preserves soil organic matter, and organic matter content actually rises in the soil near the surface.

A number of general principles apply to conservation tillage. Crop residues should not be burned, and baling or grazing of crop residues should not deplete residues below the acceptable level. Tillage and planting implements should be used properly to preserve residue cover.

Residue levels can be predicted from data on crop residues left by crops (figure 16-1) and the amount of residue remaining from each pass with an implement (figure 16-2). It can also be measured directly in the field by a number of methods.

For moisture-preserving purposes, residue levels should exceed 50 percent. For trapping snow, stubble should be left standing at least six inches tall over winter.

Differences Between Conventional and Conservation Tillage

Conventional and conservation tillage differ in more than the obvious ways. Most growers ask first about yields. Current research shows equivalent yields from conventional and conservation tillage if known technology is applied to each tillage system and the systems are matched to crops and soil types. Figure 16-15 provides some comparisons of the systems.

Tillage System	Corn Yield (bu/acre)
Conventional tillage	142
Chisel plow	138
No-till, after harvest for silage	142
No-till, after harvest for grain	144

FIGURE 16-15 Average corn yield after tillage treatment for continuous corn in Ohio. The soil is a silt loam. On this soil, no-till gave the best results. *(Source: Ohio Agricultural Research and Development Center)*

Equipment. Conservation tillage places some requirements on equipment. For example, residues should be spread evenly behind harvest equipment. Planters must penetrate the residues, place the seed, cover the seed, and ensure seed/soil contact in a rough seedbed.

Fertility. Because there is less soil mixing in conservation tillage (especially in no-till), the form and placement of fertilizers are affected. Lime, phosphates, and potash tend to stay near the soil surface. However, because residues provide a mulch, soil near the surface tends to remain moist, promoting root growth near the surface and improving uptake from that layer. Conservation tillage, especially no-till, can reduce nitrogen availability by increasing nitrogen tie-up in surface layers, increasing leaching and volatization, and by reducing average soil temperatures. The injection of nitrogen deeper into the soil and nitrification inhibitors will reduce these problems.

The pH of the top two inches of soil tends to drop rapidly, especially in no-till, affecting seed germination, crop growth, and herbicide activity. Careful testing for pH is required for this layer, followed by a topdressing of lime if needed.

Matching Tillage to Soil Type. Soils tend to be cooler and wetter in conservation tillage, especially with the no-till method. On fine-textured soils in northern states, cooler soil may delay planting and hamper seed germination. No-till is a poor choice on cold, poorly drained fine-textured soils; for these conditions, the ridge-till method is a better choice. Nor does no-till work well in highly compactable soils. On excessively drained coarse soils, no-till can improve yields by preserving moisture. Local extension agents can provide advice on the best system for each grower.

Weed Control. With less tillage, there is greater reliance on herbicides for weed control. Tillage will kill any weed seedling, but herbicides are more selective. This makes weed identification and herbicide selection more critical. Also, surface-applied chemicals are more suitable for conservation tillage than those needing to be incorporated into the soil.

Pest Control. Conservation tillage, especially no-till, alters the environment presented to pest organisms. Diseases that overwinter on crop residues, like small grain leaf diseases, can be a greater problem in conservation tillage compared with conventional tillage, where plowing buries infected residues. This factor increases the need to select resistant crop varieties and to rotate crops.

Some researchers state that insects will be a greater problem in conservation tillage, because overwintering insects are not buried by plowing. Others state that research has not proved this claim. Growers should follow the advice of local Cooperative Extension agents.

Cropping Systems

Growers employ a number of cropping systems; several are described here. A grower's choice depends on climate, economics and market demand, government programs, and grower preference. Each system requires different soil-management techniques and has different effects on the soil.

Continuous Cropping. In continuous cropping, a farmer grows the same crop each year. Continuous cropping is favored by many farmers because they can grow the most profitable crop. This method also allows the grower to specialize in the crop best suited to local soil or climate conditions. In general, however, yields often decline with continuous cropping. At the same time, expenses for fertilizer, herbicides, and pesticides tend to rise compared with expenses for a crop rotation system.

Crop Rotation. **Crop rotation** means that a series of different crops is planted on the same piece of ground in a repeating system (figure 16-16). Many farmers do not rotate because it means planting some less-profitable crops. Often a farmer has no use for certain crops in common rotations. For instance, a farmer who feeds no animals has little use for hay unless a buyer can be found for it.

FIGURE 16-16 In this typical Corn Belt rotation, there is grain stubble in the foreground, then a strip of hay, and then a strip of corn. *(Courtesy of USDA NRCS)*

However, crop rotation has important benefits for those who practice it. Crop rotation:

- Aids the control of diseases and insects that rely on one plant host, reducing a grower's pesticide bill.
- Helps control weeds. Many weed species grow best in certain crop types, so alternating crops suppresses the weeds. Some rotations suppress weeds by **allelopathy,** where one plant emits chemicals from the roots that suppress growth of other plants. For instance, soybeans planted into wheat residues suffer fewer weed problems because of allelopathic effects of the wheat.
- Supplies nitrogen if certain legumes like alfalfa are in the rotation. This can lower a farmer's fertilizer bill.
- Improves soil organic matter and tilth. Deep-rooted crops like alfalfa also improve subsoil conditions.
- Reduces erosion compared with continuous row crops, as long as the rotation includes small grains or hay. This topic is covered in more detail in chapter 18.

Generally, crop rotations involve some combination of three kinds of crops: row crops, small grains, and forages. The specific crops and crop sequence vary from place to place.

Row Crops. **Row crops,** where adapted, are usually the most profitable. Row crops are planted in wide rows and cultivated for weed control, with the help of herbicides. The crops are fertilized by broadcasting, banding, and sidedressing. Fertigation is often used in irrigated fields. Row crops usually leave the soil bare, making it erosion-prone. As a result, row crops are best suited to fairly level ground. Conservation tillage, crop rotation, and other conservation practices greatly reduce erosion from row crops (see chapter 18). Common row crops include corn, sorghum, soybeans, and cotton.

Small Grains. **Small grains,** like oats or wheat, are planted in closely spaced rows seven or eight inches apart. As a result, they quickly cover the soil surface. Land planted to small grains loses less soil to erosion. Small grains also leave a large amount of residue that controls erosion in conservation-tillage systems. The dense growth of small grains competes with weeds that infest row crops; and several suppress weeds by allelopathy.

Soil nitrogen and potash must be properly balanced for good grain yields. A good supply of nitrogen promotes growth and improves protein content. Excess nitrogen or low potassium, especially in moist soils, increases lodging. Fertilization is usually carried out by preplant broadcasting and may be followed by topdressing of the growing grain. Banding has become popular, especially in conservation-tillage systems.

Perennial Forage. Forages are harvested for their green matter and fed to animals. They may be harvested as hay or used for grazing in pasture or range. Forages improve soil tilth, add organic matter, and control erosion. Taprooted plants, like alfalfa, help break up soil pans. Legume forages also fix nitrogen that can be used by later crops. Examples include legumes such as alfalfa and a wide array of forage grasses.

Double Cropping. **Double cropping** is the practice of harvesting two crops from the same piece of ground in one year. A common example is planting soybeans into winter wheat stubble (figure 16-14). Double cropping is easiest in warm climates with long growing seasons. The use of double cropping has grown with the use of no-till systems. The second crop can be planted right behind harvest of the first crop, omitting time-consuming seedbed preparations.

Multiple cropping keeps the soil covered with vegetation for a larger part of the year. Better erosion control results. Two crops grow more green matter than one, so the practice helps maintain organic matter in the soil. Where one crop is a legume, the nitrogen addition is welcome. Two crops also draw more heavily on soil nutrients and water, so fertilizer and water must be more carefully managed.

Cover Cropping. Like double cropping, **cover cropping** involves two crops in one year on the same piece of ground. However, the cover crop is grown as a conservation tool and is usually not harvested. Several cover-cropping methods exist. In one variation, a cover crop is interseeded between rows of a taller row crop that is already growing. In a second variation, a cover crop is grown until it covers the ground, then the main crop is planted right into that first crop. The cover crop may be killed or treated in some way to reduce competition. Cover crops may consist of grasses like rye grass or legumes like various vetches. The cover crop is sometimes termed a "living mulch." A 1994 study of cover cropping in corn in mid-Atlantic states exemplifies the value of cover cropping.[1] This study concluded that of the systems studied, including clean cultivation, the most profitable was corn planted no-till in a killed winter cover of hairy vetch. The hairy vetch also contributed nitrogen and organic matter to the soil.

---

[1]Lichtenberg, E., et al. (1994). Profitability of legume cover crops in the Mid-Atlantic Region. J. Soil & Water Conservation 49(6):582-585.

Cover cropping is a potent BMP for a number of problems. The thick cover reduces runoff and erosion and loss of chemicals in that runoff. The cover crop often sops up fertilizers, especially nitrates, left over from the last crop, to lessen nitrate seepage into ground water. Cover cropping can increase soil organic matter content, lessen weed growth, and suppress some crop pests. In some areas, the standing mulch protects seedlings from blowing sand, and if a legume, adds nitrogen to the soil.

Dryland Farming

The term **dryland farming** is applied to farming in low-rainfall areas without irrigation. In the United States, dryland farming is practiced in states west of a line from western Minnesota to eastern Texas, following the ninety-sixth meridian. We will discuss two characteristic dryland farming systems: small grain-summer fallow rotation and rangeland grazing.

FIGURE 16-17 Monitoring the water table below a saline seep. Nearby plantings of alfalfa are lowering the water table. *(Courtesy of USDA ARS)*

Summer Fallow. Many dry areas lack enough water to produce good crops each year. As a result, crop rotation of small grain-summer fallow is used. In the crop year of the rotation, small grains are grown because they are relatively tolerant of low-moisture conditions. After the grain is harvested, the soil is left fallow for the next year. More complex systems include a three-year rotation of winter wheat-fallow-sorghum, common on the southern Great Plains.

Summer **fallow** is the practice of leaving the soil crop- and weed-free to store moisture. During the fallow period, weeds are controlled by cultivation or herbicides. By controlling weeds, no moisture is lost from the soil because of transpiration. Some water is lost by evaporation, but not all. After a rain, water seeps into the soil. As the soil dries, some moisture moves to the surface by capillary rise, where it evaporates. However, after the surface dries, it seals the rest of the water in the soil. After the next rain, more water is sealed in the soil. Generally about 25 percent of the rainfall on a fallow field will be stored for the following crop.

The effectiveness of summer fallow can be improved by reducing water runoff. On slopes, contour tillage helps reduce runoff. Using tillage tools that leave crop residues on the surface also helps prevent runoff. Chemical fallow, which leaves grain stubble standing and crop residues undisturbed, saves an additional one-half to two inches of moisture.

Three problems arise from the practice of summer fallow. During the fallow year, wind erosion can be quite serious. This problem will be discussed in detail in chapter 18. Second, crop-fallow rotations lead to a long-term decline in soil organic matter. The other problem is the development of **saline seeps.**

Saline seeps (figure 16-17) appear in almost two million acres of the Great Plains of the United States and Canada. Saline seeps appear where glacial till overlays an impermeable layer. During fallow, increased percolation picks up salts and carries them deeper into the soil. When salty water reaches the tight layer, it spreads out sideways, flowing downslope. Finally, the salty water seeps to the surface on a lower field. The water evaporates, leaving salt on the soil surface.

Reclaiming a saline seep begins by finding the origin of the salty water—the recharge area. Shallow ditches, land leveling, or contouring can divert excess water from the recharge area. Growers may seed the recharge area and the soil around the seep to alfalfa. With its high water demand and deep root system, alfalfa lowers the water table. During the reclamation period, salt-tolerant barley may be grown if needed.

Researchers have studied ways to reduce fallow problems by devising annual cropping systems. For instance, in the northern Great Plains about a quarter of the annual precipitation falls as snow—if all this were captured rather than allowed to blow off, it could equal the moisture saved during a fallow year. Studies using tall wheatgrass strips to capture snow—as shown in figure 8-11—have shown improved profitability for annual wheat cropping with reduced problems like wind erosion.[2]

Rangeland. Range is an uncultivated area used for livestock grazing, particularly in the western United States. **Rangeland** is particularly important because it occupies such a large proportion of the land surface of the United States: up to half may be rangeland ecosystems. Of federally owned lands in the western United States, 85 percent are grazed by livestock. This land provides food and important wildlife habitat.

Grazing is the best agricultural use of range because it is too dry, too rocky, or too infertile for other agricultural uses. Most rangeland, if cultivated, would erode badly. Generally, little is done with rangeland because water shortages make improvement unprofitable.

Care is needed, however, to keep range healthy, profitable, and acceptable as wildlife habitat. A 1994 report by the National Academy of Sciences[3] noted that because range receives few inputs like irrigation or fertilizer, the health and productivity of rangeland rely heavily on natural processes. That report goes on to say that "Rangeland health should be defined as the degree to which the integrity of the soil and ecological processes of rangeland are sustained." It also defines soil stability and natural nutrient cycles as one of the prime criteria for rangeland health.

Grazing patterns strongly affect soil properties and cover on rangeland. A North Dakota study compared long-term effects of no, moderate, and heavy grazing on native mixed-grass range and a fertilized crested wheatgrass plot.[4] The study found that soil was most compacted in the heavily grazed site, and diversity of plant cover was best maintained by moderate grazing. The authors concluded that heavy grazing degraded the soil and plant cover resource, while both moderate grazing of native grasses and wheatgrass plots could sustain grazing long-term.

Ill health of rangeland tends to result from changes in fire patterns, which alters the type of vegetation; invasion by alien weeds; and overgrazing by livestock and wild animals. Overgrazing exposes bare soil to erosion and changes the vegetation as it shifts to plants not favored by the grazers, often brush and weeds. The 1992 National Resource Inventory identified 17 percent of public rangeland with serious brush or weed problems, 23 percent with accelerated erosion, and 18 percent with multiple problems.

Controlled grazing is key to preventing erosion. The number of grazing animals should be only as large as the land can safely carry. Animals should not occupy a single range for a long period, and vegetation should be given a chance to recover before livestock is returned. Animals also should not feed on specific rangeland during periods of slow plant growth.

Sustainable Agriculture

Increasing concern for long-term farm productivity and the effect of agricultural practices on the environment led to the concept of **sustainable agriculture.** The American Society of Agronomy in 1989 declared that "a sustainable agriculture is one that, over the long-term, enhances environmental quality and the resource base on which agriculture depends; provides for basic human food and fiber needs; is economically viable; and enhances the quality of life for farmers and society as a whole."[5]

Those who research or practice sustainable agriculture have several concerns. There is concern that agriculture's

---

[2]B. G. McConkey et al. (1990). "Perennial Grass Windbreaks for Continuous Wheat Production on the Canadian Prairies." Journal of Soil and Water Conservation, *45, (4): 482–485.*

[3]National Academy of Sciences. 1994. Rangeland health: New methods to classify, inventory, and monitor rangelands. National Academy Press.

[4]*ASA (1989). "Decisions Reached on Sustainable Agriculture.* Agron. News, *Jan 1996, p. 10.*

[5]Wienhold, B., Hendrickson, J., & Karn, J. 2001. Pasture management influences on soil properties in the northern Great Plains. J. Soil Water Cons. 56(1): 27-31.

resource base is being depleted: declining soil productivity due to erosion and loss of organic matter and nutrients; depletion of fertilizer sources like phosphate rock; and cost and availability of energy. A feared consequence of agriculture is a degraded environment: pollution of water by agricultural chemicals, nutrients, and siltation. Stability of the farm economy and community further motivates sustainable agriculture.

Sustainable agriculture, then, is a philosophy and collection of practices that seeks to protect resources while ensuring adequate productivity. It strives to minimize off-farm inputs like fertilizers and pesticides and to maximize on-farm resources like nitrogen fixation by legumes. Top yields are less a goal than optimum and profitable yields based on reduced input costs.

Soil and water management are central components of sustainable agriculture. Techniques include crop rotation, conservation tillage, cover cropping, nutrient management, and others. A recent mid-Atlantic study[6,7] comparing plastic mulch and vegetative mulch, (produced by shredding a standing cover crop of hairy vetch) on a tomato crop illustrates research on sustainable methods. Researchers found the vetch mulch greatly reduced runoff, erosion, and sediment and pesticide transport off the field compared to polyethylene mulch, a standard practice for growing tomatoes. The runoff from polyethylene mulched plots was found to be more toxic to aquatic organisms. Half the nitrogen was used in the vetch plots, of interest in light of fertilizer nitrogen issues discussed in chapter 15. The study suggests that the killed vetch mulch was a more sustainable practice than conventional plastic mulch for tomato production.

Organic Farming. **Organic farming** is a type of sustainable agriculture that also prohibits the use of synthetic substances, including inorganic fertilizers and pesticides. A major theme shared by organic farms is promoting a healthy soil by controlling erosion and keeping organic matter levels high. Buyers believe organic products to be safer, more nutritious or flavorful, or support the process of organic farming.

Organic growers add nitrogen by the use of manures, composts, legumes, and organic nitrogen fertilizers. Phosphorus and potassium come from manures and mineral fertilizers such as rock phosphate. Organic farms tend to rely more on natural nutrient cycles than do conventional farms. Crop rotations figure centrally in many organic operations. Weed control tends to rely on crop rotation, cultivation, and sometimes flaming or mulches.

According to a 1980 United States Department of Agriculture study, successful organic farms come in all sizes and crops. This and other studies point to soil benefits including reduced erosion, increased soil organic matter content, higher populations of earthworms, richer soil flora, and others. In a 2000 review of studies comparing conventional, sustainable, and organic systems in horticultural crops, their comparative production and profit were highly variable; results depended greatly on the specific sites and practices. Profitability for organic production tended to rely on the higher prices offered for organic produce.[8]

State-sponsored programs to certify organic production have grown in recent years, and in 1990 the Organic Foods Production Act directed the USDA to set up a federal program. The final rules for that program were published in 2000. The rules set certain production standards, prohibit the use of many substances on organic land including sewage sludge, and provide a list of allowed and disallowed synthetic materials. It also sets labeling requirements. These and state standards define what can be sold as organic and help the consumer purchase organically grown foods.

---

[6] Rice, P. et al. (2001). Runoff loss of pesticides and soil: A comparison between vegetative mulch and plastic mulch in vegetable production systems. J. Environmental Quality 30(5): 1808–1821.

[7] Rice, P. et al. (2002). Comparison of copper levels in runoff from freshmarket vegetable production using polyethylene mulch or a vegetative mulch. Environmental Toxicology and Chemistry 21(1): 24–30.

[8] Brumfield, R. 2000. An examination of the economics of sustainable and conventional horticulture. HortTechnology 20(4):687–691.

SUMMARY

Tillage has three goals: weed control, alteration of physical soil conditions, and management of crop residues. Tillage also has a number of side effects, however, especially an increase in erosion, compaction, and reduced soil permeability.

Conventional tillage buries crop residues to produce a smooth, residue-free seedbed. Conservation tillage leaves residues on the soil surface to prevent erosion and preserve soil water.

Three cropping systems are used by growers: continuous cropping, crop rotation, and multiple cropping. Continuous cropping (or a simple corn-soybean rotation) allows a farmer to grow the most profitable crops yearly. Crop rotation, on the other hand, is better for the soil and helps control erosion.

In low-rainfall areas, small grains are grown in rotation with summer fallow. During fallow, weeds are controlled by cultivation or weed killers to store moisture for the following crop. Problems with summer fallow include erosion and saline seeps.

Animal grazing is the most practical agricultural use of dry, steep, or rocky land in the West. Controlled grazing, seeding of improved grasses, and sometimes fertilization and water management keep range in good condition.

Organic farming replaces chemical fertilizers and pesticides with crop rotation, manuring, cultivation, and mineral fertilizers. Organic farmers focus on having a "healthy" soil. Sustainable agriculture aims to reduce some of the problems of standard agriculture by using techniques that lower off-farm inputs, increase use of resources found on the farm, and the use of Best Management Practices.

REVIEW

1. What is conservation tillage and what are criteria for determining if a practice qualifies as conservation tillage?
2. Compare conservation-tillage methods most suitable for areas with soils that tend to be cold and damp in the spring to those that would be warmer and drier.
3. Some people think of sustainable agriculture as a return to "old-time" methods. Do you agree or disagree? Explain your answer.
4. What influence do you think widespread adoption of conservation tillage could have on global climate change?
5. Describe the purposes of tillage.
6. After the last cultivation, we plant a fast-growing legume between the rows of a corn crop. What do we call this practice? What might be benefits and drawbacks?
7. Why does summer fallow store some water in the soil for next year's crop? Review the discussion of capillary water movement in chapter 7 if necessary. How efficient is this practice?
8. Discuss the degree to which rangeland grazing, conventional row-crop agriculture, and organic farming utilize natural nutrient cycles.
9. Would you guess that organic farms consume directly and indirectly less or more energy than conventional ones? Think about fertilizer sources as well as other factors.
10. Using charts in this chapter, estimate the amount of residue left on the soil surface in these three practices:
    a. 150 bushel/acre corn after moldboard plow and discing to six inches.
    b. 150 bushel/acre corn after a single chisel plow with spear point.
    c. 50 bushel/acre soybeans treated as in a and b above.

    Which crop leaves the most residue? Which system is likely to have the least erosion?

ENRICHMENT ACTIVITIES

1. If you are not already familiar with the tillage tools described in this chapter, visit an equipment dealer to look at them.
2. Survey and visit local farms to find out what tillage and cropping systems they use.
3. Study organic production certification rules for your state or the rules for the federal program (Federal Register, 7 CFR Part 205: National Organic Program; Final Rule. December 21, 2000). Typing "organic certification and [name of state]" into your browser should find information.
4. An excellent source of information on cover crops is at <http://www.attra.org/attra-pub/covercrop.html>.

CHAPTER 17

Horticultural Uses of Soil

OBJECTIVES

After completing this chapter, you should be able to:

- state how to select soils for horticultural crops
- describe fertilization practices for horticultural crops
- describe how growers manage their soils
- solve the special problems of container soils
- describe how soil influences landscaping

TERMS TO KNOW

alkalinity
coarse aggregates
conductivity meter
media
perched water table
perforation
perlite
soil-based potting mix
soil-less potting mix
stem-girdling roots
vermiculite
vertical mulching
xeriscaping

Merriam-Webster's Collegiate Dictionary (10th edition, 1993) defines horticulture as the "science and art of growing fruits, vegetables, flowers, or ornamental plants." The information presented so far in this text about using soils applies to horticultural crops as well as to other crops. Let's see how soil is used by growers of different crops, starting with vegetable growers.

Vegetable Culture

Vegetables are the most important horticultural crop in terms of total value. They make an important contribution to the human diet, supplying starch, fiber, and minerals and vitamins missing in grains and meats. There is also increasing interest in vegetables as a source of antioxidant chemicals shown to be important for human health. Vegetables are grown throughout the United States, but growing areas are concentrated in regions best suited to the economic production of vegetables, like California.

Soil Selection. Vegetable growers often select a specific soil type that suits the needs of the crop to be grown and marketing needs. Many growers in northern areas choose coarse soils because they warm up rapidly in the spring, allowing early planting and early harvest when prices are best. In general, vegetable growers favor coarser soils than do other farmers. Following are the soil types and their uses:

- Coarse-textured soils are best for early crop growth, especially for cool-season crops like lettuce or carrots. Several crops, like melons, grow best on sandy loams. For best yields, these soils are usually irrigated.
- Medium-textured soils are good for all crops. Where yields are more important than an early harvest, medium soils are better than coarse ones.
- Fine-textured soils are less desirable for vegetables. They tend to stay wet too long, hampering field operations, and crusting can inhibit germination of fine seeds. Heavy soils are very poor for root crops.
- Organic soils, especially mucks, are favorites for cool-season and root crops (figure 17-1), such as carrots, onions, and celery. Highly loose, porous mucks are especially favorable for growth of popular long-tapered carrots. Warm-season crops like tomatoes seldom thrive in organic soils because they tend to be cold and are often in "frost pockets."

Soils selected for growing vegetables should be loose, friable, and high in humus. For most crops, a slightly acid soil is best (except for potatoes, which suffer less disease between pH 4.8 and 5.4). Many vegetables are also sensitive to high soil salts.

FIGURE 17-1 Radishes growing on raised beds in an organic soil. *(Courtesy of USDA NRCS)*

The most essential factor for success is good drainage. Poorly drained soil warms up slowly and cannot be planted early. Vegetable crops keep their quality for a very short time after they are ready for harvest. Thus, they must be picked regardless of soil conditions. In this case, growers cannot afford a muddy soil.

Soil Management. Vegetable growers prepare their fields much as do other growers. If needed, drainage is installed. Most growers use conventional tillage; that is, the field is plowed and harrowed before planting, or the field is prepared for furrow irrigation. There is some trend by vegetable growers toward conservation tillage. However, many vegetables have very fine seeds that are not well suited to the rough seedbed and high residue levels of conservation tillage. Also, live plants are usually planted for tomatoes, cabbages, and other crops, and transplanters do not handle the thick residues well. However, some strip-till planters have been adapted for use in vegetables. Herbicides and/or frequent tillage controls weeds.

Even when not furrow irrigating, vegetable growers often bed or list their fields (figure 17-1). Raised beds improve drainage and because the soil is warmer, promote rapid early growth. The loosened, deeper rooting zone that beds provide is also ideal for a well-formed root crop.

It can be difficult for vegetable growers to keep organic matter levels high, because many vegetables simply do not grow the bulk of green matter that field crops do. A large part of the plant may even be harvested, as in lettuce or cabbage. Growers may harvest both roots and tops, such as in green-top carrots or beets. To make up for the lack of crop residues, many growers use animal or green manures as often as possible.

Not only do many vegetable crops leave little residue, but many are also inefficient fertilizer users. This raises the potential for percolation of nitrates into groundwater. Cover cropping, as described in chapter 16, presents great potential for preserving organic matter levels, reducing runoff and erosion, and for sopping up leftover nutrients in vegetable crops.

Plant nutrition presents problems because of the unusual needs of many vegetables and the soils they are grown in. The irrigated sands favored by many growers leach readily and are often short of growth elements. Mucks are generally rich in nitrogen, but phosphate, potash, lime, copper, and others may be in short supply. The edible part of a vegetable is usually an enlargement of normal plant tissue—like the enlarged leaf stalk that is celery—and movement of nutrients into these tissues may be slow. Calcium never moves readily in plants and is often a problem in vegetables. Blossom-end rot of tomatoes, cracked stems in celery, and other problems result. Thus, fertilization with secondary and trace elements is common in vegetable production.

In general, growers feed vegetable crops by the same methods as those described in chapter 14. One difference is the use of starter solutions on vegetable transplants. Crops like tomatoes and peppers are transplanted into the field rather than seeded. Transplanting equipment has a tank to hold a weak fertilizer solution that soaks the planting hole of each plant as it is transplanted. Starter solutions are high in phosphate to overcome phosphate immobility and stimulate rapid rooting. Starter fertilizers typically have ratios of 1-2-1.

Fruit Culture

Fruits are important in the human diet because they are rich in vitamins, particularly ones missing from many other foods, and because they are a rich source of antioxidants. Unlike most crops, fruit are usually long-lived woody plants that remain in place for many years; this influences their soil management.

Soil Selection. Tree fruits require soil well-drained to at least three or four feet deep. Shallow soils cannot support a tree crop during dry periods, unless irrigation is provided. Soils with a high water table also restrict the deep rooting of fruit trees. In poorly drained soils, fruit trees may experience a serious health decline as soil-borne fungi attack weakened roots.

Fruit plants tolerate a wide pH range. The recommended pH range is 6.0 to 6.5, because it suits both the fruit plant and any cover crops that may be grown with the fruit. Blueberries, however, need a pH of 4.3 to 4.8.

Many fruits are clean cultivated to remove competition from weeds or sod. The plants must be grown on fairly level ground that is unlikely to erode, unless a slope is terraced. Apples, when grown in sod, can be grown on rather steep land. In fact, orchardists favor hillsides for growing apples because heavier cold air drains off hills on frosty nights.

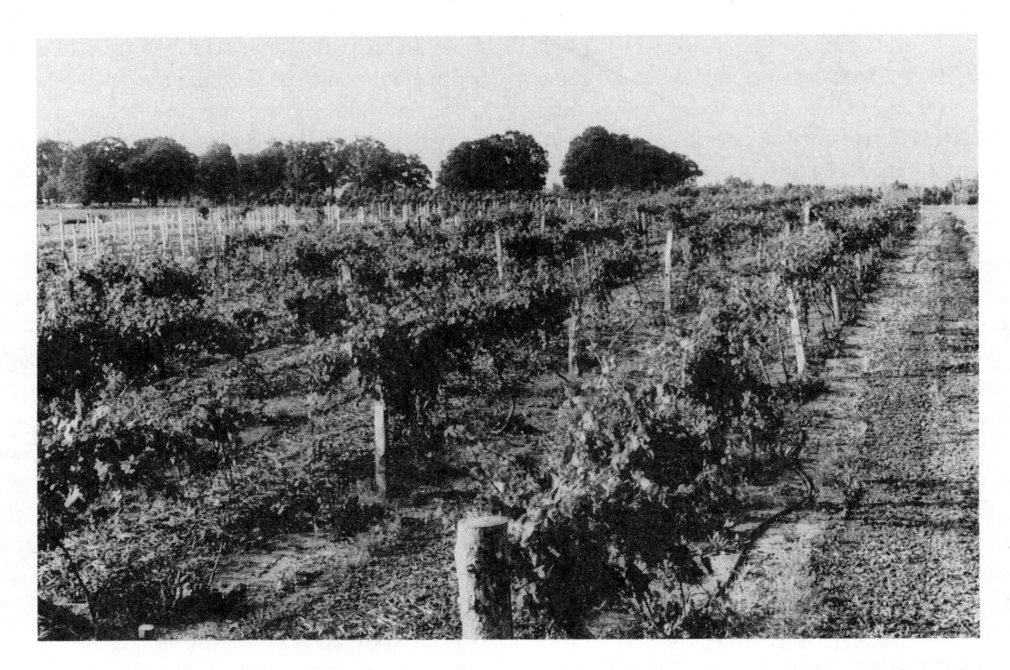

FIGURE 17-2 Grape vineyard operators often favor a coarse, gravelly soil because it produces a high-quality, sweet grape.

Fruits also tolerate a wide range of soil textures. Apples and pears do best on moderately fine-textured soils, while stone fruits, such as plums and peaches, prefer a coarser texture. Grapes grow on any soil, but the sweetest grapes are grown on sandy or gravelly soil (figure 17-2). Berry plants do best on moderately coarse soil.

Soil Management. All fruit crops are perennials and occupy the same ground for ten to twenty years. Therefore, the soil must be properly prepared before planting—there is no way to try again without a major financial loss. The site should be carefully selected. Any major soil changes like terracing, drainage, or leveling should be completed. Based upon soil-test results, potash, phosphate, and lime are broadcast and plowed into the soil.

Nitrogen is the most important nutrient for the established fruit crop. However, excess nitrogen or shortages of other elements causes poor fruiting, poor fruit quality, and late ripening. In addition, the tree is more likely to be injured by diseases or cold. Growers usually fertilize by topdressing in the late fall or early spring with a high-nitrogen fertilizer. Plant tissue testing can be useful in determining if elements other than nitrogen are needed on an established crop.

Many fruits are sensitive to low levels of trace elements. The best method of determining crop needs is to perform a soil test followed by a tissue test. Trace element problems can be solved by mixing trace elements into the base NPK fertilizer or by foliar feeding with chelated elements. Chapters 12 and 14 suggest some trace element fertilizers. In some cases, liming or acidifying the soil solves micronutrient problems.

Many fruit crops are clean-cultivated, which results in a loss of organic matter, poor tilth, and erosion. To overcome the problem, growers use manures, cover crops, mulches, and sod. Figure 17-3 shows an example of how these treatments are used. The figure shows part of a dwarf apple tree planting. The grower wants to ensure that apples do not compete with weeds or sod for nutrients and water. One approach keeps the soil surface bare, or clean-cultivated. Another way, which causes less soil damage, is to keep a strip of ground in the tree row free of weeds or grass by

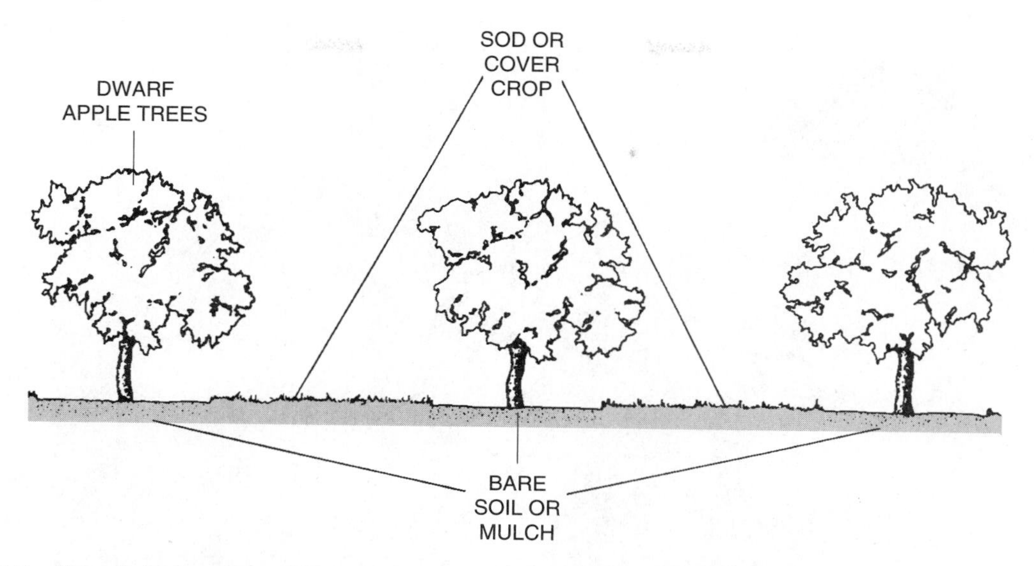

FIGURE 17-3 Many fruit plants grow best in bare soil. To prevent soil damage, growers may plant cover crops or sod between the rows of fruit trees.

mulching, cultivation, or herbicides while the row middles are sodded. The vegetative cover between the rows protects the soil and is a source of organic matter. Mulching in the row also provides benefits like saving moisture, adding organic matter, and protecting the soil and fruit.

Mulching can also be useful for small fruits (figure 17-4) to suppress weeds and save moisture. By mulching, pickers can work in the field after rain or irrigation; mulch also keeps fruit off the ground.

FIGURE 17-4 These strawberries are mulched to protect both the soil and the fruit and to conserve moisture.

Nursery Field Culture

Soil management is a constant challenge for the nursery grower. The very process of growing nursery stock is very hard on the soil. The soil is clean-cultivated, and the only crop residues are a few leaves (figure 17-5). The soil itself is dug up and hauled away when trees and evergreens are balled and burlapped. For example, using the figures for root ball sizes from the American Standards for Nursery Stock and average soil bulk densities, it can be determined that digging up an acre of five-year-old evergreens removes 165 tons of soil.

FIGURE 17-5 A nursery soil is bare and exposed to rainfall. Little organic matter returns to the soil from the crop. It is difficult to preserve soil structure and control erosion in this situation.

Site Selection. Soil is one of several factors to consider in selecting a nursery site. The land should be level, so erosion can be controlled. Sloping land may be used if it is terraced and rows are planted across the slope. It should be well drained, because wet soils hamper nursery operations. A pH between 5.5 and 6.5 is preferred for most nursery crops.

Soil texture is an important concern for nursery crops. Plants to be harvested bare-root (soil shaken off the roots) should be grown on sandy or sandy loam soil. Coarse soils most easily shake off plant roots without damaging them. In nurseries where stock is balled and burlapped by hand (soil ball stays on roots), finer-textured soils are better. Soils high in sand do not stick together, so it is hard to dig up an intact soil ball in sandy soils (figure 17-6). Silt loam or clay loam soils are best for balling and burlapping. Using machines to dig soil balls allows a wider range of soil textures.

Soil Management. Nurseries, perhaps more than any other operation, need to make a serious effort to maintain organic matter levels. It is typical to rotate nursery crops with green manure. Sudangrass, alfalfa, or other vegetation is used to build the soil. Many growers apply animal manures and/or sewage sludge. A few growers plant cover crops between nursery rows or mulch in the rows with chopped hay or wood chips. Some growers are beginning to sod row middles.

A new option could be called a "living mulch." Winter rye is planted in the nursery in early fall, allowed to grow overwinter, and develop in the spring. When the trees begin putting on growth, the rye is killed with an herbicide. The stand of dead rye acts like a mulch, protecting soil and moisture. Even more, chemicals given off by the rye suppress weed growth (allelopathy).

Before planting nursery stock, a soil sample should be taken from the top two feet of soil. Based upon the

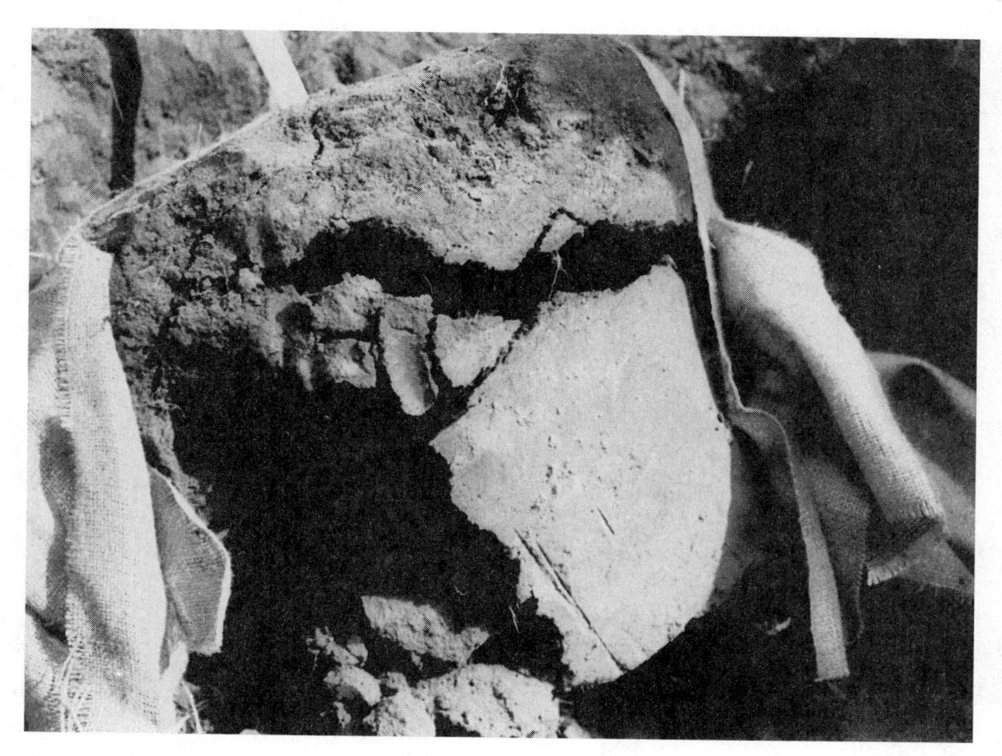

FIGURE 17-6 It is difficult to dig up an intact soil ball around plant roots in a coarse soil.

results of the soil test, the correct amount of potash, phosphate, and lime (or sulfur if the pH is too high) is mixed into the soil. Each year thereafter, high nitrogen fertilizers with ratios of 4-1-2 or 3-1-1 may be top-dressed in the fall or spring. Figure 17-7 suggests how much nitrogen should be applied.

Crop	Lb N/1000 Ft^2	Lb N/Acre
Deciduous plants	5	225
Narrowleaf evergreens	4	175
Broadleaf evergreens	3	125

FIGURE 17-7 Sample guidelines for fertilizing nursery stock. The best practice is to follow local guidelines.

Container Growing

One of the most demanding ways to grow a plant is to grow it in a pot. A containerized plant requires constant attention to watering, fertilizing, and other practices. Despite this, more and more plants are being grown in containers. Not only are greenhouse growers growing flowers in pots, more and more nurseries grow shrubs, evergreens, and trees in containers (figure 17-8). The container grower has complete control over soil conditions, making it easier to grow a large, uniform crop of quality plants. More recently, the business of landscaping the interiors of buildings with potted plants has grown rapidly. Apartment dwellers and even homeowners now garden in containers.

Growing plants in containers differs from growing plants in the ground in one key way: the plant's root system is confined to a small soil volume that must supply all the plant's water and nutrient needs. This means the container grower waters and fertilizes far more than those who till the ground. This section examines six

FIGURE 17-8 Growing nursery stock in a container presents special challenges. One key to successful growing is the use of highly porous potting mixes.

major topics: the naturally poor drainage of potted soil, types of potting soil, soil sterilization, soluble salts and alkalinity, soil temperature, and water pollution.

Container Drainage. A pot of soil is by nature poorly drained because of the shallow soil profile. Compare soil in a pot with soil in the ground. Recall from the discussion in chapter 7 that capillary action "pulls" water into the drier soil below a wetting front. In a deep soil profile, then, lower layers of soil pull water downward.

In a pot, the soil column ends abruptly at the bottom of the pot. The bottom layer of soil has no capillary connection to deeper soil and the last bit of water cannot drain away after watering. Thus, in spite of drainage holes, a layer of soil on the bottom of the pot remains saturated after drainage ceases. This layer is called a **perched water table** (figure 17-9). As a result, potted soil is wetter and has less air after drainage than the same soil in the ground.

The difficulty is the short water column—no "depth" to pull water down, no heavy mass of water pushing down by gravity. Therefore, the taller the pot, the less severe the problem (figure 17-9). A six-inch pot filled with a standard greenhouse mix has an air-filled porosity of

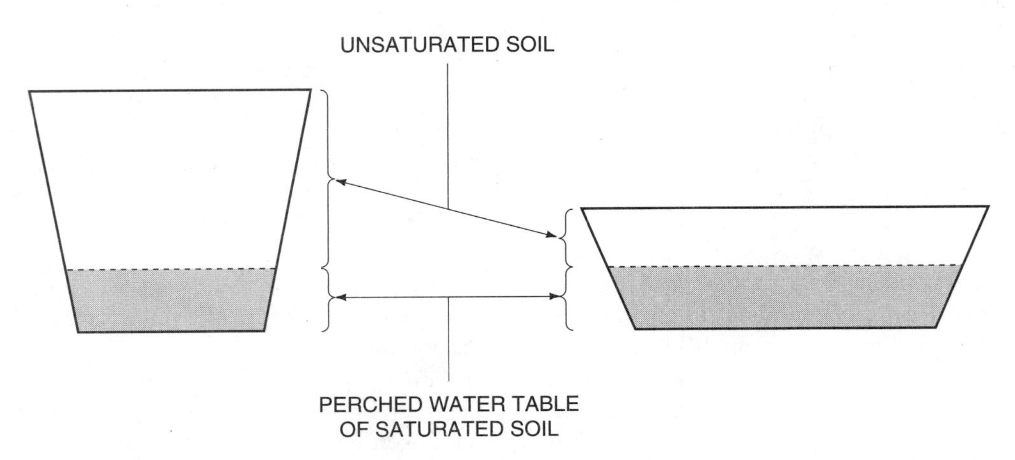

FIGURE 17-9 A perched water table in a pot causes poor drainage. Highly porous soils and fairly tall pots help solve the problem.

about 20 percent. Some greenhouse containers are no more than an inch deep. With the same mix, such a pot has a porosity of perhaps *2 percent.*

Because of poor drainage, a potting mix must be highly porous, with very large pore spaces. Large pore spaces apply less capillary force to hold the water in the pot. There are two approaches to making a porous mix. One is to mix materials into a field soil to make a porous **soil-based potting mix.** The second method omits soil altogether, making a **soil-less potting mix** of other materials.

The key to making potting mixes is to use large particles to create large pore spaces. For example, one simple soil-less mix is half sand and half peat. When fine sand is used to prepare the mix, the percentage of total soil volume filled with air after watering is about 5 percent. The same mix with coarse sand has an air capacity of 16 percent. Thus, the mix with coarse sand holds more than three times as much air as the fine sand mix.

Potting Mixes. Good potting mixes, or **media,** have a high holding capacity for both air and water. To accomplish this, the mix needs large particles that can absorb water. Picture a pot full of shredded sponge. Each piece of sponge can soak up water, and the empty spaces between the pieces can hold air. Growers may not use shredded sponge, but they produce media that behave the same way. The mixes contain varying amounts of three materials:

- Soil is the main part of soil-based mixes. It cannot be used alone, but must be mixed with the other two materials. Soil helps a mix hold water and nutrients, and helps buffer the medium from rapid changes. However, the fine particles retard drainage and lower aeration. Avoid soil in shallow pots, or use only a small percentage of soil. In pots large enough to have a deep soil column, soil may be used more freely. Soils used in potting mixes should be loamy with good structure and be free of pesticide residues.
- **Coarse aggregates** are large, inorganic particles used to create large pores in the mix. Coarse aggregates include coarse sand, **perlite** (large granules of lightweight expanded volcanic glass), **vermiculite** (expanded mica), shredded plastics, and other materials.
- Organic amendments hold water and may themselves help porosity. Shredded sphagnum peat, mined from peat bogs, is most common. Many growers shred and compost tree bark or sawdust. Rice hulls, shredded coconut hulls, and wood chip/sludge compost are also being used.

Growers combine these materials into various mixes to suit their needs. Standard soil-based mixes follow the model set by the John Innes mixes developed in England in the 1930s. These mixes consist of loam, peat, a coarse aggregate, and fertilizers. Later, soil-less mixes based on mixing peat with a coarse aggregate were developed at the University of California (UC mixes) and Cornell University (Peat-lite mixes). Figure 17-10 briefly summarizes these mixes. Many growers now use mixes based on composted hardwood bark chips or pine bark. Bark

	Common Potting Mixes				
	John Innes	*UC Mix D*	*Cornell Peat-Lite A*	*Cornell Peat-Lite B*	*Bark Mix*
Loam	7	—	—	—	—
Sand	2	1	—	—	1
Peat	3	3	1	1	—
Perlite	—	—	—	1	—
Vermiculite	—	—	1	—	—
Composted bark	—	—	—	—	2

FIGURE 17-10 For these common potting mixes, the numbers represent parts. For example, Peat-lite A consists of one part peat and one part vermiculite. These mixes also contain some combination of lime, gypsum, fritted trace elements, superphosphate, and other fertilizers. These mixes are only a sample of the large number of mixes in common use.

chip mixes have very high porosity and seem to suppress many harmful soil organisms.

Soil Sterilization. Soil-based mixes contain weed seeds, insects, nematodes, and parasitic fungi and bacteria. Of special concern are fungi that destroy young seedlings or cause root rots. To kill these organisms, some growers treat the soil with chemicals. The soil must "air out" for several weeks after such a treatment to avoid injuring crops planted in the mix. More commonly, soil is sterilized by heat. Many growers use steam—normally a temperature of 212°F. However, this high temperature may cause three problems:

- High heat kills the bacteria that convert ammonia to nitrates (nitrification). This break in the nitrogen cycle can cause a buildup of ammonia to harmful levels.
- High heat raises the solubility of manganese, which can result in toxic levels.
- High heat creates a "biological vacuum," destroying organisms that could compete with pathogens if the soil were reinfected.

Several methods avoid these problems. First, lime can be used in a potting mix to maintain the pH between 6.0 and 7.0. At this level, manganese is insoluble, reducing toxicity problems. Second, potting mixes should be used soon after treatment, before high levels of ammonia can build up. Third, "live steam" (temperature 212°F) can be replaced by "aerated steam" (temperature 160°F to 180°F). Special devices inject air into live steam to lower the temperature. This temperature reduces ammonia and manganese problems, and allows some helpful organisms to survive while killing pathogens.

Soluble Salts and pH. To grow potted plants successfully, a great deal of fertilizer (fed through irrigation water or incorporated into the potting mix) must be poured into a small soil volume. Most irrigation water contains dissolved salts, and most fertilizers are salts. Therefore, a troublesome problem of growing in pots is the buildup of soluble salts.

Controlling soluble salts in potted plants means frequent testing of both irrigation water and the medium itself. Growers send samples of potting media to testing labs, as described in chapter 12. However, they also monitor it

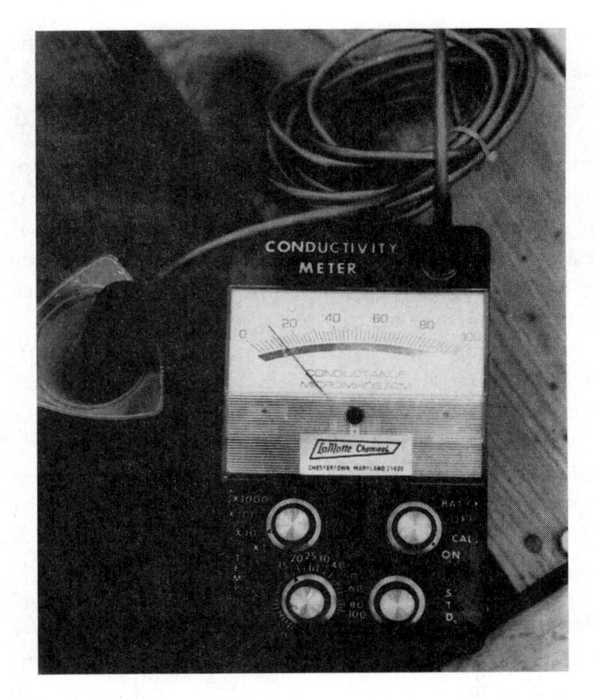

FIGURE 17-11 A conductivity meter can be used to measure the soluble salt content in greenhouse soils.

themselves with a **conductivity meter** (figure 17-11). The more ions dissolved in water, the more easily an electrical current will pass. Therefore, measuring the conductance of a standardized mixture of medium and water also measures its soluble salt concentration. Conductivity meters provide growers with critical and timely information needed to manage soluble salts and fertility. These other practices also help growers avoid problems:

- Saline water can be treated with special devices, but the process can be expensive for large amounts of water. A common water softener merely replaces calcium ions with sodium ions and does not lower salinity. Some California nurseries, contending with salty irrigation water, have installed very large—and expensive—systems to remove salts from their water supply.
- All pots should have good drainage. When watering, enough water is added so that some water leaks from the drainage holes. Drainage water leaches out excess salts.

A major difficulty encountered by container growers is dramatic changes in soil pH. In many areas of the country, high levels of dissolved carbonates—lime—in irrigation water raise pH of the potting mix. In other areas, the water has so little dissolved lime that leaching calcium from the potting mix during watering lowers pH. For most containerized crops, pH should be between 5.8 and 6.5.

The carbonate content of water, termed **alkalinity,** should be measured by a testing laboratory. If it is too high, the most common treatment injects acid into the water supply to neutralize the alkalinity. Water could also be treated to remove the ions, and some have even suggested mixing irrigation water with rainwater collected from the greenhouse roof. Crops irrigated with low alkalinity water may need to be treated with lime to counteract falling pH.

Soil Temperature. Plants growing above ground in containers, especially when exposed to hot sunlight, suffer large swings in soil temperature that can damage roots.

More severe is the difficulty of overwintering containerized plants in cold climates. Plant roots are less hardy than plant tops, and a potted root system exposed to subzero temperatures will be damaged. Northern nurseries must protect container plants over winter by covering them with straw or some protective structure. Similarly, landscape plants in containers, like urns by the doorstep or trees in planters on city streets, experience the same freezing. Northern gardeners and landscapers should use only the root-hardiest plants for this purpose or plant only annuals that do not need to survive the winter. Insulating containers with rigid insulation also helps.

Water Pollution. The public is concerned about water pollution; this issue strongly affects nurseries and greenhouses. For example, container nurseries have pots sitting on the ground, and fertilize through overhead sprinklers. Most of this water—as much as 70 percent—lands between, rather than in, the pots. Because a container field is watered and fertilized daily, the potential for water pollution is great.

One answer, where feasible, is to use drip irrigation (figure 17-12). Little water or nutrients are wasted this way. Second, greater use of slow-release fertilizers added to the pots could reduce the amounts added to irrigation water. Third, pots should be set on a sealed surface that stops leaching into the soil and the land surface graded so water runs off into sealed holding ponds, where the water can be pumped for reuse.

FIGURE 17-12 Drip irrigation on boxed stock in a California nursery. Drip irrigation can reduce water use and fertilizer pollution.

Landscaping

Landscaping makes our surroundings more beautiful and more livable. While most of us cannot live in a setting like the one shown in figure 17-13, even modest landscapes improve our lives. Landscapers must know about soil so that landscape plants will remain healthy, attractive, and easy to care for.

Most transplanted shrubs and trees survive—yet many fail to thrive, and too many actually die. Some have estimated that the average life of a new urban tree

FIGURE 17-13 Cheekwood Gardens in Nashville, Tennessee. Knowledge of soil is important to landscaping success, whether the project is an elaborate estate or a backyard garden.

is less than ten years. These failures cost money, in replacements and damaged customer relations. Often building contractors are responsible, sometimes homeowners fail to provide proper care, and other times a landscape professional is at fault.

When plants fail to thrive, more often than not, roots hold the answer. That means an understanding of soils is essential. Here are just two examples:

- A few inches of soil are added to the yard to raise the grade. The blanket of new soil cuts off oxygen to tree roots, trees decline for several years, then die. Many trees are very sensitive to raised grades.
- Many soils are not acid enough for the trees being planted in them. In the Midwest, pin oaks are planted by the thousands, yet many suffer iron chlorosis from unsuitable soil pH.

More than most, landscapers must understand how soils, roots, and water interact. Landscape soils are complex compared to field soils, because they are radically altered by construction and landscaping. Soils of different textures are placed together, and as chapter 7 points out, that affects water movement.

Now let's see how our knowledge of soil can help us understand proper landscape practices.

Site and Plant Selection. Too often, landscape designers pick a plant only because it would look pleasing in a certain spot, not because it would grow well in that spot. Mismatches between plants and site lead to unhealthy plants and endless maintenance. Landscape designers should know which plants grow well in their area and select plants that match soil drainage, pH, salinity, and degree of compaction on the landscape site. Appendix 6 lists soil preferences for selected trees.

A thorough designer will actually examine the soil on a site. Salinity and pH can be checked with a soil test or portable testing device. Compaction can be checked with a penetrometer or by seeing how hard it is to shove a screwdriver into the ground, taking soil type and moisture into account. Soil color provides guidance on drainage, or a very simple, crude percolation test can be done. A foot-deep hole is filled with water, and after it drains empty, filled again. Drainage can be inferred from the amount of water that drains in an hour. Experience makes these simple tests most valid.

Avoiding Compaction. Most landscape sites are slightly to severely compacted during construction. Studies on the effects of compaction on tree growth showed interesting results. Compaction caused shallow root systems. Often, roots from deep in the planting hole actually grew upward along the sides of the hole toward the surface, rather than outward into the surrounding soil. Failure to explore the native soil results in drought damage, nutrient shortages, and poor anchorage.

Tillage helps tear up the compaction, but often little can really be done about it. However, landscapers can avoid adding to compaction while working. Landscapers who drive on a site with trucks full of rock mulch only compound the problem. Landscapers should especially take care with wet, fine-textured soils.

While no method exists to completely relieve compaction in established landscapes, techniques are available that can be helpful. In **vertical mulching,** many holes are drilled in the ground and filled with organic matter or coarse material like sand. This allows some movement of air and moisture into the root zone of trees. Devices have also been developed to inject air or water into the soil under very high pressure, fracturing the compacted soil. The fractures are then filled with a coarse material. The long-term effectiveness of these treatments has not yet been fully determined.

Deep Planting. Recent work at the University of Minnesota has uncovered long-term problems associated with deeply planted trees. Trees planted too deep often fail because roots are deprived of oxygen, but those that survive face a longer-term problem called **stem-girdling roots.** When planted too deep, even by a few inches, tree roots often develop in ways that cause some to grow across the stem. As the tree ages, those roots prevent proper development of the trunk and compress the tree's vascular system. Eventually the tree begins to weaken and die. More dramatically, the tree may snap off at the soil line in high winds.

Stem-girdling roots can be avoided by planting trees at the proper depth. Established trees may be inspected for the condition—often visible by a flattened side on a tree trunk—and girdling roots removed.

Transplanting. The key to successful transplanting is rapid root growth. After all, for plants dug out of a field nursery, up to 98 percent of the root system is lost. Even containerized stock must grow new roots quickly, because it can't survive on just the soil mass it was planted with. How do landscapers ensure that the roots of newly planted trees and shrubs will grow quickly?

Research shows that plastic mulch, commonly used in shrub beds to keep out weeds, restricts root growth by keeping oxygen out of the soil, as pictured in figure 1-10. Organic mulches, like wood chips, keep down weeds and conserve moisture without robbing the soil of oxygen. Below the wood chips, one could place landscape fabric rather than plastic sheeting. This woven material is porous, allowing movement of air and moisture but restricting weed growth.

There is disagreement over the need for fertilization of newly planted trees and shrubs. Some experts claim little need; that trees come already supplied from the nursery with all the nutrition needed at first. The author, however, has observed that plants being held for sale often appear depleted by frequent watering without fertilization. However, high fertilization can actually retard new root development. The author supports light fertilization with a complete slow-release fertilizer at the time of planting if the soil tests low in nutrients.

There has been increasing interest in inoculating plants with mycorrhizae before planting them. This may improve transplant success and later plant growth. Preparations of inoculum are now available.

A recommended procedure for transplanting is as follows:

1. Check the drainage of the site. If it is poor, drainage systems may need to be installed. If that is impractical, then create planting berms—attractive, rounded, raised beds that will get the plants above the wet soil.

2. In a balled and burlapped tree, use a probe like a screwdriver to find the uppermost root in the soil ball. This is necessary because trees may come from the nursery with roots buried deep in the ball. When planted, this root should be at the soil surface or even slightly above in heavy soils. Remove soil above that root.
3. Prepare a planting hole much wider than the tree roots but never deeper. If dug deeper and then partially filled back in, the plant will settle and end up too deep. To improve air movement to roots and to lessen the chance of soil remaining water-soaked around the roots, make the hole even shallower in heavy, less well-drained soil. A large hole will provide loose soil for the roots to grow into.
4. Carefully place the root system of the plant in the hole. If a tree, the top major root should be at the soil surface.
5. Backfill the hole with the same, unamended soil that came out of the hole. Research indicates that amending the soil is not helpful and may create more problems with sharp interfaces. If the tree was planted high in heavy soil, create a gentle mound to cover the whole root system. A slow-release fertilizer may be included in the backfill if it seems called for.
6. Water the plant carefully to remove air pockets and thoroughly soak the soil.
7. Mulch around the tree with about three inches of coarse organic mulch (figure 17-14). Avoid fine mulches or more depth to avoid retarding aeration. Do not create a volcano-looking pile, and it is best if the mulch not actually touch the trunk.

Amending Soil pH. All too often, the native pH of a soil is not proper for every plant one wants to use. Obviously, this calls for designers and others to check for pH, an easy enough task with a soil test or even a piece of pH paper. Once the pH is known, a designer can make informed plant selections.

Nevertheless, there are times when soil pH should be altered. Because many landscape plants prefer acid soil, acidification is common. One common suggestion for acid-loving plants is to dig an especially large planting hole and then put sphagnum peat moss (half by volume) into the backfill. This helps temporarily, but as the peat decays, the pH returns to normal.

Three chemicals are used to acidify soils more permanently: sulfur, aluminum sulfate, and iron sulfate. There are disagreements about which is best, but the author prefers sulfur as being inexpensive, easy to use, and because of concern with aluminum toxicity. Figure 11-19 suggests amounts to be used. Mix the sulfur into the soil. The pH of calcareous or very alkaline soils may be very difficult to change even by this means.

Landscapers should be aware of the many sources of calcium that can raise pH in a landscape site, including lime in a concrete foundation, limestone rock mulches, and alkaline irrigation water. Annual use of acid-forming fertilizers like ammonium sulfate or special acid preparations may maintain acidity. Sulfur-coated urea, a slow-release fertilizer, will also enhance soil acidity.

Trees that do begin to suffer chlorosis can be treated by sulfur spread over the soil surface. While this changes the pH of only the top inch or so, there may be enough roots there to provide adequate iron for the whole tree. However, this and other treatments do not always work well. Foliar feeding with chelated trace elements can provide a temporary green-up in some cases. One may also use devices that inject trace elements into the trunk or root flare, but may be risky to tree health drilling holes in the trunk.

Fertilizing Established Trees. Experts disagree on the value of fertilizing landscape trees and shrubs, because there is so much variation between different trees and sites that experimental evidence is mixed. It is generally agreed that young, growing trees benefit from fertilization. Some feel that older established trees benefit little except to relieve an actual deficiency. Stressed trees should only be fertilized lightly because fertilizers may add to stress by increasing soil salinity or by supplying nitrogen the tree cannot handle. Nevertheless, fertilizing trees remains a common practice.

Timing strongly affects landscape plant response to fertilization. The best time to fertilize woody plants is when they go dormant. Nutrients absorbed in the fall will promote rapid growth the following spring. Early spring is also a good time to fertilize landscape plants. In cold climates, woody plants should not be fertilized from midsummer to early fall because it could spur late growth and keep the plant from hardening off for winter.

FIGURE 17-14 Landscape tree mulched with pine straw in Georgia. Commonly used in parts of the southeast, these pine needles are helpful to tree establishment and can reduce the chances of "lawnmower blight."

Nitrogen is the most important element for trees. To understand how to fertilize trees, one must first know where tree roots are (figure 17-15). Trees may root deeply, but 80 percent of a tree's feeder roots are in the top foot of soil. These feeder roots reach far from the trunk of the tree. One misconception is that tree roots all grow within the dripline of the tree. In fact, they often extend triple that distance from the tree trunk.

Shade trees may be fertilized by broadcasting over the whole root system of the tree. To be most accurate, dig a few holes to find how far the root system reaches. If this is not practical, assume the root system extends 100 percent beyond the dripline. Then spread a high-nitrogen fertilizer (2-1-1, etc.) over the entire root system at the rate of 3 1/2 pounds of nitrogen per 1,000 square feet. Nitrogen will, of course, leach into the root zone. If the shade tree grows in a lawn, split the fertilizer into two applications to avoid overfeeding the turf. One application should be in the late fall, the other in spring.

Another method of feeding shade trees is called **perforation.** This method is more difficult but is better if the tree lacks phosphorus or potassium. The amount of nitrogen to be applied can be calculated by the method described previously. Another method, simpler but less accurate, is based on the trunk diameter (caliper) measured twelve inches above the ground. If the caliper is less than six inches in diameter, one-quarter pound of nitrogen is added per inch of caliper. For trees more than six inches in diameter, a half pound of nitrogen per inch is applied. A complete fertilizer high in nitrogen, such as a 2-1-1 ratio fertilizer, should be used. This fertilizer is applied by drilling holes into the ground in a grid pattern, on twenty-four-inch centers, under the tree, and dividing the fertilizer amongst the holes. An eighteen-inch depth is often recommended, but this places the fertilizer below most tree roots, where it will be less available and subject to leaching. An eight- to twelve-inch depth is preferred. Figure 17-16 shows how this is done.

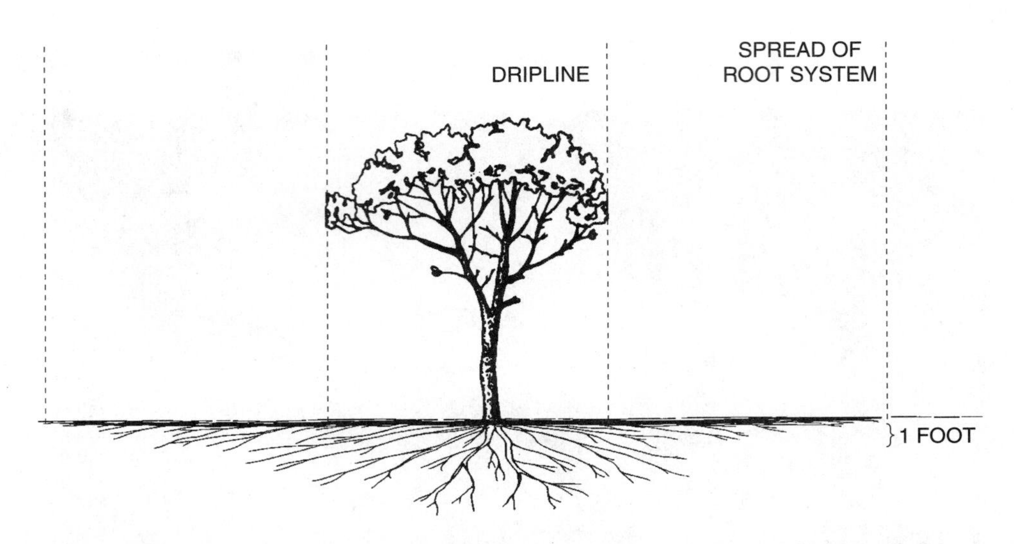

FIGURE 17-15 Tree roots cover a very large area and extend far beyond the dripline of the tree. Most feeder roots grow in the top foot of soil.

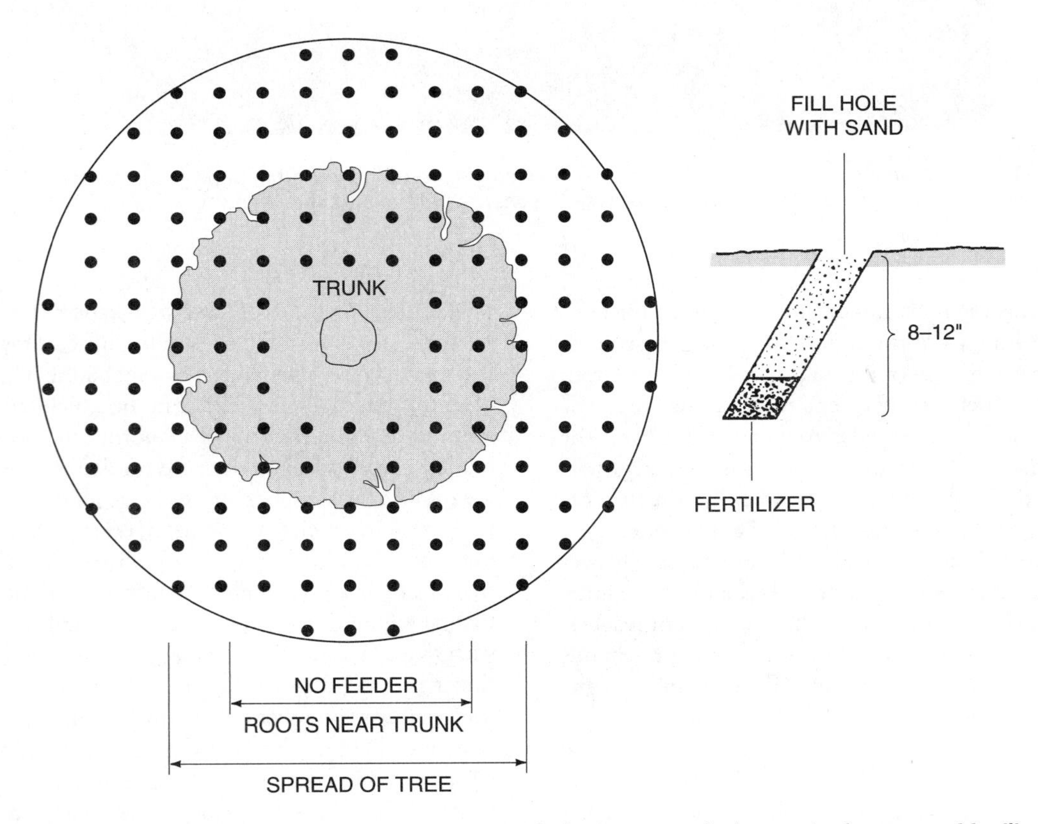

FIGURE 17-16 To feed trees by perforation, a grid pattern of holes is prepared. The required amount of fertilizer is divided by the number of holes to be filled. Perforation is useful for getting phosphate into the root zone.

Some workers feed trees by injecting a solution through a root feeder. In this system, a solution is pumped out of a tank under pressure and forced through a tube inserted into the soil. The pattern for injection would be the same as in figure 17-16, except that it should be on thirty-inch centers. Inexpensive versions are sold in most garden centers.

Turf. Except in arid regions, turf is the most common element of a landscape. Turf not only looks nice, it also protects the ground from erosion. Turf may be planted by seed, sod, or in some cases, sprigs and plugs. Whatever method, good soil preparation is important to success.

Many turf specialists no longer recommend covering the existing soil with "black dirt" before planting turf, because the interface is a problem. If any is added, it should be tilled into the existing soil to avoid the sharp interface. It is beneficial to till a two-inch deep layer of compost into the soil.

Chemical amendments should be tilled into the soil before planting, based on good soil tests. Phosphorus should be added now because of its immobility and importance for good root growth and turf spread. Potassium, which promotes good wear tolerance and resistance to disease, drought, and cold, also would be best incorporated now. Lime or sulfur may also be incorporated to adjust pH to 6 to 7.

After the soil is tilled to incorporate amendments and to loosen the soil, it is carefully raked or harrowed to make a fine seedbed free of bumps or pits. Turf may then be planted by seed, sod, or sprigs, according to local practice. Frequent irrigation is needed while the turf is being established.

Turf is fed by broadcasting a high-nitrogen fertilizer. Often a fertilizer with a ratio of 20-1-3 is suggested, but seek out local recommendations. Fertilizers with a high water-insoluble nitrogen (WIN) content are best. In areas where soil salinity is a problem, turf fertilizers should be of low salinity, and care must be taken to avoid overfertilization.

Fertilizer recommendations vary widely depending on the region of the country, the type of turfgrass, the level of maintenance, and the use of the turf. The annual rate of nitrogen application varies from as little as one pound nitrogen per thousand square feet for low-input, low-maintenance turf with tolerant turf types to as much as six pounds. The larger amounts are split into several applications—often spring, early fall, and late fall. Fall applications are the most beneficial.

There has been concern that nitrates may leach into the ground water below turf, or that phosphates will run off into surface waters and induce eutrophication. There is good evidence that turf roots are very efficient at scavenging nitrates out of the soil, and phosphates tend to tie-up in the thatch. Thus, properly applied fertilizers at moderate rates, following soil testing, should not create serious problems. One should take care to sweep up fertilizers and clippings that fall on surfaces like sidewalks.

Core aeration is a most useful turf soil-management practice. To aerate, a machine, preferably one with hollow tines, creates small pits in the lawn. This breaks up compaction and increases aeration. Aerate only when the soil is moist; wet soil will compact worse, while turf in dry soil is under stress. Schedule aeration when the turf is growing actively. This means spring and early fall for cool-season grasses and late spring and early summer for warm-season grasses.

For season-long green color, irrigation of turf is usually needed. One expert suggests adding about an inch of water when 30 to 50 percent of the lawn shows signs of water stress—a slight rolling of the older leaf blades. Some grasses can be allowed to go dormant in a dry season, but excessive dryness can damage turf.

Xeriscaping. Landscapers—and their customers—of the more arid regions of the United States face the problem of water shortages. The problem has been dramatically compounded by the planting of landscapes adapted to the more moist eastern and northern states. An answer to excess water use is xeriscaping (from the Greek word *xeros,* or dry). **Xeriscaping** is landscaping adapted to dry climates.

A wide range of plants are available that thrive under low-moisture conditions. The most dramatic of these include cacti, succulents, and yuccas (figure 17-17). However, xeriscaping need not mean only a mixture of sand and cacti. A number of shrubs, trees, and flowers can tolerate dryness. Examples of nonthirsty plants include palo verde (*Cercidium sp.*), rose periwinkle (*Catharanthus roseum*) and several salvias (*Salvia sp.*). Some short-grass prairie grasses can be grown for lawn with minimal watering, such as blue gramma grass, the state grass of Colorado.

FIGURE 17-17 A xeriscape at an old mission church in California. The planting consists of cactus and succulents. A xeriscape need not be this severe.

For even lower water usage, some plantings can be replaced by "hard" features. Turf areas may be replaced by paving, such as brick patios, or by mulching with pebbles. Some shrubs can be replaced by boulders, pole tips set upright in the ground, or similar features.

Lastly, the xeric landscape strives to conserve water. Plants can be installed into beds recessed below grade several inches to catch water, then mulched to slow moisture loss. Also, plants tend to be planted less densely, reducing competition for water. Drip irrigation greatly reduces water consumption. Irrigation systems should be well designed, zoned for different plant needs, and be well maintained and audited.

Xeriscapes may not always function as planned. In sixteen months of monitoring water use in Phoenix, Arizona, one study found that homeowner study subjects applied *more* water to xeriscapes than to conventional landscapes.[1] These results suggest there may be some challenge to making xeriscapes successful.

Arid regions that might practice xeriscaping contend with soil problems other than dryness. These can include alkaline soil pH, and saline or sodic soils. These problems call for selection of tolerant plants, or treatment as described in chapter 11. Incorporation of peat and sulfur into beds is helpful. These soils often suffer from the presence of a lime-cemented hardpan

[1]Peterson, K., Stabler, L, & Martin, C. 2000. Irrigation application volumes in urban residential landscapes. 2000 Symposium, Central Arizona—Phoenix Long Term Ecological Research.

called caliche. If severe, the caliche may need to be broken up before planting.

SUMMARY

This chapter discussed how soils are managed to grow fruits, vegetables, and ornamental plants. An irrigated, coarse soil allows early production of vegetables, while a medium soil gives the best production. Vegetable soils must be very well drained. Many growers make good use of animal and green manures to make up for the small amount of organic matter most vegetable crops produce.

Fruits and field nurseries need a deep, well-drained soil. Clean-cultivated fruits and nurseries need level land and close attention to erosion control. Because the crops stay in the ground for many years, the land must be carefully prepared before planting, including drainage, terracing, irrigation installation, leveling, and plowing lime and nutrients into the soil. In the years after planting, nitrogen is the most important nutrient. Maintenance of soil organic matter is a challenge for these growers.

Soil in pots is poorly drained because the soil column ends at the bottom of the pot. To overcome this, potting mixes are very porous and consist of coarse aggregates such as sand, an organic material like peat moss, and sometimes soil. Soil-based mixes must be sterilized by heat or chemicals to kill weed seeds and soil pathogens.

For landscape plantings to grow well, the soil must be handled properly. Plants are chosen that will thrive in the soil on a landscape site. Landscapers must be aware of the effects that textural interfaces have on the movement of water and the growth of roots. Proper soil preparation for planting turf and good transplant practices are needed for healthy plants. Fertilizing established turf and trees also keeps them growing well. Xeriscaping is landscaping adapted to dry climates.

Chapter 19 of this text contains information related to this chapter, particularly material on structured soils in urban landscapes.

REVIEW

1. Why is preliminary soil testing and preparation more important for fruit crops than for most other crops? What are examples of preparations?
2. Why are vegetables often grown on sandier soils in northern states? Provide a full explanation.
3. Why is nursery culture hard on soil? What practices do nurseries do to compensate?
4. Gardeners often put a layer of rocks in the bottom of a pot to improve drainage. Does it? Explain your answer.
5. Why is it critical for health of plants growing in containers that there be drain holes in the pots?
6. We plant a tree in heavy clay soil. To make sure it does not fall over, we bury a few inches of the stem. To make it easier to water, we leave a deep depression under the tree. What are possible consequences of such actions?
7. As a landscape designer, you like balsam fir (*Abies balsamea*). What kind of soils should be avoided? What soil tests might you ask for to evaluate a yard for suitability for balsam fir? See appendix 6.
8. Describe a soil we would consider good for both a home landscape and a nursery. See appendix 5.
9. What are elements of a xeriscape that help save water?
10. A case study: In 2002 the author's home state passed a law restricting phosphorus in lawn fertilizers in its largest city to preserve water quality in city lakes. The city is sometimes known as the City of Lakes. Describe the problem of phosphorus in surface water. Knowing what you have learned in your soils class, and following an Internet search on "lawn fertilizer and phosphorus," write a paragraph about your reaction to this ordinance.

ENRICHMENT ACTIVITIES

1. This little experiment will demonstrate the effect of the depth of a soil column on drainage. Soak a common kitchen sponge until it is saturated. Now hold it up horizontally (but not flat). When it stops dripping, turn to a vertical position. What happens now that the column has become deeper?
2. Check out numerous sites about xeriscaping on the Internet. An example is:
   <http://www.pan-tex.net/usr/a/aggie/ag05055.htm>. Compare several sites and judge them on reliability. Are they from a knowledgeable source? What is the authority or credentials of the source? Do they seem to have a biased viewpoint? How old is the site?
3. Study the Web site on stem-girdling roots by the man who did most of the research, Dr. Gary Johnson, at:
   <http://www.extension.umn.edu/distribution/naturalresources/DD7501.html>.
   If you type "stem-girdling roots" into an Internet search engine, you will also find other pages on the topic.

CHAPTER 18

Soil Conservation

OBJECTIVES

After completing this chapter, you should be able to:

- list the effects of soil erosion
- describe how soil erosion occurs
- list the types of water and wind erosion
- calculate soil loss from water erosion on a field
- describe ways to prevent erosion

TERMS TO KNOW

cross-wind ridges
cross-wind trap strips
ephemeral gullies
filter strips
gully erosion
Revised Universal Soil Loss Equation (RUSLE)
rill erosion
saltation
sheet erosion
splash erosion
surface creep
suspension
Universal Soil Loss Equation (USLE)
Wind Erosion Equation (WEQ)

Erosion

Each year almost two billion tons of soil wash or blow from farmlands of the United States. This quantity is equivalent to losing the full plow layer from two million acres of farmland. Most of the loss—about a billion tons—results from water erosion. The remaining .8 billion tons are lost in wind erosion.

Soil scientists follow the rule of thumb that one acre of land can afford to lose at most one ton to five tons of soil each year. The average soil loss to water erosion on cropland is thought to be 3.1 tons per acre per year, close to the limit. Added to this amount, however, is an average soil loss of 2.0 tons per acre per year to wind erosion (figure 18-1). As noted in chapter 3, most erosion occurs on erosion-prone land. About 21 percent of American cultivated cropland suffers soil losses greater than the acceptable limit for water erosion. Figure 18-2 shows where water erosion takes place in the United States.

Erosion data are from the National Resources Inventories conducted by the USDA in 1982, 1987, 1992, and 1997. As shown in figure 18-1, erosion slowed greatly during those years. While two billion tons of soil lost annually may seem large, it is much less than the five billion tons estimated in 1982. A dramatic decline in airborne dust has also been reported in several western states due to a decline in wind erosion.

Some decline in erosion rates was certainly due to a voluntary increase in erosion-control practices, especially conservation tillage. Some was due to conservation compliance measures required of growers by the 1985 and 1996 Farm Bills for participation in other USDA programs. Much improvement is attributed to removal

	Sheet and Rill Erosion		Wind Erosion (tons/acre/yr)	
	1982	1997	1982	1997
Cultivated cropland	4.5	3.5	3.7	2.9
Pastureland	1.1	1.0	0.1	0.1
Rangeland	1.2	1.2	4.7	4.4

FIGURE 18-1 Estimated average annual loss of topsoil in tons/acre to erosion in nonfederal lands of the United States. *(Source: USDA, 1997 National Resource Inventory)*

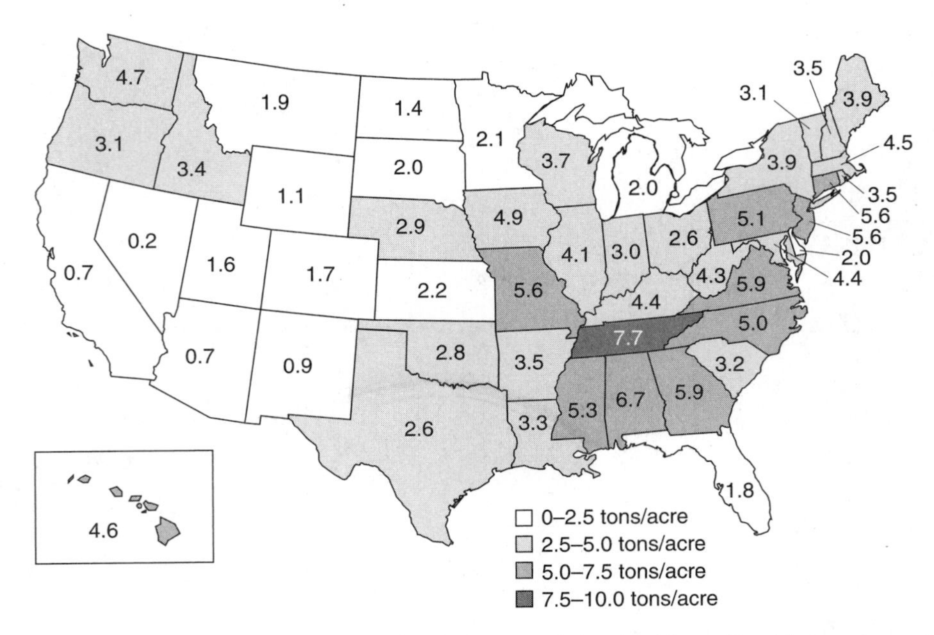

FIGURE 18-2 Average annual sheet and rill erosion by state on nonfederal cultivated land in tons/acre. *(Source: USDA, 1997 National Resource Inventory)*

of over thirty-six million acres of erodible land from cultivation, entered into the Conservation Reserve Program begun in 1985. Chapter 20 describes these programs in greater detail.

What damage is caused by this amount of erosion? Some of the damage occurs on a grower's fields; this is considered on-site damage. Some occurs elsewhere; this is off-site damage. Examples of on-site damage include these:

- Erosion removes topsoil. Topsoil affords the best root environment by providing the best structure, the most air, and an active population of organisms. Once the topsoil is lost, only the less productive subsoil remains.
- Topsoil contains most of the soil's organic matter and plant nutrients. Erosion carries away nitrogen, phosphorus, and any nutrient stored mostly in organic matter. Large amounts of fertilizer paid for and applied by the grower are wasted.
- Erosion thins the soil profile, decreasing the root zone. This is a particular problem on already shallow soils.
- Thinning topsoil and declining organic matter greatly reduces total plant-available water in the root zone. This has been considered the most serious effect of erosion on productivity.
- Gullies cut up fields into odd-shaped pieces and make it very difficult to operate farm equipment.

Examples of off-site problems include these:

- Eroded soil contains nutrients and pesticides that pollute lakes and streams. For instance, large fish kills have occurred in streams fed by runoff water from fields treated with soil insecticides.
- Soil washed away by erosion settles in streams, lakes, harbors, and reservoirs. The sediment fills in lakes and reservoirs (figure 18-3), creating a need for expensive dredging. It reduces the ability of streams to carry water, resulting in an increase in flooding.

FIGURE 18-3 When sediments eroded off farm fields filled one end of this Iowa lake, there was an economic loss as well as reduced recreational and ecological value. *(Courtesy of USDA)*

Cost of Erosion. It is only fairly recently that data have become available (such as National Resource Inventories) to attach values to the cost of erosion. There are actually two separate sets of costs: (1) costs to the farmer and consumer of production losses, and (2) costs to the public of pollution and sedimentation.

Productivity losses to the grower remain difficult to pin down precisely and are not usually severe. Estimates vary widely. Crosson and Stout [1] in 1984 estimated that each year an average corn farmer loses about $0.33 per acre more than lost the year before; that is, losses increase over time. In a 1994 Indiana study [2] of soybeans, yield losses varied between 17 percent and 24 percent on a severely eroded soil phase compared to a slightly eroded phase of the same soil. The authors impute this loss primarily to loss of available water and nutrients.

Estimates of total productivity losses in the United States vary widely. Crosson [3] estimated a value of $40 million per year at that time. Other estimates vary between $83 million [4] all the way up to a value of $27 billion.[5]

---

[1] Crosson and Stout. 1984. Productivity Effects of Cropland Erosion in the United States, John Hopkins Press.
[2] Weesies, G., et al. 1994. Effect of soil erosion on crop yield in Indiana: Results of a 10-year study. J. Soil & Water Cons. 49 (6): 597–600.
[3] Crosson, P. 1994. New perspectives on soil conservation policy. J. Soil & Water Conservation 39 (4): 222–225.
[4] Den Biggelar, et al. 2001. Soil erosion impacts on crop yields in North America. Advances in Agronomy 72 (1): 1–52.
[5] Pimental, D., et al. 1995. Environmental and economic costs of soil erosion and conservation benefits. Science 267: 1117–1123.

Off-site damage estimates for the nation as a whole also vary widely and may be greater than for on-site values. Estimates range from $3.1 billion [3] to $17 billion [5] per year. Dredging costs alone were estimated in 2002 to be about $257 per year in the United States .[6]

The nation has been battling erosion since the 1930s, yet it remains a major problem. This data indicates why. The short-range cost to individual growers is not great, and the cost of installing conservation measures can be high. Surveys of farmers also show that most underestimate how easily their fields can erode. Further, over the past forty years, a stream of technological improvements, including fertilizers and improved crop varieties, has masked the effects of erosion. Growers have every reason to assume such advances will continue. But when the soil becomes thin enough, technology will not save yields.

However, some scientists suggest that direct costs to growers have been set too low. Neglected costs include lost fertilizer, lime, and pesticides, additional equipment needs to work harder subsoils, deposition in drainage and irrigation structures, and many others. To these scientists, direct costs are high enough to economically justify erosion-control expenses for the grower.

How Water Erosion Occurs

Erosion follows three steps. First, the impact of raindrops shatters surface aggregates and loosens soil particles (figure 18-4). Some of these particles float into soil voids, sealing the soil surface so water cannot readily infiltrate the soil. The scouring action of running water also detaches some soil particles. Second, detached soil grains move in flowing water and are carried down slopes. Finally, the soil is deposited when the water slows down. These three steps are known as detachment, transport, and deposition (figure 18-5).

Erosion is a form of work, and work takes energy. The energy for water erosion comes from the energy of a falling raindrop or running water. The amount of energy in a moving object is the product of the mass of the object and its velocity (speed) squared. Expressed as an equation,

$$E = 1/2mv^2$$

The energy of a falling raindrop relates to its size and especially to its speed. A two-inch per hour rainfall has the same energy as a one-pound object falling forty-seven feet onto one square foot of soil. The erosive energy of running water depends on its volume and speed of flow.

With high erosive energy, water can detach and move larger soil particles. It can also move more soil particles. Thus, erosive energy relates directly to the amount of soil carried off a field. Deposition occurs when the energy of running water decreases—such as when it slows down at the foot of a slope. This energy concept will help in the understanding of four erosion factors: (1) soil texture and structure, (2) slope, (3) soil cover, and (4) roughness of soil surface.

Texture and Structure. Texture has two effects on soil erosion. First, texture influences the infiltration rate of water. If rainwater infiltrates soil quickly, less water runs off. With a lower volume of running water, less soil can be transported. Second, particles of different sizes vary in how easily they can be detached. Silt particles are most easily detached, so silty soils are liable to water erosion.

Structure also influences infiltration—good structural grades like granules reduce runoff. The strength of soil aggregates is important too, since strong peds better resist the impact of raindrops. Because of the importance of organic matter to structure, soil organic matter content has a strong bearing on erodability. Compaction, loss of organic matter, and destruction of soil peds by tillage all reduce infiltration and increase the volume of water available to transport eroded soil. The combined effects of organic matter content, texture, and structure are called the erodability of a soil.

Slope. Slope has two components—length and grade. On a steep slope, water achieves a high runoff velocity, increasing its erosive energy. On a long slope, a greater surface area is collecting water, increasing flow volume. On a longer slope, running water can also pick

[6] Hansen, L., et al. 2002. The cost of soil erosion to downstream navigation. J. Soil & Water Conservation 54(4): 205–212.

FIGURE 18-4 Raindrop impact on the soil. *(Courtesy of USDA NRCS)*

FIGURE 18-5 Soil erosion on this hill occurred in three steps: raindrops and running water detached soil particles, transported them down the hill, and deposited them at the bottom. *(Courtesy of USDA NRCS)*

Slope	Slope Length *(feet)*
4%	1,000
6%	200
8%	100
10%	50
12%	30
14%	20

FIGURE 18-6 Both slope grade and length affect soil loss. All slopes listed have equivalent soil loss.

FIGURE 18-7 A complete vegetative cover almost eliminates soil loss. Here highly erodible land in Iowa is protected by a solid cover of native switchgrass. *(Courtesy of USDA)*

up speed. Figure 18-6 shows that long, gentle slopes can have the same erosive potential as short, steep slopes. This fact helps explain why many growers underestimate how erodible their land is.

Surface Roughness. A rough soil surface impedes the downhill flow of water, slowing its velocity. If the roughness takes the form of ridges across a slope, water ponds behind the ridges, decreasing the volume of runoff water. However, if enough water collects behind a ridge, it overflows the ridge and wears it away. Thus, roughness can fail to stop erosion during very heavy rains, on long slopes, or if sealing of the soil surface stops infiltration.

Surface roughness depends largely on tillage practices. The seedbed resulting from conventional tillage is smooth, while that from chisel plowing is rough. Tillage across slopes acts to impede downhill flow; tillage up and down the slope promotes downhill flow.

Soil Cover. Bare soil is fully exposed to the erosive forces of raindrop impact and the scouring of running water. Soil cover reduces the energy available to cause erosion. A mulch or cover of crop residues absorbs the energy of a falling raindrop, lessens detachment, and reduces sealing of the soil surface. Mulches also slow down runoff water. A complete crop cover like turf or hay has the same effect, plus plant roots hold soil in place (figure 18-7).

Crops that are less close growing, such as row crops or nursery stock, have a slightly different effect. As these crops close in between the rows, they form a canopy over the soil. This canopy intercepts rainfall and absorbs most of its impact. When water drips off plant leaves, it again gains velocity, and thus energy. The energy of these drops depends on the height of the crop canopy but is not as great as free-falling raindrops. Thus, it is important for erosion control that crops cover the soil surface as quickly and completely as possible. Unlike mulches, crop canopies have no effect on runoff speed or volume, so have a less protective effect.

Types of Water Erosion. A raindrop strikes the soil surface forcefully. The impact shatters soil aggregates and throws soil grains into the air. On a slope, water begins to flow downhill, carrying detached soil grains with it. This water joins other flowing water, increasing in speed, volume, and soil-carrying capacity. This order of events leads to five types of erosion. All five types can occur at the same time on any given slope.

- **Splash erosion** is the direct movement of soil by splashing. A soil grain can be thrown as far as five feet by raindrop splash.
- **Sheet erosion** is the removal of a thin layer of soil in a sheet. On gentle slopes, or near the tops of steeper

FIGURE 18-8 Sheet erosion probably went unnoticed by the farmer until the subsoil appeared on these eroded knolls. *(Courtesy of USDA NRCS)*

FIGURE 18-9 Rills, ephemeral gullies, and gullies. Rills can be filled in by tillage, ephemeral gullies can be crossed but not completely filled in, and gullies cannot be crossed by farm equipment. *(Courtesy of USDA NRCS)*

slopes, water moves in tiny streams too small to be noticed. This gives the impression of losing soil in a thin sheet. The eroded knolls shown in figure 18-8 are an example of sheet erosion. Sheet erosion may go unnoticed until the subsoil appears.

- **Rill erosion** is visible as a series of many small channels on a slope. Water tends to collect in channels, picking up energy as it runs down the slope. As a result, running water carves out small but visible channels called rills. A rill is small enough to be filled in by tillage. Figure 18-9 shows rill erosion on a roadside embankment.
- **Ephemeral gullies** are large rills. The channel is small enough that tillage equipment can cross it and largely, but not completely, fill it in by tillage. During another heavy rainfall, water will collect in the old channel, and erosion will begin here.
- **Gully erosion** is the most highly visible erosion. Gullies are so large that equipment cannot cross them (figure 18-10). Gullies usually begin to form near the bottom of a slope or on steep slopes, where running water has enough force to carve a deep channel. Gully heads may back up the hill as water running into the gully collapses the sides.

FIGURE 18-10 Gullies are large channels that form where water collects as it runs down a hillside. *(Courtesy of USDA NRCS)*

Each type of erosion is important to understand for different reasons. Sheet erosion is a hidden soil loss, since there are no visible signs until the subsoil appears. Rill erosion can also be hidden, because each tillage causes the rills to disappear. The amount of the hidden erosion can be easily underestimated by a grower.

The distinction between regular rill and ephemeral erosion is also important. Tillage does not fill in the ephemeral channel, thus it can act as a "seed" for gully formation. Some soil conservation measures must treat these two forms of erosion differently. The **Universal Soil Loss Equation,** the main tool for estimating erosion rates, predicts only sheet and rill erosion, not ephemeral erosion. Thus, the equation can seriously underestimate soil loss on fields with a great deal of ephemeral erosion.

Gullies are the most dramatic image of soil erosion. However, much more soil is lost over most fields by sheet and rill erosion. Gullies chop a field into inconvenient shapes and add to the sediment load of streams.

FIGURE 18-11 A rainfall simulator being used by research scientists in South Dakota. The booms imitate rainfall and various treatments are used to model soil loss. *(Courtesy of USDA)*

Predicting Soil Loss: The Universal Soil Loss Equation

Using the soil loss factors described, an equation has been developed to predict the average soil loss from sheet and rill erosion on a specific site. The Universal Soil Loss Equation (USLE) was developed over several years from some 10,000 plot-years data from forty-nine test sites around the country (figure 18-11). A grower can use this equation to decide what conservation practices are needed to keep soil losses within acceptable levels. The USLE also helps growers determine the most economical way to preserve the soil.

The USLE does have weaknesses. The equation predicts sheet and rill erosion only. If a slope shows signs of ephemeral or gully erosion, the results from the equation will understate soil loss. The equation predicts the average soil loss over time. A single, highly erosive rainfall may cause far more erosion than is predicted by the equation. Nor does the USLE provide an accurate estimate of soil losses from snowmelt runoff. Also, applications of the equation to some crops, rangeland, and to farmland in some parts of the country have not been fully reliable.

Tolerable Soil Loss. At the heart of the USLE is the assumption that a certain soil loss can be tolerated because soil-forming processes will replace lost soil. The equation can then be used to determine if soil loss exceeds this amount. In virgin grasslands, erosion may amount to an inch every five thousand years or about three pounds/acre/year. In the Palouse area of the Pacific Northwest, erosion levels sometimes reach 50 to 100 tons per acre per year. This is a loss of an inch of soil every one and a half to two years. Between these two extremes, what soil loss can be tolerated?

Soil scientists have decided that soil losses between one ton and five tons per acre per year can be tolerated, depending on quality and depth of a soil. It is assumed that erosion damages a deep soil least. The amount that can be tolerated is symbolized by the letter "T." Each soil series has its own value of T, which is noted on a soil survey report. However, T probably exceeds normal soil-formation rates. That is, even at T, soil is being lost more rapidly than it is being replaced, so many soil scientists maintain that T is not a sustainable loss over the long term and not a reliable guide to the need for erosion-control measures.

The Universal Soil Loss Equation. The USLE is based on a standard test plot representing an average eroded site. This plot has a 9 percent slope 72.6 feet long. The slope is kept in clean-tilled fallow, using conventional tillage up and down the slope. The equation works by comparing a specific spot to this test plot.

The equation reads as follows:

$$A = R\,K\,LS\,C\,P$$

A is the tons of soil lost per acre each year. Obviously, *A* should be less than *T.* To solve for *A,* values are inserted for the six variables and are multiplied. The variables are

- *R*—rainfall and runoff factor. *R* is based on the total erosive power of storms during an average year. *R* depends on local weather conditions. The isoerodent map of figure 18-12 shows *R* values for the United States.
- *K*—soil erodability factor. *K* depends on texture, structure, and organic matter content. Soil survey reports give the value of *K* for mapped soils. They may also be calculated.
- *LS*—slope factor. *L* compares the slope length and *S* compares the grade with the standard plot. *L* and *S* are separate factors, but they can be treated as one variable, *LS. LS* values can be determined from the chart in figure 18-13.
- *C*—cover and management factor. *C* compares cropping practices, residue management, and soil cover to the standard clean-fallow plot. *C* values are calculated from detailed tables and are valid only within the area for which they are calculated. NRCS offices prepare simplified tables for use in the field, and some have computerized the computations. Figure 18-14 is a sample of such a table, and is included here for use in the solution of USLE problems presented in this text. Check local NRCS offices for local charts. If necessary,

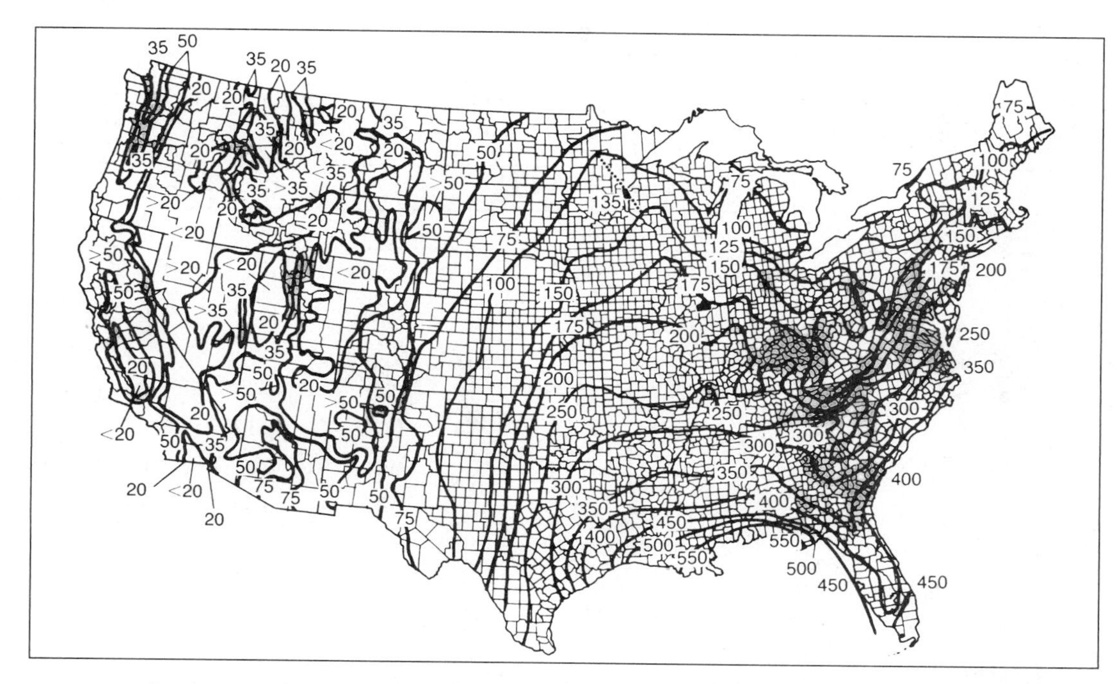

FIGURE 18-12 The isoerodent map gives the average yearly values of the rainfall erosion index, factor *R.* The dotted line is used in examples in this text. *(Source: USDA Handbook 537)*

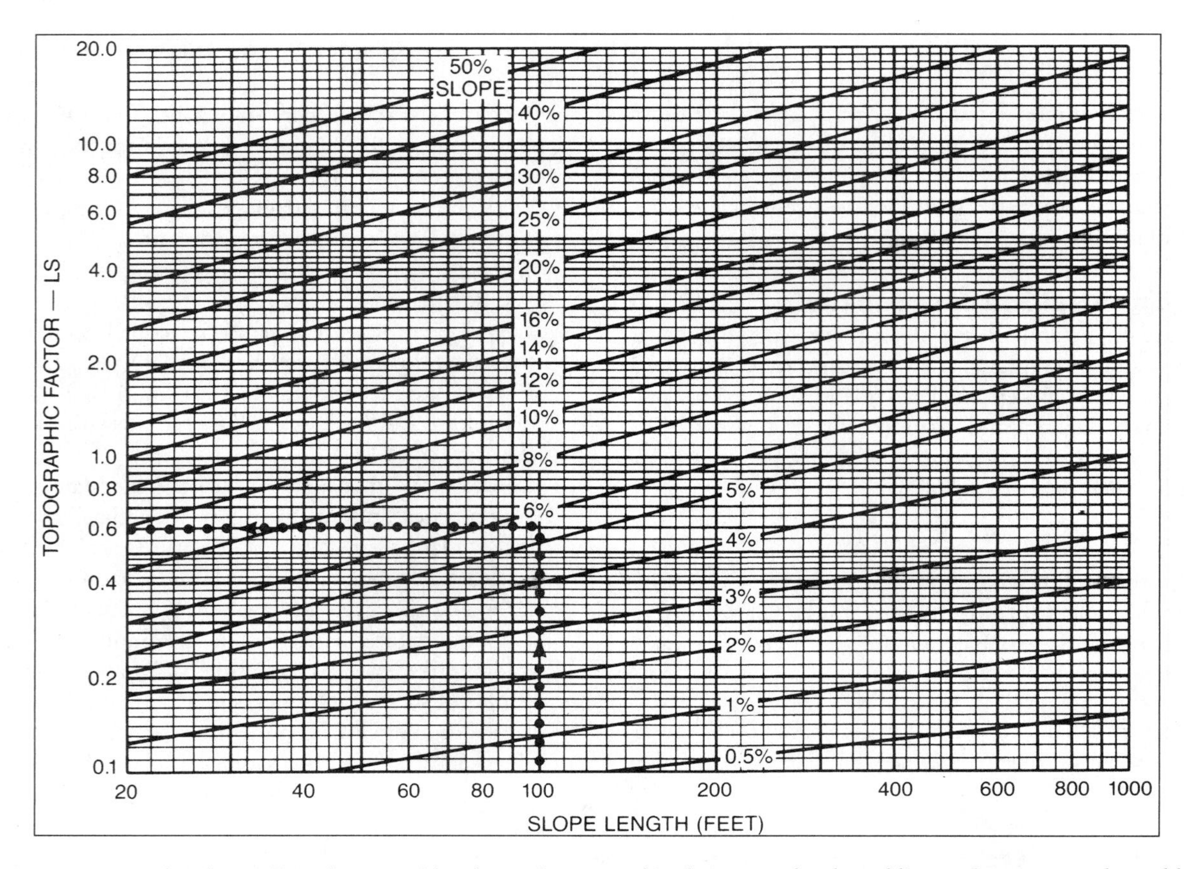

FIGURE 18-13 The slope effect chart provides the *LS* factor used in the USLE. The dotted lines refer to a sample problem in this text. *(Source: USDA Handbook 537)*

	Conventional Plowing		Conservation Tillage		
Crop Sequence	**Fall**	**Spring**	**30%**	**40%**	**50%**
Continuous soybeans	0.48	0.45	0.30	0.24	0.20
Corn, soybeans	0.41	0.37	0.24	0.19	0.16
Continuous corn	0.37	0.36	0.19	0.15	0.12
C, SB, SG	0.32	0.29	0.17	0.14	0.11
C, C, SG	0.30	0.27	0.14	0.12	0.09
C, SB, C, O, M, M	0.20	0.18	0.12	0.11	0.09
C, C, C, O, M, M	0.17	0.15	0.10	0.09	0.08
C, C, O, M, M, M	0.11	0.10	0.08	0.07	0.06
Permanent grass	0.003–0.013				
Grazed forest	0.01–0.04				
Ungrazed forest	0.001–0.003				

FIGURE 18-14 Some crop management factors for southern Minnesota. These numbers apply to the averages of crop rotations using corn (C), soybeans (SB), small grains (SG), oats (O), and meadow (M). The percentages are residue coverage. *(Source: NRCS Minnesota office)*

the values can be calculated (see the reference listed in the enrichment activities section of this chapter).

- P—support practice factor. P compares the effect of contour tillage and contour strip-cropping with the test plot. Figures 18-15 and 18-16 give P values.

Sample Solution. This sample solution of the USLE considers a certain field in mapping unit 158C in figure 3-5. The slope is 2 percent to 6 percent and about 100 feet long. According to the soil survey for this area, $K = 0.18$ and $T = 5.0$. Assume the grower uses conventional tillage to grow continuous corn and plows up and down the slope in the fall. The farm is in east-central Minnesota—the county is shaded on the isoerodent map.

We begin with the formula:

$$A = R\,K\,LS\,C\,P$$

The value of R is read from the isoerodent map. Since the farm location does not lie on a curve, draw a line between the two curves it lies between and estimate the value of its place on that line (see figure 18-12). In this case, $R = 135$:

$$A = (135)\,K\,LS\,C\,P$$

The value of K is given in the soil survey report.

$$A = (135)\,(0.18)\,LS\,C\,P$$

The value of LS can be obtained from figure 18-13. The slope is somewhere between 2 percent and 6 percent; let's use 5.5 percent. Find the slope length on the bottom axis (100) and follow up the chart until it touches the 5 percent slope line. Now go up to where the 5.5 percent line would be if there were one. From this point, move left to the vertical axis. The LS factor is shown as $LS = 0.6$.

$$A = (135)\,(0.18)\,(0.6)\,C\,P$$

The first three factors are fixed for the site—they are not changed by anything the farmer does, short of land

Land Slope Percent	*P* Value	Maximum Slope Length
1 to 2	0.60	400
3 to 5	0.50	300
6 to 8	0.50	200
9 to 12	0.60	120
13 to 16	0.70	80
17 to 20	0.80	60
21 to 25	0.90	50

FIGURE 18-15 Support practice factor P for fields that are contour tilled. The figures are unreliable if slopes are longer than indicated. *(Source: USDA Handbook 537)*

Land Slope Percent	*P* Value			Strip Width	Maximum Slope Length
	A	*B*	*C*		
1 to 2	0.30	0.45	0.60	130	800
3 to 5	0.25	0.38	0.50	100	600
6 to 8	0.25	0.38	0.50	100	400
9 to 12	0.30	0.45	0.60	80	240
13 to 16	0.35	0.52	0.70	80	160
17 to 20	0.40	0.60	0.80	60	120
21 to 25	0.45	0.68	0.90	50	100

FIGURE 18-16 Sample support practice factors for contour strip-cropping. Column A is a four-year rotation of row crop, small grain, meadow, meadow. Column B is row crop, row crop, winter grain, meadow. Column C is alternative strips of row crops and small grain. *(Source: USDA Handbook 537)*

leveling, terracing, or incorporating organic matter to lower K. Simplify the equation by multiplying the three factors:

$$A = (135)\,(0.18)\,(0.6)\,C\,P = (14.58)\,C\,P$$

Refer to figure 18-14 to find the C value under conventional fall plowing in continuous corn, $C = 0.37$.

$$A = (14.58)\,(0.37)\,P$$

The farmer used no support practices, so $P = 1$, meaning it is the same as in the test plot.

$$A = (14.58)\,(0.37)\,(1) = 5.39 \text{ tons/acre/year}$$

The computed soil loss is slightly greater than the acceptable level for this soil. However, since the equation, if anything, understates losses, the farmer should curtail some of the erosion. What practices will lower the erosion level? What if the farmer changes to contour plowing rather than up-and-down plowing? For contour plowing, P changes from 1.0 to 0.5 (figure 18-15); the other factors remain the same:

$$A = (14.58)\,(0.37)\,(0.5) = 2.70 \text{ tons/acre/year}$$

The equation says that changing to contour tillage will cut erosion in half, well below five tons per acre. Changing from moldboard plowing to chisel plowing, leaving a 30 percent cover, would save even more soil:

$$A = (14.58)\,(0.19)\,(0.5) = 1.39 \text{ tons/acre/year}$$

Applications of the USLE. The most obvious way to use the USLE is to predict erosion from a certain field and to help select the best control measures. The example showed how this can be done.

The USLE can also be used to identify erodible lands. Land planners could use this information to prepare use maps based on erodibility. Or officials could identify lands most in need of assistance.

One proposed scheme defines an "erosion index," or EI, as follows:

$$EI = \frac{RKLS}{T}$$

The formula calculates the "native" erodibility of land, such as slope and texture, and excludes management practices. This is divided by T to get a multiple of the acceptable soil loss if farmed without protection. Any value of EI greater than one means this land, if tilled without erosion-control measures, will lose soil more rapidly than T. This index could be used to identify the soils most in need of good management practices.

The Revised Universal Soil Loss Equation. Needed improvements, described earlier in this chapter, in the USLE are incorporated into the **Revised Universal Soil Loss Equation (RUSLE),** first released in 1992 and still being refined. RUSLE follows the basic USLE equation, but has refined factor values from improved data and computer modeling. The new model allows soil-loss calculations on some crops not covered by the USLE, like horticultural crops such as vegetables. The revised equation was designed primarily for computer use.

While RUSLE is currently replacing the USLE for erosion prediction, the older version is retained here because it is more easily described and practiced in a text such as this. The RUSLE requires more field data, like information about crop canopy and surface roughness. However, most basic procedures remain the same.

Even more complex erosion-prediction tools based on extensive computer modeling of erosion processes, such as the Water Erosion Prediction Project (WEPP), are being prepared. These may be very potent tools if validated by experimental data.

Controlling Water Erosion

All methods of controlling erosion are based on one of the following three actions:

- Reducing raindrop impact to lessen detachment. This can be done by growing vigorous crops that fill in the canopy quickly, by leaving crop residues on the surface, by mulching, or by growing a total vegetative cover.
- Reducing or slowing runoff. This lessens detachment by scouring and reduces the amount of soil that can be transported. Avoiding compaction, maintaining organic matter levels, and subsoiling help water infiltrate the soil. Contour practices and conservation tillage both reduce runoff.
- Carrying excess water off the field safely by use of grass waterways or tile outlets.

Conservation Tillage. Conservation tillage sharply reduces sheet and rill erosion. It is the lowest cost conservation method per ton of soil saved and carries other benefits as described in chapter 16. For these reasons, conservation tillage is the most widely accepted Best Management Practice for controlling soil losses.

The effectiveness of conservation tillage depends on a rough soil surface and surface residues, especially during the critical period before the crop canopy fills in. The definition of conservation tillage states that residue coverage be adequate to bring erosion below *T*—30 percent coverage is often taken as a minimum amount. Coverage can be measured directly by stretching a fifty-foot cord, marked every six inches, diagonally across several rows. The number of marks touching a piece of crop residue is the percentage of coverage (figure 18-17). For instance, if thirty-five marks touch crop residues, the coverage is 35 percent. This procedure should be repeated in several parts of the field and the results averaged. Figure 18-18 shows the effect of several residue levels according to the USLE.

Soil Cover. In all situations where soil is worked or exposed, covering the surface with mulch or vegetation reduces erosion. Examples include cover crops between rows of nursery stock, sodding new roadsides or building sites, winter cover crops in row crops, and numerous others.

Crop Rotation. Crop rotation reduces erosion if a close-growing crop like small grains or forages is included. These close-growing crops reduce the detachment and transport energy of water. Also, they improve the soil's physical properties so that water seeps into the soil better. Figure 18-19 shows the effect of several rotations.

Grassed Waterway. A grassed waterway is a shallow, sodded, wide ditch that runs down a slope

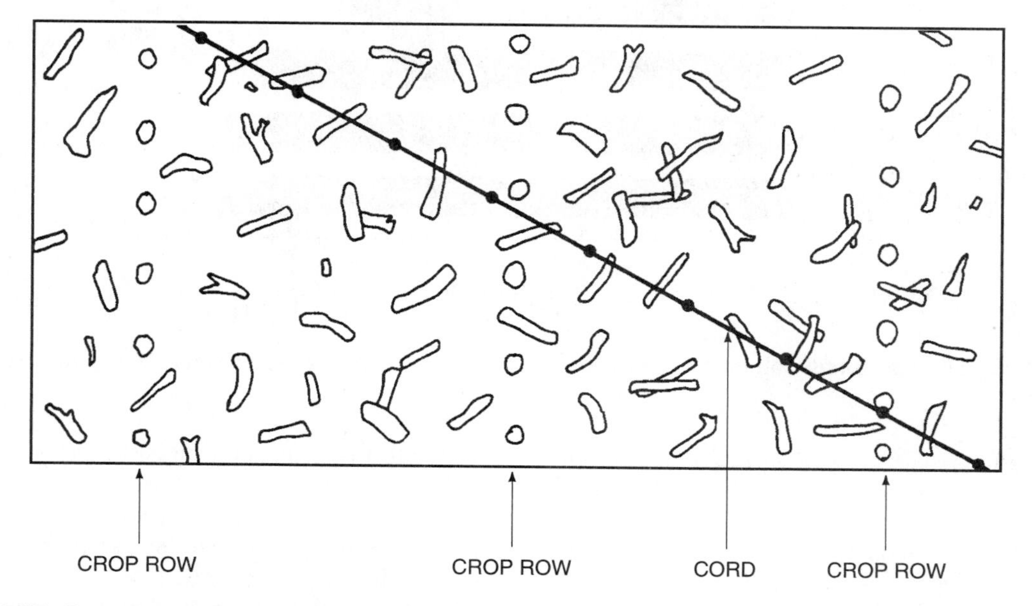

FIGURE 18-17 To measure percent residue cover, a fifty-foot cord marked every six inches is stretched across crop rows. The number of pieces residue touched by a mark is the percent coverage.

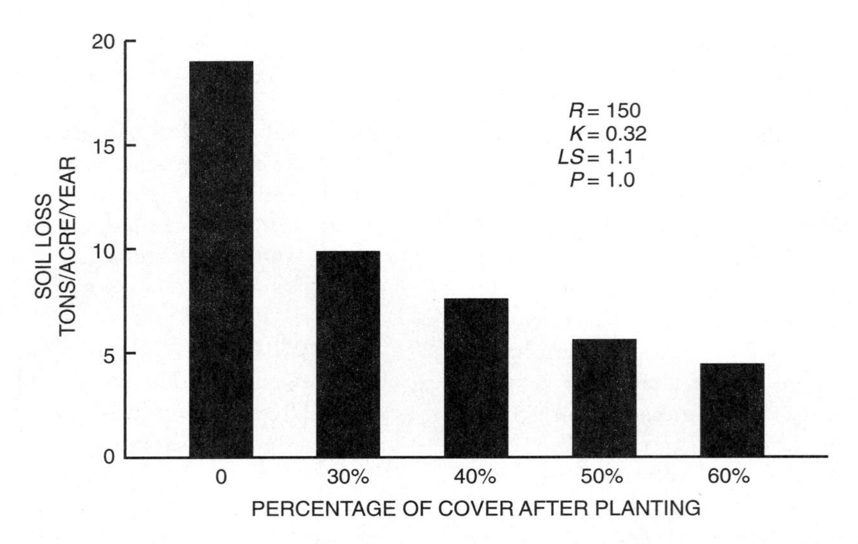

FIGURE 18-18 Effect of several crop residue levels on erosion according to the USLE for continuous corn. *(Source: USDA NRCS, Minnesota office)*

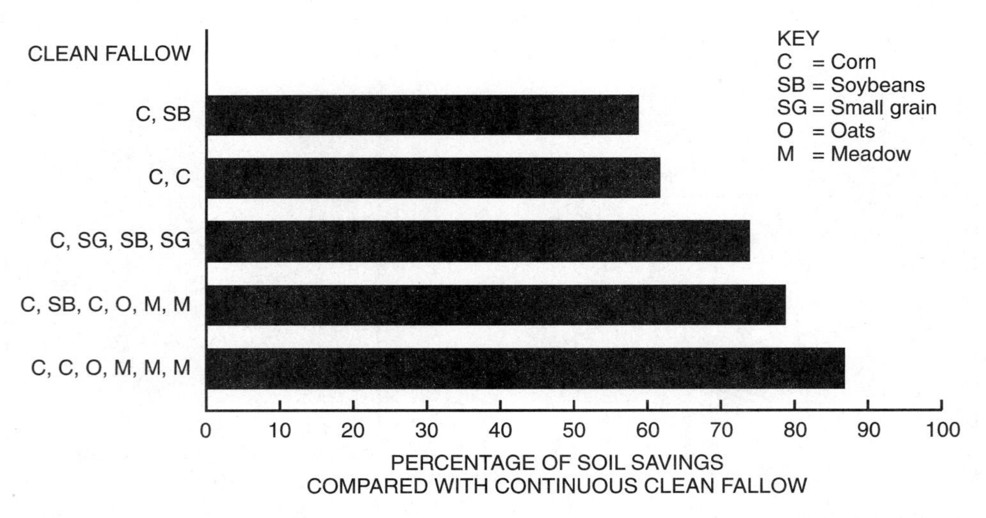

FIGURE 18-19 Effect of several crop rotations on soil erosion. Other USLE factors remain constant. *(Source: USDA NRCS Minnesota office)*

(figure 18-20). It is designed to carry excess water off the field safely. Grassed waterways serve several purposes:

- Waterways prevent gullying where water naturally gathers on a slope.
- Waterways can be used to collect excess water from tillage contours.
- Waterways serve as outlets for terraces.

Small waterways may be built with grading tools mounted on farm equipment. Larger ones require construction equipment to grade the shallow ditch. Grassed waterways must be carefully maintained and fertilized. Equipment should not be turned on a waterway, and tillage tools should not be dragged across it. Any damage should be repaired immediately with sod.

FIGURE 18-20 This grassed waterway collects runoff water and carries it safely into a farm pond. Contour strips empty into the waterway. *(Courtesy of USDA NRCS)*

Contour Tillage. Contour tillage works best on permeable soils in areas of low intensity rainfall. Generally, moderate slopes suffering from rill erosion benefit most from contour tillage (figure 18-21). Simple cross tillage is inadequate if the slope has much ephemeral erosion. Cross tillage does not erase ephemeral rills, and the remaining depressions act as channels for water flow. Therefore, land with ephemeral erosion may need to be smoothed by more than simple tillage before it can be contoured.

Where runoff is not too great, contour tillage can stand alone. It is often helpful, however, to gently slope contours toward a waterway. Water that might otherwise overtop the ridges can flow toward the waterway and be carried off the field.

To establish contours, one surveys a guideline across the slope. The line is either parallel to the slope or dips gently toward a waterway. A new guideline is added when the slope changes. When plowing the field, begin plowing at the guideline and plow parallel to it.

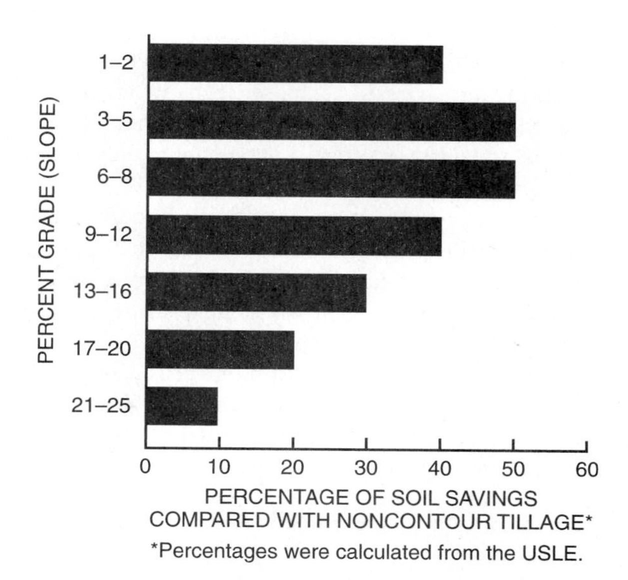

FIGURE 18-21 Effect of contour tillage on soil erosion for several grades. Other USLE factors remain constant. Contour tillage works best on moderate slopes of 3 percent to 8 percent. On steep or long slopes, water overflows the ridges and wears them away.

Strip-Cropping. Strip-cropping (figure 18-20) can be used in all conditions along with contour tillage.

FIGURE 18-22 Contour buffer strips and grass waterways on an Iowa farm. *(Courtesy of USDA)*

FIGURE 18-23 Terraces and no-till farming work to control erosion on this farm in Iowa. *(Courtesy of USDA)*

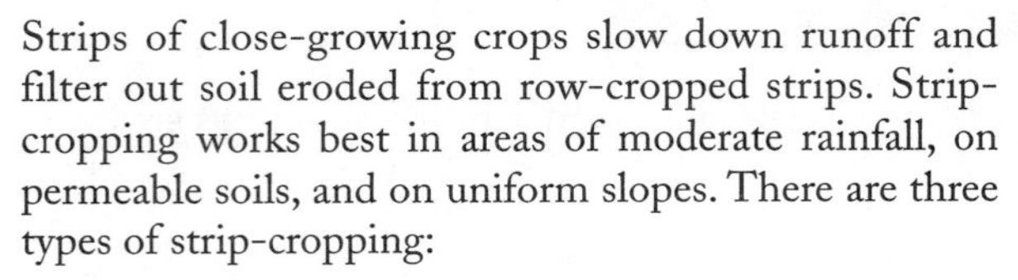

Strips of close-growing crops slow down runoff and filter out soil eroded from row-cropped strips. Strip-cropping works best in areas of moderate rainfall, on permeable soils, and on uniform slopes. There are three types of strip-cropping:

- Buffer strips (figure 18-22) correct an eroded area by planting it in grass. Grass buffer strips are much narrower than the cultivated strips between them. Buffer strips, as narrow as fifteen feet, are not as effective as the other types.
- Field strips are placed straight across a slope, but they may not follow the actual contour. Where slopes are irregular, it is difficult to design strips that follow the contour exactly. In such cases, field strips may be more practical than contour strips.
- Contour strips follow the contour and may empty into a waterway. Where the slope changes often, contour strips are more difficult to establish than field strips, but they are more effective.

As in contour tillage, guidelines are needed to plow for strip-cropping. Strips of equal width are placed across the slope and are planted to alternating row crops, small grains, and meadow. Strips vary in width from 50 feet for steep slopes to 130 feet for gentle slopes. Nurseries often alternate strips of stock and cover crop.

Improving Organic Matter. Improving organic matter can greatly reduce erosion because moisture will seep into the soil more quickly. Growers who make an effort to increase soil humus benefit by reduced erosion. Review chapters 6 and 15 for details on improving organic matter by such methods as cover cropping, manuring, conservation tillage, and others.

Terraces. Where strip-cropping cannot halt erosion, terraces may be built (figure 18-23). Long or steep slopes on impermeable soil, for example, require terraces. Terraces are costly to install and are used most commonly for valuable crops like the nursery in figure 1-12, or where there is a shortage of good land. In general, two kinds of terraces are used:

- Level terraces parallel the slope and do not empty into a waterway. This type of terrace is used where soil is permeable enough so water can seep in once captured in a terrace.
- Graded terraces are needed where water cannot soak in enough. These may slope gently towards a waterway or be drained by an underground tile outlet.

Several terrace designs are shown in figure 18-24. Of these designs, the broad-based terrace is most common. Terrace construction begins by designing them to fit

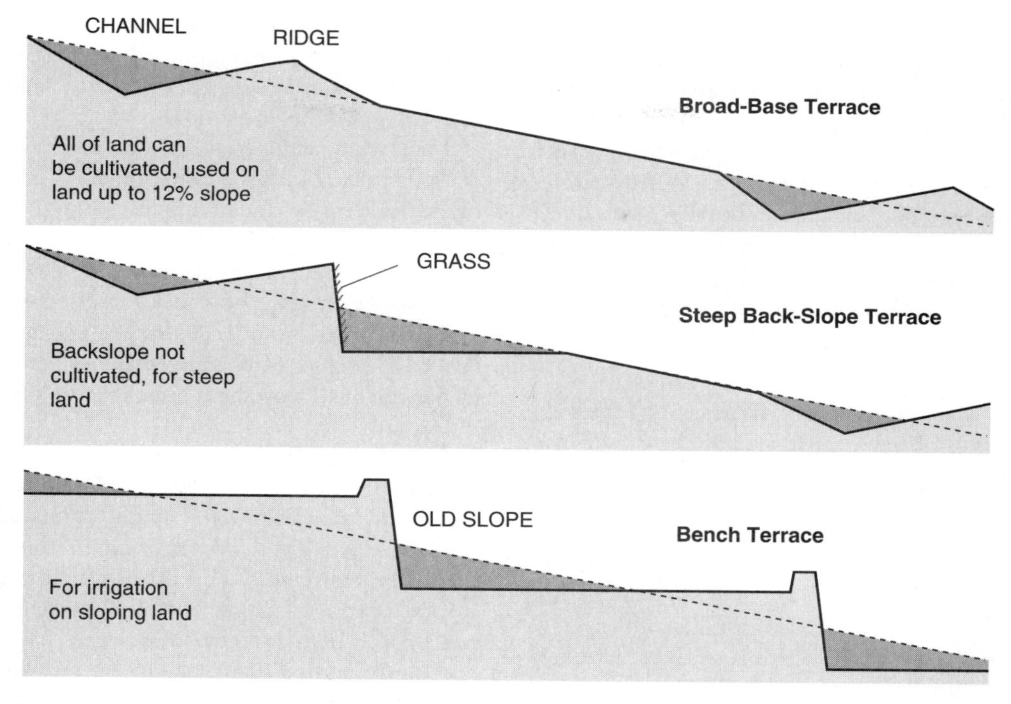

FIGURE 18-24 Each terrace design serves special purposes. The broad-based design is most common. Scale is exaggerated in these drawings. Broken lines show the old grade.

conservation needs without overly hampering farming. The land is surveyed and the terraces are marked on the slope.

Growers must be careful to maintain terraces. Obviously, tillage follows the terrace. If growers use a moldboard plow, they must plow correctly. The moldboard throws soil off to the side, so can change the terrace. Some growers who purchase new, larger equipment fail to maintain terraces properly because the terraces are in the way. Without proper care, terraces cannot be effective.

Terraces are accounted for in the USLE by the *LS* factor. Terraces break up a slope into several shorter slopes. In solving the USLE, the length of a single terrace is taken as the length of the entire hill. Depending on terrace design, the slope factor may or may not change. It is assumed that most of the eroded soil within the cropped part of a terrace is deposited in the terrace channel, and that a smaller amount leaves the field in waterways or outlets. For the *P* factor, use the contour tillage or strip-cropping value.

Diversions. Diversions are large-capacity terraces that divert runoff from higher elevations. Diversions are not farmed but are covered with grass. Their uses include:

- protecting fields from runoff flowing from higher elevations.
- diverting water away from active gully heads.
- diverting water from feed lots, farmsteads, or other sensitive areas.
- in residential housing, diverting water away from homes.

Filter Strips. **Filter strips** are zones of dense vegetation between a source of runoff, like a farm field, and a receiving body of water, like a stream or lake. It primarily traps sediments and farm chemicals from runoff before they can enter the body of water and so addresses off-site damage of erosion.

Filter strips, an important BMP for protecting surface waters, range from ten to forty feet wide, with the larger widths needed for steeper slopes, higher soil clay contents, and larger areas collecting water. They should be used with other in-field erosion control measures like conservation tillage. Cool-season sod-forming grasses perform the best, where adaptable, because they form a dense cover early in the year. The grass is not mown. Warm-season grasses, native vegetation, and dense shrubbery are variations. Lakes may also be protected from lawn runoff by filter strips along the shoreline, often decorative masses of native or nonnative flowers and ornamental grasses.

Wind Erosion

Wind erosion accounts for about 40 percent of the soil loss in the United States, mostly in the Great Plains states. Other areas with wind erosion problems include muck and sandy soils of the Great Lakes states and Atlantic seaboards. Figure 18-25 shows amount of wind erosion for each state in average tons per acre per year for cultivated land.

Dry areas with high winds are most likely to experience wind erosion. At greatest risk is soil kept bare by clean-till summer fallow.

Cause of Wind Erosion. Figure 18-26 shows the effect of wind blowing across a bare soil. A very thin layer of still air covers the soil surface, but larger soil particles stick up above the layer. When wind reaches 10 to 13 miles an hour at a height of one foot above the surface, soil grains begin to move.

First, wind begins to roll soil grains in the size range of 0.004 to 0.02 inches (0.1 to 0.5 mm), fine to medium sands. Suddenly a sand grain jumps straight into the air, rising as high as twelve inches. Wind blows the sand grain several feet. The grain strikes the ground, where it may bounce up again or knock loose some other particles. This process is called **saltation** and causes 50 percent to 75 percent of all wind erosion. In fact, more than 90 percent of all movement occurs within one foot of the soil surface.

Very fine silt and clay particles are too small to be picked up by the wind. However, the impact of a sand grain moving by saltation may knock dust into the air. Once the wind hits it, dust rises high into the air and is carried long distances. This process, called **suspension,** accounts for about 3 percent to 40 percent of all wind erosion. Silt particles move most easily.

Coarse sand particles, ranging in size between 0.02 and 0.04 inches (0.05 to 1.0 mm), are too large to be kicked into the air. However, under the impact of saltating sand grains, they can roll along the ground. This is

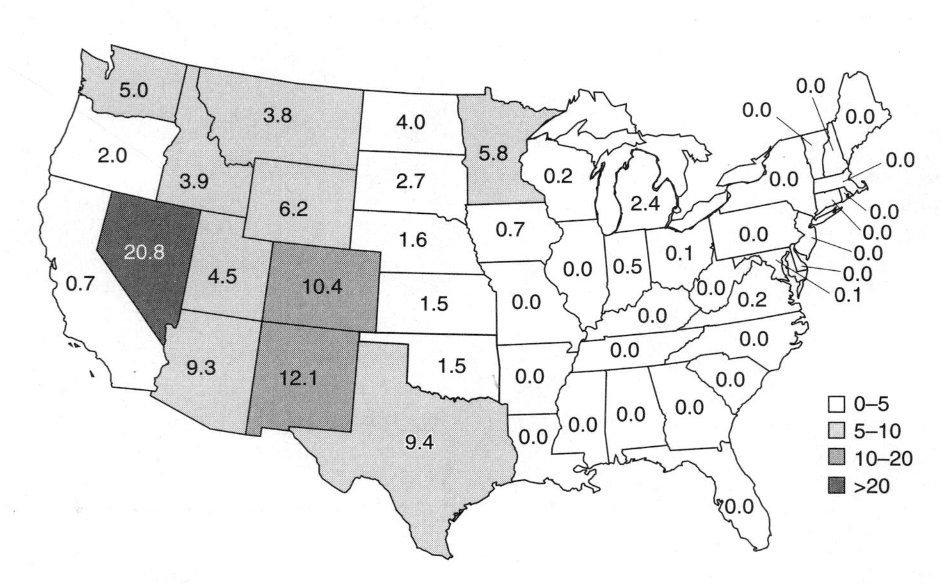

FIGURE 18-25 Average annual wind erosion by state on nonfederal cultivated land, tons/acre/year. *(Source: USDA National Resource Inventory, 1997)*

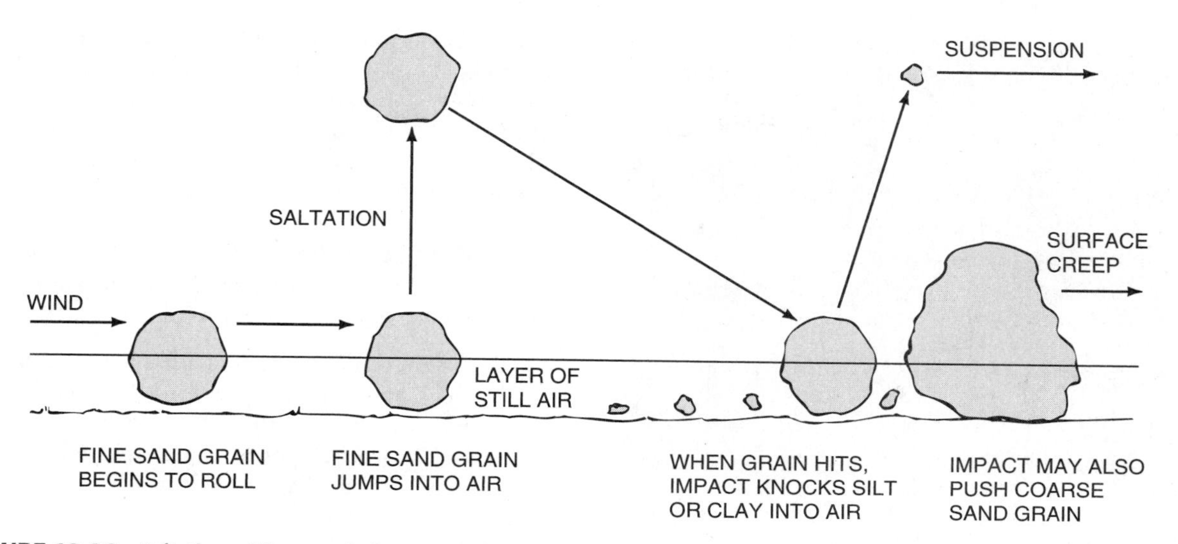

FIGURE 18-26 Saltation of fine sand triggers wind erosion. Fine sands are large enough to protrude above the layer of still air but small enough to be picked up by the wind. Surface creep and suspension both depend upon the impact of fine sand in saltation.

known as **surface creep.** It accounts for 5 percent to 75 percent of wind erosion.

Like water erosion, the detachment and transport of soil particles by wind are functions of energy. The higher the wind velocity, the greater the energy. Blowing soil itself contributes to erosive energy. A soil grain has greater mass than air, thus the impact of a soil grain carries more energy. As a result, wind erosion has an avalanche effect—as wind blows across an open field, more and more soil is picked up as the erosive energy increases.

When wind velocity dies down, so does its energy. Soil particles in saltation or suspension come to rest when the wind dies down. They may also fall out on the lee (downwind) side of an obstruction, the way snow gathers on the lee side of a snow fence (figure 18-27).

FIGURE 18-27 Windblown soil is deposited on downwind side of obstacles. Therefore, it will pile behind windbreaks, snowfences, soil ridges, crop strips, and in ditches such as here in Iowa. *(Courtesy of USDA)*

Effects of Wind Erosion. Like water erosion, wind erosion removes the best soil first—the topsoil. It carries off fine soil particles, especially silt and organic matter. The shift toward a coarser soil texture reduces nutrient-holding and water-holding capacity. Further, windblown soil particles "sandblast" young plants (figure 18-28), tattering leaves and tearing away plant cells. Young plants can easily lose half their dry weight when exposed to sand-laden winds. Off-site damages can also be severe—and expensive. Blowing soil can fill road or drainage ditches (figure 18-27), affect the respiratory health of animals and people, increase cleaning and laundry costs, and wear at paint and other surfaces.

FIGURE 18-28 Windblown soil "sandblasts" young crop plants. *(Courtesy of USDA NRCS)*

Factors in Wind Erosion. The following factors determine the amount of wind erosion.

- Soil erodability relates mainly to texture and structure. Soils high in fine sand and coarse silt are most liable to wind erosion; soils high in clay least liable. Drained organic soils are also easily eroded by wind. Soil grains cemented into soil aggregates are less likely to be blown away.
- Soil roughness makes a larger still air layer at the soil surface. Each clod or ridge also acts like a tiny windbreak to slow wind and capture blowing soil.
- Climatic conditions that promote wind erosion include low rainfall, low humidity, high temperatures, and high winds. Dry, windy conditions cause faster soil drying, and dry soil is more erodible than moist soil. Dry soil also supports a thinner vegetative cover.
- Length of field affects erosion. On the leading edge of a field, there is no wind erosion. As the wind travels across the field, it picks up more and more soil grains, like an avalanche.
- Vegetative cover protects the soil, as does a mulch. Bare soil, on the other hand, is fully exposed to the erosive force of wind.

These factors together can be arranged to create a soil loss formula similar to the USLE called the **Wind Erosion Equation,** or WEQ. Since WEQ is more complex to apply than the USLE and generally considered less reliable, it will not be detailed in this text. Like the USLE its limitations are being addressed in a new computer-driven model, the Wind Erosion Prediction System (WEPS).

Preventing Wind Erosion. Preventing wind erosion means reducing the factors listed previously. The following practices help control wind erosion:

- Till at right angles to the wind, leaving the soil surface rough and cloddy (figure 18-29). Lister furrows are very useful because they act as small windbreaks, called **cross-wind ridges,** to capture blowing soil.
- Use conservation tillage or subsurface tillage tools, such as rodweeders and subsurface sweeps, to leave crop residues on the soil surface.
- Keep the soil covered with vegetation as much as possible. Cover crops of winter grains work very well to protect the soil over winter.
- Plant crops in strips at right angles to the wind. Strips of soil in summer fallow may be partially protected by alternate strips of small grains or row crops.
- Plant windbreaks of trees and large shrubs. Windbreaks shorten the field, reduce wind velocity, and capture blowing soil (figure 18-30). Windbreaks should not be solid; they should block about 50 percent of the wind.
- Plant buffer strips as temporary windbreaks. For instance, tall wheatgrass barriers planted north-south, about 50 feet apart, reduce erosion by about 93 percent.[7] The strips also capture blowing soil, so are called **cross-wind trap strips.**
- Plant the most critical areas to permanent grasslands or other vegetative cover.

Critical Periods. Two periods are critical for wind erosion. The first is when the soil is fallowed. Large clean-fallow fields in the Great Plains are ideal grounds for wind erosion. Subsurface tillage tools that leave crop residues on the surface reduce this problem. Instead of controlling weeds by tillage, herbicides can be used to kill weeds while leaving crop stubble standing to protect the soil from wind.

[7] Aase, J. et al. 1985. Effectiveness of Grass Burness for Reducing Grass Erosion. J. Soil Water Cons. 40(4): 354–357.

FIGURE 18-29 Creating surface roughness and cross-wind ridges in Texas. This will help reduce wind erosion. *(Courtesy of USDA ARS)*

FIGURE 18-30 Field windbreaks in North Dakota protect the soil against wind erosion. *(Courtesy of USDA)*

The period between harvest and the following crop cover is the critical period in many areas. A good snow cover protects soil, but if the snow blows off the field, so will soil. Again, conservation tillage and winter cover crops protect soil during the critical period.

SUMMARY

About three-fifths of the soil lost to erosion in the United States washes away in running water. Water erosion strips the topsoil, reduces yields, and deposits sediments in streams, lakes, and reservoirs. About two-fifths of erosion in the United States results from wind. Wind also strips the topsoil, blows away the smallest soil particles, and buries ditches and other structures.

Falling raindrops and running water detach soil particles from the soil surface and carry them away. Depending on the slope, erosion removes soil as a sheet or creates rills and gullies. Water erosion is promoted by bare and erodible soil, long or steep slopes, and the lack of conservation practices.

Soil scientists use the Universal Soil Loss Equation to compute soil water erosion loss. The USLE accounts only for losses from sheet and rill erosion and will understate soil loss where there is ephemeral or gully erosion. Using the USLE, a specialist can suggest practices to keep a farm productive.

Growing vigorous crops, maintaining organic matter, and avoiding overtillage and compaction help to control erosion. Both conservation tillage and crop rotation sharply curb erosion. Contour tillage, contour strip-cropping, and terraces are effective ways to slow runoff. Where these are not enough to stop runoff, they may be combined with grassed waterways or outlets to carry the excess off the field without erosion.

Wind blows soil off fields by saltation, suspension, and surface creep. Wind erosion is most likely to occur on soils that are high in fine sand or on organic soils, in hot, dry, windy climates, and where soil is kept bare, especially for summer fallow. Control practices include breaking the wind, keeping the soil rough, and planting at right angles to the wind. The most effective method is keeping the soil covered by vegetation or crop residues.

REVIEW

1. Discuss the types of water erosion.
2. Discuss the types of wind erosion and the types of soil particles associated with each.
3. Speculate how climate change predicted by many scientists might affect erosion rates (hint: consider the USLE and WEQ).
4. You operate a farm in Maclean County, Illinois (indicated in figure 18-12). For the soil, a Saybrook, K = 0.32, T = 5.0. One field has a slope of 4 percent and is 400 feet long. You grow a corn-soybean rotation. Use the C factors of figure 18-14. Calculate average annual soil loss for these two practices and compare that rate to T:
   a. fall-plow with a moldboard plow (conventional plowing) up and down the slope.
   b. contour chisel, leaving 40 percent cover.
5. Land in the Conservation Reserve Program is usually planted to permanent grass cover. What would be the average annual erosion rate on the field in question 4 under permanent grass? How does this compare to the above?
6. Calculate the erosion index (EI) for the soil in question 4. How would you rate this soil's erodibility?
7. How might you prepare an interpretive soil map rating soils as to their water-erosion hazard?
8. Describe practices that might keep sediments and chemicals from farm fields out of nearby bodies of water.
9. Surface roughness affects both water and wind erosion. How does it affect each? How might we use roughness to reduce erosion?
10. How much sediment is delivered to the rivers in your part of the country? Knowing your area and the information in this chapter, explain this amount in broad terms. What part of the country has the biggest problem with sedimentation? What major river basin is this? Can you guess why this area is so prone to erosion and sedimentation? This question requires viewing the Internet site <http://www.nrcs.usda.gov/technical/land/meta/m2087.html>.

ENRICHMENT ACTIVITIES

1. Visit growers in your area who use different conservation measures.
2. Practice the use of the USLE to land in your area using charts in this chapter and local soil survey reports.
3. This chapter concentrates on erosion problems in the United States. Investigate erosion in some other country by typing "soil erosion and [name of country]" into your browser. Write a short report.

4. For thousands of years, in many parts of the world like Chile, the Philippines, and Nepal, steep mountainsides have been made usable for agriculture by terracing. Look for images of these mountainsides by searching on the Internet. Many such images are travelogue photographs. Describe the terraces.
5. Research one soil conservation practice you think most useful in your area using the Internet or other resources. Prepare a short report that provides more detail than provided in this chapter.
6. There are excellent images and descriptions of types of soil water erosion and erosion-control practices at: <http://topsoil.nserl.purdue.edu/nserlweb/weppmain/overview/intro.html>.
7. A variety of useful maps and tables on soil erosion is available from the USDA at: <http://www.nrcs.usda.gov/technical/land/>.

CHAPTER 19

Urban Soil

OBJECTIVES

After completing this chapter, you should be able to:

- list characteristics of many urban soils
- describe ways of dealing with urban soils
- describe modified and structured soils
- describe erosion control on urban sites
- discuss zero-runoff concepts

TERMS TO KNOW

bioswales
bioremediation
brown fields
debris basins
hydroseeding
retention ponds
recharge basins
rip-rap
silt fences

Urban soils, sometimes identified as "developed land," are those soils found within a city, town, or metropolitan area. The USDA reports that there were 98 million acres of developed land in 1997, an increase of 20 million acres since 1982. The problems and traits of urban soils demand a discussion in a soil text because they continue to grow as a fraction of our land surface and because most Americans—indeed, most of the world's population—inhabit such land.

If we add people to the standard list of soil-formation factors, as discussed in chapter 2, then the most heavily human-influenced soils are these urban soils. This chapter describes characteristics of urban soils and makes suggestions for managing them.

Characteristics of Urban Soils

Urban soils present certain difficulties to landscapers, gardeners, building contractors, and others who use them. Compared to rural soils, urban soils have been greatly altered by construction, excavation, contamination, and other activities. Earth has been moved from site to site, grades and drainage patterns changed, foundations dug, and debris and contaminants left behind.

Several characteristics of urban soils have been proposed by researchers. These include:

- Great soil variability, both vertically in the soil profile and across the urban landscape.
- Severely modified soil structure from soil moving and stockpiling, often with a very high degree of soil compaction.
- Modified soil pH caused by buried debris like concrete (which raises pH) and contaminants.
- Restricted aeration and drainage. Sometimes urban soils are also hydrophobic—that is, they repel water so it runs off the surface instead of being absorbed.
- Interrupted nutrient cycling because contamination and other changes alter biological processes. Mycorrhiza levels may be reduced, and in some urban soils, very high earthworm populations alter recycling processes in the soil.
- Elevated soil temperatures from the "heat island" effect of cities.
- Presence of substances added by humans, including solid debris and chemical contaminants.

A few of these will be examined in detail next.

Soil Variability and Classification. Much of the surface of an urban area has experienced earthmoving and "cut and fill" operations. In construction of typical suburban tracts, topsoil is removed and stockpiled, soil is moved about by bulldozers, roadbeds and foundations built, and finally the topsoil may be returned as a thin topdressing on the yards. Large-scale commercial building causes even more severe changes.

As a result, urban landscapes show extreme soil variability. Across the landscape, soils may vary greatly over a space of a few feet. Strange soil profiles also cause vertical variation, as shown in figure 19-1. This variation complicates the use of urban soils; even soil testing presents a problem when one corner of a yard may be wholly different than another.

Trying to survey and classify such variable soil presents great challenges. Many soil surveys label areas of developed soils simply as "Urban complexes." In soil taxonomy,

FIGURE 19-1 "Soil profile" of a city parking lot shows pavement over sand. After the pavement was removed, three inches of "black dirt" were spread over the sand and sod was planted.

urban soils may be classified as Entisols, being, in a sense, new soils in which soil-forming processes are starting over.

In the 1990s, the New York City Soil Survey Program examined 320 acres of parkland with the detail needed for such a variable landscape. To do this, five new soil series had to be named to identify human-constructed landforms. Such a detailed survey will provide valuable data for the proper use of that parkland.

Buried Debris. Urban soil usually has a lot of debris buried in it. During construction of a building, contractors may bury wood or masonry scraps on the site, rather than haul them away. Buried rubble may retard drainage or even cause excessive drainage. It can be a physical barrier to root growth. Buried masonry, which contains lime, can raise pH to unacceptable levels. The debris is also a constant source of frustration to those trying to work the soil. Recognizing this problem, many suburbs now require builders to remove debris from construction sites.

Compaction. Urban soils are usually moderately to severely compacted from the use of heavy equipment on the soil during construction, vehicle parking, and maintenance equipment. Countless footsteps on yards and parks also cause compaction. Even vibration from nearby roadways settles soil. Compaction from construction may kill sensitive trees like oaks, cause greater erosion, and make it more difficult to establish a landscape. Compaction promotes the growth of compaction-tolerant weeds in yards and makes it difficult to establish good turf.

Compaction can be measured by bulk density. Chapter 1 mentioned that an "ideal soil" is about 50 percent solid particles and 50 percent porous space. The bulk density of such a soil is about 1.3 grams per cubic centimeter. As bulk density rises above 1.4, root growth begins to suffer from lack of air and direct physical resistance. At a bulk density of 1.7, roots cannot penetrate the soil. The people in charge of landscaping around the nation's Capitol measured some bulk densities on the Capitol Mall ranging between 1.8 and 2.2.

Many fine upland trees like sugar maple thrive only on soils of high air-filled porosity. Many of our best urban trees hale from floodplains, where trees are adapted to soils often low in oxygen. Appendix 6 lists responses of many tree species to soil compaction.

Planning can avoid some compaction. For instance, on new construction sites, limit the area driven on by construction equipment. Landscape architects and designers can help control pedestrian compaction by remembering that people walk the shortest route between two points. Designers should carefully analyze expected foot traffic patterns on a new site, and install sidewalks or mulched paths where people can be expected to walk.

Although difficult, it is possible to break up compaction if no plants are in the way. Deep tillage will break up compaction. The soil can then be heavily amended to stop further compaction. Digging large, solid particles like calcined clay into the soil helps by creating a "skeleton" that resists compaction. Large amounts of organic matter like wood chips or leaf mold also help. Where heavy foot traffic is expected, a deep layer of wood chips cushions the soil.

Another method of dealing with foot traffic is to pave the soil with special pavers and grids that have large holes built into them (figure 19-2). Grass can grow in the spaces, giving the impression of turf, yet the grid protects the soil. Water and air can also move through the spaces.

If trees or turf already occupy a site, compaction is more difficult to repair without hurting roots. Machines called aerators remove vertical cores from the soil. This process helps break up compaction. For turf, vertical coring to six inches breaks up the soil and makes passages for air and water movement. Machines that remove a core should be used, and the cores should be left on the surface as a topdressing. For trees suffering from the effects of compaction, coring to eighteen inches is needed.

Soil Contamination. Soil contamination may severely impact the usability of some urban soils. These contaminants may include heavy metals like cadmium or lead, deicing salts, or industrial and home wastes like paint thinner and solvents. Severely affected land may require massive reclamation before use, such as stripping and replacing soil.

DEICING SALTS. In northern states, roadway deicing salts may be sprayed some distance by passing vehicles and snowplows. The most common deicer, common salt (sodium chloride), is used in huge amounts. Sodium chloride creates saline or even sodic soil conditions, and

FIGURE 19-2 This porous pavement system protects soil from compaction while allowing the planting of turfgrass. *(Courtesy of Presto Products Company)*

chloride may reach toxic levels. Salt spray also damages plant tissue directly, especially evergreens, by desiccation. The author has noted flower buds of lilacs killed by salt spray and observed thousands of junipers killed in a new roadside planting.

The degree of damage from road salts is proportional to the distance from the roadway and traffic levels on that roadway. Thus plantings can be designed to minimize problems. One answer is to plant salt-tolerant species. Appendix 6 provides some guidance.

Common roadsalts may be replaced by less harmful materials like calcium chloride or calcium magnesium acetate (CMA), but these may be less effective in the coldest zones and more costly. Urea and sand mixtures are also used in warmer weather. In a homeowner's yard, heavy irrigation may leach salts from the root zone.

HEAVY-METAL CONTAMINATION. Heavy metals, like cadmium or lead, may be present in some soils from the parent materials. In urban settings, they tend to rise to high levels because of settling of urban dust and air pollution, as well as contaminants that enter soil like paint chips or industrial waste. Many soil amendments and fertilizers contain low amounts of heavy metals.

Heavy-metal soil contamination is an increasing problem worldwide and is most severe in urban settings. Heavy metals are toxic to plant roots, animals, and people.

Lead contamination is the most severe example in urban soils. In the soil survey of a New York park mentioned earlier, soil levels ranged from 6 parts per million to 2,003 parts per million, showing lead distribution is quite patchy. Lead is a highly toxic metal that was once added to paints and to leaded gasoline. Such uses have been phased out in recent years, but lead remains in urban soils. It primarily affects children, causing learning, memory, behavioral, and other problems. It has been shown that lead blood level in urban children correlates with lead levels in surrounding soils.

Much lower lead levels are measured in the blood of inner-city children since lead was removed from gasoline in the 1970s and 1980s. However, urban children still acquire lead by ingesting contaminated soil or paint chips or by breathing airborne dust. In old, especially run-down neighborhoods, chips of lead-based paint may be eaten by infants and young children. Paint chips may also contaminate soil around a building.

The following suggestions may be useful in controlling health problems from lead:

- Keep yards covered by a good stand of turf to prevent children from playing in contaminated soil to avoid ingestion of lead on dirty hands. Sandbox sand can be changed yearly.
- Cleaning children's hands, keeping the house free of dirt, and other cleanliness measures lower the amount of contaminated soil ingested or breathed by children as dust in the air.
- Remove chipping lead-based paint from old homes, clean up thoroughly, and repaint with new paints.
- Have a child's blood tested for lead levels in high-risk neighborhoods. These include areas with old, run-down housing and areas near heavy traffic.

Many home gardeners are concerned about lead in their garden soil. The biggest risk is the direct ingestion of soil by children. The other, much lesser risk, is absorption of lead by garden crops, which are then eaten by the family. In some cities, it may be possible to have garden soil tested for lead.

If a gardener suspects a lead problem, a number of measures may be taken. One could dig out the old soil to a depth of six inches, then build an elevated bed with landscape timbers or by other means. If the bed were raised six inches above the grade, this would give a total of twelve inches fresh soil. This depth would contain most plant roots. A second measure is to keep the soil pH near neutral, since lead is much more soluble in acid soils. It is also possible that high levels of organic matter from compost might form a complex with lead, tying it up.

Brown Fields. Brown fields are so severely damaged by human abuse that the land is unusable without costly abatement efforts. The Environmental Protection Agency defines **brown fields** as "abandoned, idled, or underused industrial or commercial facilities where expansion or redevelopment is complicated by real or perceived contamination." Examples include land polluted by industrial chemicals, numerous piles of waste, or extensive debris burial.

Brown fields present chances for redevelopment in cities and for slowing of urban sprawl by opening more land in older parts of metropolitan areas. However, costs may be high, and cleanup is unlikely to occur without government aid. Abatement begins with a site assessment, looking at historical records and sampling the site. Teams of soil scientists, chemists, geologists, and engineers may be involved. A site plan is then produced, and the plan executed. Economic development can then follow.

A site plan could include **bioremediation**—the use of living things to reduce pollution. For example, the area could be seeded with certain soil bacteria that break down organic pollutants like spilled oil. Some plants take up high amounts of heavy metals or other pollutants; they can be grown to remove the pollutants from the soil, then removed and disposed of.

Modified and Structured Soils

While in residential neighborhoods soil is healthy enough to support the growth of trees, turf, and other landscape plants, the same is not true in the urban center. The "downtown tree" has an average life expectancy of ten years because of soil and environmental extremes.

The downtown tree is often planted in a small soil pit or elevated planter (figure 19-3), with an extremely restricted root zone and surroundings covered by pavement.

FIGURE 19-3 A downtown tree in St. Paul, Minnesota. The only place for roots to grow is in the compacted base under the pavers.

FIGURE 19-4 These street trees in downtown St. Paul were planted at the same time. On the street side, the trees are stressed by growing in small tree pits. On the other side, trees thrive in some open soil.

Below the pavement is a "soil" base severely compacted to support the pavement. This treatment severely restricts aeration and raises mechanical resistance to root growth. Trees that do survive often do so by finding better soil on the other side of the pavement, but in so doing, roots may buckle the pavement and so are cut off. The difference between trees growing in tree pits and open soil is dramatic (figure 19-4). Similarly, turf and other landscape ornamentals perform better in improved soils.

These problems can be answered by either heavy modification of existing soil or by creation of a wholly new structured soil, depending on the situation. In areas of relatively open soil exposed to extreme compaction, like boulevards, a modified soil may suffice. In paved areas, a structured soil is required.

Modified soils use regular soil with additional treatments. While actual projects need a lot of expertise, we can make suggestions here. Usually subsoil compaction should be broken up, then an A and B horizon built with modified soil, and finally the soil slightly compacted to prevent later settling.

A modified soil consists of soil with a large amount of inorganic or organic amendments mixed in. Inorganic materials include coarse sand larger than 0.25 mm in size (never fine sand), perlite, cindered fly ash, and others. These amendments are added in enough amounts that the grains touch each other to form little bridges that prevent compaction and to create large numbers of large pores. This could be as much as 60 percent to 75 percent of soil volume.

Organic amendments may also be used, but tend to be temporary. Such materials should be composted to prevent nitrogen-tie and settling because of rapid decay. Examples of good materials include composted milled pine bark or redwood bark and sphagnum peat.

The designed soil should be added to create A and B horizons to a minimum depth of eighteen inches, an optimum depth of twenty-four to thirty-six inches, and a maximum of forty-eight inches. Otherwise, a thin layer of four- to six-inch topdressing to rebuild an A horizon may be useful, but only if the subsoil is not compacted, is adequately drained, and can qualify as a B horizon. Some topdressing should first be applied, mixed into existing soil to prevent a sharp interface, then additional topdressing added to the four- to six-inch depth.

Structured soils are radically different than existing soil. These were devised by the Urban Horticulture Institute at Cornell University. Structured soils create a rooting zone under pavement while preserving a stable base for that pavement. The primary component of structured soils is crushed stone, into which is mixed a certain percentage of clay loam soil and a water-absorbing gel that helps hold water and "glues" the whole thing together. This mixture will meet the load-bearing needs of pavement, but creates a rigid lattice that contains large voids for air exchange that can be explored by roots. The soil occupying a portion of those voids then supports root growth.

Urban Erosion

Once urban sites are well established, with turf and trees growing, little erosion occurs, but during the construction phases of urban or road development, erosion can be 10 to 100 times greater than on similar farmland (figure 19-5). Large developments leave soil bare for long periods of time. Topsoil removal exposes the subsoil, which is often more erodable than the topsoil. Areas that normally absorb water, like woods, are often graded. Once stripped, these areas absorb less moisture, causing more runoff. Even more water runs off roads, parking lots, and roofs. When this extra runoff flows over bare soil produced during construction, massive erosion results.

FIGURE 19-5 Severe rilling on a roadside embankment caused the ditch to fill with sediment. Much of the soil reached a nearby lake. *(Courtesy of NRCS USDA)*

The impact of construction site erosion differs slightly from farmland erosion. Generally, loss of plant nutrients and growth media is a secondary concern. A developer is concerned with erosion damage—like rilled and gullied land that must be repaired before a site can be sold. In extreme cases, erosion can undercut and collapse roads and foundations. And, of course, urban erosion contributes to off-site problems. About 5 percent of the sediment load annually reaching surface waters in the United States comes from urban sources.

Controlling Erosion. Controlling erosion on construction sites requires careful planning, using soil surveys, topographic maps, and other tools. There are five general principles for controlling runoff, erosion, and sedimentation:

KEEP DISTURBED AREAS SMALL. During the planning stage, identify critically erodable areas and plan to avoid disturbing them. Such areas can be left as "green" areas. Disturb only that land being actively built on, rather than stripping the soil over a large area. Where possible, retain natural soil cover.

PROTECT DISTURBED AREAS. As soon as possible, vegetation should be planted on disturbed sites. Such vegetation could include a temporary cover of annual grasses, permanent turf, shrubs, or ground covers. On very steep slopes, a layer of rock, called **rip-rap,** can control erosion. Retaining walls made of concrete, railroad ties, or stone can be used to terrace slopes.

KEEP RUNOFF VELOCITIES LOW. When grading land, keep slopes short and gradients as low as possible. Keep vegetative cover intact to slow runoff where feasible. Various permanent or temporary control structures may be used. For instance, highway construction crews often make temporary dams of hay bales across ditches on long sloping roadsides to keep running water from picking up speed downhill.

DIVERT RUNOFF AWAY FROM DISTURBED AREAS. Land or pavement at a high elevation can act as a small watershed, collecting rainwater and directing it onto bare soil at a lower elevation. These bare areas should be protected by diversions, and the water led into waterways or outlets. Waterways can be covered by turf (grassed waterways), rip-rap, or even concrete, depending on the amount of waterflow they must handle.

RETAIN SEDIMENT ON-SITE. It is impossible to stop all erosion during construction, but sediments can be captured and retained on the site. This avoids pollution and sediment damage in other areas. Grass strips filter soil out of runoff water, as do **silt fences** (figure 19-6). Most modern developments include **debris basins** or **retention ponds** to capture sediments. These are small ponds (figure 19-7) built in the path of drainage patterns. Sediment-laden water flows into the basin, and soil settles out. Later, these ponds can provide an attractive amenity to local housing.

Establishing Vegetation. Growing vegetation on denuded areas is essential. Permanent turf is the best cover, but mowing can present a problem on steep

FIGURE 19-6 Silt fences, made of tightly woven plastic fabric, filter sediments from construction sites. They protect nearby ponds, streams, wetlands, or other sensitive areas. *(Courtesy of NRCS USDA)*

FIGURE 19-7 Erosion control at this housing development consists of a retention pond (background) to capture runoff and trap sediments. In the foreground, soil has been seeded to grass and a mulch applied. *(Courtesy of USDA NRCS)*

banks. In recent years native plants like prairie grasses have come into use. Very closely spaced shrubs, especially those that spread by suckering, may also replace turf. Using turf, groundcovers, native grasses, shrubs, and trees properly can bring beauty and a wildlife habitat as an added benefit.

It is often difficult to establish plants on critical erosion areas, because soil may be poor and both seeds and soil may wash away. Before planting, make sure all runoff is diverted from the site. The soil should be free of rills and gullies, compaction broken up, and lime and fertilizer added as needed. If groundcovers or shrubs are being planted, wood chip mulches can be used to control erosion.

When an erodable slope is seeded to grasses or legumes, a light mulch is needed to keep soil and seeds from washing away. About eighty pounds of clean hay or straw per 1,000 square feet ($1\frac{1}{2}$ tons per acre) provides a satisfactory cover. This mulch should be held in place by netting (such as jute netting), spraying with special asphalt materials, or punched into the soil with a modified disc. Sometimes, heavy jute netting (a fiber mesh) replaces the mulch. In very difficult areas, seed and mulch may be applied at the same time by **hydroseeding.** In this method, a mixture of water, seed, and chopped hay is blown on a slope from the side of the road.

Sodding provides a quick cover of critical spots like steep slopes (figure 19-8). Sodding, if properly done, stops erosion quickly, but is a costly solution. Because a steep slope is a problem site, careful attention must be paid to the proper sodding techniques. The sod should be staked.

FIGURE 19-8 Sod eliminates erosion on disturbed slopes. The unsodded portion of the slope is severely rilled and gullied. *(Courtesy of Turfgrass Producers International)*

CALCULATING SOIL LOSS. The Universal Soil Loss Equation (USLE) can be used to estimate soil losses from construction sites and to identify critical erosion sites. The results are not precise, but they do offer some guidance. The USLE is written as:

$$A = R\,K\,LS\,C\,P$$

A is the soil loss in tons per acre. Refer to chapter 18 for instructions in how to solve the equation. The following values can be used to solve the USLE for construction sites:

- R. The rainfall factor can be obtained from figure 18-12.
- K. The erodibility factor can be obtained from soil survey data or calculated from USDA Handbook 537 (listed in the activities section of chapter 18). If the topsoil has been removed, then the K value for the exposed subsoil must be used.
- LS. The slope factor can be obtained from figure 18-13. Slope factors are reliable only up to a 20 percent slope, and so are not reliable for very steep roadside embankments.
- C. The cover and management factor for bare, stripped soil is the same as a clean-fallow plot, so C equals 1.0. If a mulch covers the soil, C is much lower. Figure 19-9 gives some C values for construction sites.
- P. The support practice factor is usually 1.0, because few of the support practices are applicable to construction sites.

Zero-Runoff Urban Sites

Standard practice for design of urban sites is the rapid removal of stormwater by grading the land to move it quickly to storm sewers or culverts and discharging it

Type of Mulch	Mulch Rate	Land Slope	Factor *C*	Length Limit
	Tons per acre	*Percent*		*Feet*
None	0	all	1.0	—
Straw or hay,	1.0	1–5	0.20	200
tied down by	1.0	6–10	0.20	100
anchoring and				
tacking equipment				
	1.5	1–5	0.12	300
	1.5	6–10	0.12	150
	2.0	1–5	0.06	400
	2.0	6–10	0.06	200
	2.0	11–15	0.07	150
	2.0	16–20	0.11	100
	2.0	21–25	0.14	75
	2.0	26–33	0.17	50
	2.0	34–50	0.20	35
Crushed stone,	125	<16	0.05	200
1/4 to 1 1/2 in.	135	16–20	0.05	150
	135	21–33	0.05	100
	135	34–50	0.05	75
	240	<21	0.02	300
	240	21–33	0.02	200
	240	34–50	0.02	150
Wood chips	7.0	<16	0.08	75
	7.0	16–20	0.08	50
	12	<16	0.05	150
	12	16–20	0.05	100
	12	21–33	0.05	75
	25	<16	0.02	200
	25	16–20	0.02	150
	25	21–33	0.02	100
	25	34–50	0.02	75

FIGURE 19-9 Cover and management factor (*C*) and length limits for construction slopes. If slope length is exceeded, higher mulch-application rates or some means of shortening the slope is needed. *(Source: USDA Handbook 537)*

into surface waters. This reduces flooding and wet basements in developed sites. However, it may merely move the flooding downstream. Further, urban runoff often carries pollutants and sediments to streams and rivers. In the natural cycle of things, some of this stormwater supplies needed water to vegetation and recharges groundwater. The flow of many streams depends on groundwater, so a failure of groundwater recharge reduces streamflow. In areas with cold-water streams that support trout or salmon, runoff water raises stream temperature, making them less fit for fish habitat.

A different option is to design developed areas to retain water on-site and allow it to filter into the soil, using the water as a resource for irrigation and recharge rather than a disposal problem. This is called zero-runoff design.

A number of BMPs can be employed for keeping most water on-site. Some are presented in the previous section of this chapter. Lawns, usually being on compacted soils, can be planted to be more absorptive or even replaced with other types of vegetation that absorb water better. Hard surfaces like parking lots can be made of porous concrete that actually absorbs water rapidly. Wetlands on site can be retained, or new ones created. Most retention ponds develop a sealed bottom that stops infiltration; these can be designed to have porous bottoms to become **recharge basins.**

Water is often carried off yards by swales, shallow ditches. These can be designed with deep, organic, porous soils and planted to dense vegetation that slows down runoff speed, promotes infiltration, and traps sediments. These **bioswales** can lead to wetlands or recharge basins for capturing any water that exits the swale.

SUMMARY

Urban soils exhibit great variation, poor health, buried debris, compaction, and contamination. They are difficult to classify and map because of the extreme variation. Examples of contamination include road deicing salts and lead, which leads to health problems in children. Severely damaged brown fields can be repaired and be used for redevelopment.

In urban centers, soil can severely limit plant growth. Here, modified and structured soils can improve growth of trees and other plants to improve the environment for city inhabitants.

During construction of roads or housing, soil is fully exposed to the forces of water erosion. Keeping disturbed areas small, establishing vegetative cover, and diverting runoff from soil can prevent erosion. Sediments can be captured by grass filter strips or catch basins, and zero-runoff techniques can safely retain stormwater on-site.

REVIEW

1. Describe activities that might have damaged soil structure in some urban soil.
2. Why might the existing county soil survey in a new suburban area not accurately describe the soil in someone's yard? Give several specific examples.
3. Using appendix 6, identify five trees you might conclude would make good urban trees in highly disturbed sites. Explain your reasoning.
4. Describe strategies for limiting the amount of sediment that leaves a new construction site.
5. Assume that behind a new home is a slope fifty feet long and forty feet wide with a gradient of 10 percent. The home is in central Illinois, marked on the map of figure 18-12. The subsoil, exposed after grading, has a K value of 0.20. How much total soil will be carried off this slope in an average year if it is left bare? How much if it is mulched with 1.5 tons/acre of hay? One acre equals 43,560 square feet.
6. Distinguish modified soil from structured soil.

7. A case study: Known for its pristine clear waters, Lake Tahoe is beginning to suffer from pollutants running off development around the lake. A rule has been passed requiring zero runoff from properties in the Lake Tahoe watershed. Speculate as to courses of action that could be taken by various property owners.
8. A case study: In the soil survey of a park in New York described in this chapter, the surveyors described several new series. One of them, the Greenbelt, has a soil profile listed as A (0–3 inches), Bw (3–13 inches), C1 (13–27 inches), C2 (27–57 inches), Ab (57–58 inches), Bwb (58–60 inches). Create a story describing a possible history of this soil during historical times.

ENRICHMENT ACTIVITIES

1. If any large-scale housing projects are being built nearby, visit them and identify erosion- and sediment-control practices.
2. For study of urban soils from an ecological perspective, read an interesting series of studies on changing soil conditions along an urban-rural transect beginning in New York City and ending in rural Connecticut. For example, read McDonnel, M. et. al. 1997. Ecosystem processes along an urban-rural gradient. Urban Ecosystems 1:21–36.
3. Find out about soil lead problems in your state, and how you get soil tested for lead, using your favorite Internet browser. Entering search terms like "lead testing and [Name of state]" should work.
4. For more detailed information about lead in gardens, try: <http://www.extension.umn.edu/distribution/horticulture/DG2543.html>.
5. The United States Geologic Survey has an Internet site on bioremediation at <http://water.usgs.gov/wid/html/bioremed.html>.

CHAPTER 20

Government Agencies and Programs

OBJECTIVES

After completing this chapter, you should be able to:

- list federal agencies that assist growers
- describe some of the soil programs of these agencies

TERMS TO KNOW

Agriculture Experiment Stations
Agricultural Research Service (ARS)
conservation compliance
Conservation Reserve Program (CRP)
Consolidated Farm Service Agency (CFSA)
Cooperative State Research, Education, and Extension Service (CSREES)
cost-sharing
Natural Resources Conservation Service (NRCS)
Soil and Water Conservation Districts (SWCD)

FIGURE 20-1 The provision of advice is an important service of agencies described in this chapter. Here a NRCS soil conservationist greets a South Carolina farmer. *(Courtesy of USDA)*

Research yearly comes up with new methods that can help growers, but growers cannot spend much of their time in school learning new methods. Market forces pressure growers to change, not always in ways that are best for the soil. As businesspeople, growers must often put their money where the financial return is best—which is not always for long-term benefit. A number of programs support growers with technical assistance and have information to answer questions such as: How do growers keep up with change? How can they afford soil improvements?

A network of agencies and regulations helps the grower in several ways.

- Education provides information on new and old methods in the form of publications, workshops, and advice (figure 20-1).
- Technical assistance (figure 20-2) advises growers how to complete specific projects like irrigation or grassed waterways.
- Financial assistance helps growers pay for projects.
- Research helps by creating new and better techniques.

Government programs operate through a complex web of federal, state, and local agencies. In addition to the government, there are private sources of information and help. We will concentrate here on government programs, especially those in the area of soil and water conservation.

FIGURE 20-2 Erosion in a streambed caused damage to this cornfield. Growers can obtain help from various agencies to solve this type of problem. *(Courtesy of USDA NRCS)*

USDA Agencies

Help comes to growers and other soil users from many sources. One major source is the United States Department of Agriculture (USDA), established in 1862. The USDA includes several agencies that help farmers and others with soil and water management as well as many other facets of agriculture. Many USDA programs work through state or local groups.

The USDA was dramatically reorganized and streamlined in 1994, with fourteen of forty-three agencies eliminated or merged with other agencies. This chapter reflects those changes.

Agricultural Research Service. The main research arm of the USDA is the **Agricultural Research Service (ARS).** Established in 1953 to do basic and applied research in agriculture, the ARS does a variety of research, some with soils. For example, the ARS participated in the development of the Universal Soil Loss Equation and Revised Universal Soil Loss Equation. They publish an interesting nontechnical magazine reporting on their current projects titled *Agriculture Research.*

Cooperative State Research, Education, and Extension Service. The **Cooperative State Research, Education, and Extension Service (CSREES)** replaces the older Cooperative Extension Service, and merges extension and research activities in CSREES programs at state **Agriculture Experiment Stations.** Experiment Stations, and the CSREES programs housed in them, are located at land-grant universities and are jointly funded by both state and federal dollars.

Personnel funded by CSREES may have extension appointments, research appointments, teaching positions at the college, or some combination of the three. Some positions are in soil science specifically, while others are in related agriculture areas. Research positions, of course, conduct basic research in agriculture.

Extension positions aim to provide information and assistance to a variety of audiences and to spread information obtained by research at the Experiment Stations. Many extension personnel act as county agents and serve the needs of people in their home county. They may be helpful in a wide variety of subject areas. Others may be specialists, either attached to counties or located at the Agriculture Experiment Stations. Examples could include nursery, greenhouse, or urban horticulture specialists. Extension services publish informational bulletins, run workshops, and provide expert advice that is specific to the state they serve.

Consolidated Farm Service Agency. The **Consolidated Farm Service Agency (CFSA)** was created by a merger of several earlier USDA agencies, including the Agricultural Stabilization and Conservation Service (ASCS) and the Farmer's Home Administration (FmHA). CFSA administers a number of funding programs and two major soil conservation programs (described later). Many CFSA activities are administered at the state and local level by committees of local growers in **Soil and Water Conservation Districts,** or SWCD (described later). In carrying out its conservation programs, CFSA utilizes NRCS technical expertise.

Natural Resources Conservation Service. The **Natural Resources Conservation Service (NRCS)** replaces the older Soil Conservation Service, established by Congress in 1935 to carry out a national program of soil and water conservation. The NRCS provides a variety of assistance, including technical support to CFSA programs. These include:

- aiding the CFSA in establishing policies for CFSA conservation programs.
- conducting soil surveys in a joint effort with Agriculture Experiment Stations.
- helping city and county officials with land-use planning (figure 20-3).

FIGURE 20-3 The NRCS helps to ensure that land is used properly when being developed for urban uses. Here, a soil scientist surveys a development site. *(Courtesy of USDA NRCS)*

FIGURE 20-4 A newly installed grassed waterway installed with the help of the NRCS.

- administering conservation programs except those assigned to the CFSA (figure 20-4).
- developing useful plants for conservation.
- measuring winter snowpack on mountains in the West to predict the amount of water available the following year.
- providing technical assistance in resource management issues such as fish and wildlife habitat, pasture, and range.
- conducting a National Resource Inventory of the status and trends of the nation's soil and water every five years. The NRIs serve as major databases for public land and water planning and are the source of many statistics in this text. The most recent NRI was conducted in 1997.

Soil and Water Conservation Districts. While not federal agencies, many USDA programs operate through Soil and Water Conservation Districts. Feeling that federal conservation programs should operate through local authority, then-President Franklin Roosevelt proposed to all state governors in 1937 a model for creating Soil and Water Conservation Districts.

Almost all states now have such districts. Generally, the boundaries of a SWCD follow county boundaries (figure 20-5). Each district is governed by locally elected or appointed boards of growers or ranchers. The districts plan for and carry out programs they feel have a priority in their district. The actual role and duties of the SWCD vary from state to state, since they are state, not federal, agencies.

All districts have formal agreements with the U.S. Secretary of Agriculture to set up programs. The secretary, in return, agrees to help the districts. Many districts have NRCS soil scientists assigned to them for technical aid.

USDA Conservation Programs

The National Resources Conservation Service and the Consolidated Farm Service Agency, among other duties, operate several programs aiming to conserve the nation's soil and water resources. These can be characterized as cost-sharing programs, conservation compliance

FIGURE 20-5 Most of the United States is divided into Soil and Water Conservation Districts that set local conservation priorities. It is through these committees that many programs operate. Here SWCD Commissioners and a NRCS agent discuss local conservation issues in Iowa. *(Courtesy of USDA NRCS)*

programs, and conservation reserve programs. The following are among the most important. It should be noted that the 2002 Farm Bill was under consideration at the time this edition was being prepared, and some of the following could change.

Environmental Quality Incentives Program. EQIP, authorized by the 1996 Farm Bill (known officially as the Federal Agriculture Improvement Act of 1996), combined several earlier cost-sharing programs. EQIP, administered by CFSA and NRCS, helps growers in soil and water conservation efforts by providing money on a cost-share basis. **Cost-sharing** means the USDA shares part of the cost for conservation efforts with a grower. Growers can enter into five- and ten-year agreements with the USDA for technical, educational, and cost-share assistance.

Conservation and Wetland Reserve Programs. The Food Security Act (or Farm Bill) of 1985 created the **Conservation Reserve Program (CRP).** CRP purchases ten-year conservation easements from growers. In exchange for payments, growers plant the land to permanent cover such as grasses or trees. The CRP targets highly erodable land or land that if used improperly might degrade water quality, and is administered by the CFSA with NRCS technical help.

In 1997 about thirty-three million acres were enrolled in CRP, mostly in the middle sections of the country. The USDA estimates that CRP land experiences an average reduction in soil loss of nineteen tons per acre per year, and much of the national reduction in erosion in recent years has been attributed to CRP. CRP also improves wildlife habitat and large increases in pheasant, duck, and other game populations have been attributed to CRP. The 1996 Farm Bill shifted program priorities slightly to emphasize protecting environmentally sensitive areas, like streams, from the effects of erosion.

The 1996 Farm Bill also extended the Wetland Reserve Program, an effort to preserve wetlands in farm areas by long-term easements. In 2001 about a million acres were enrolled, many in the lower Mississippi watershed. Cost-sharing pays for the easements and helps pay for wetland restoration efforts.

Conservation Compliance Programs. **Conservation compliance,** sometimes called "Sodbuster," requires growers to take certain conservation measures to remain eligible for federal price-support programs. For instance, growers are asked to submit and execute soil erosion-control plans for certain highly erodable lands. According to the NRCS, during the decade 1985–1995, 1.7 million such plans were developed and implemented on 143 million acres of highly erodable land.

Wetlands in agricultural land may also be protected by "Swampbuster" provisions of the Farm Bill. These provisions deny eligibility for other USDA programs to growers who drain and farm certain wetlands.

Clean Water Act. Section 404 of the Clean Water Act protects many of the nation's wetlands by requiring permits to discharge dredged or fill materials in wetlands, streams, rivers, and other waters of the United States. Most activities of growers are exempt from Section 404 permitting, including maintenance of existing drainage structures or cultivation of "prior converted croplands." However, conversion of wetlands to new use, such as crop cultivation, filling for parking lots, or other development usually requires a permit. Permits usually call for applicants to minimize impacts on or replace affected wetlands.

The U.S. Army Corps of Engineers is the lead agency enforcing Section 404 and issuing permits, with strong involvement by the Environmental Protection Agency. The Natural Resources Conservation Service and U.S. Fish and Wildlife Service also have input into Section 404 policies and activities. However, early in 2001 the U.S. Supreme Court removed Corps jurisdiction over certain wetlands that lie away from what are known as "navigable waters," so one would have to check on the application of Section 404 for any particular wetland.

State and Local Efforts

In addition to federal programs, states and localities also have laws and programs. Foremost of these, of course, are the Soil and Water Districts, Agricultural Experiment Stations, and Extension Services. Obviously, this text cannot list all the state and local efforts.

One example is the Reinvest in Minnesota program, or RIM, enjoyed by citizens and sportsmen of Minnesota. Enrolled land is planted to wildlife cover, especially to native vegetation. Wetlands have also been restored. RIM is administered through the Soil and Water Conservation Districts. The intent is to protect erodable land and water quality, and to promote wildlife habitat.

Many local and state laws involve controlling land use, such as zoning laws. For instance, many outer suburban areas limit how far land can be subdivided. To save farmland, many states have "green acre" laws, which give tax breaks to land in developing areas kept as farmland. Some of these and other efforts aim to slow the loss of farmland to urban use (figure 20-6).

FIGURE 20-6 A shopping center is being planned to replace this farmland. In the future, we may see greater efforts to slow the loss of farmland to nonfarm uses. *(Courtesy of USDA NRCS)*

SUMMARY

Several groups assist growers with education, technical help, financial assistance, and research. Many of these groups are part of the United States Department of Agriculture.

Agricultural research is carried out by the Agriculture Research Service. The Cooperative State Research, Education, and Extension Service, attached to State Agriculture Experiment Stations, conducts research and provides education through extension services.

The Consolidated Farm Service Agency conducts a number of funding programs, as well as administering the Environmental Quality Incentives and the Conservation Reserve Programs. The former supplies cost-sharing for conservation activities, while the latter puts highly erodable land into ten-year easements.

The Natural Resources Conservation Service administers several other conservation programs and provides a wide range of technical assistance in conservation.

Many federal conservation programs operate through Soil and Water Conservation Districts, state funded and administered by local grower committees.

While CRP and many other programs are voluntary, several other laws require some protection of our soil and water resources. These include Section 404 and Swampbuster provisions to protect wetlands, Sodbuster provisions to keep uncultivated highly erodable land from being plowed, and conservation compliance measures to increase the number of soil conservation projects undertaken on farms. These programs are largely responsible for erosion and wetland loss reductions over the past decade, as noted in earlier chapters.

REVIEW

1. Discuss the National Resource Inventories. Find at least three examples of data in this book supplied by the NRIs. (Hint: look at the captions of various charts in the text that are included to give a picture of soil use and conditions in the United States.)
2. The informal terms for a couple of conservation programs involve the use of the word "buster." What are those programs and what do they do?
3. What is the CRP? What are its benefits? What percentage of U.S. nonpublic land is under CRP (see a chart in chapter 1).
4. Discuss the cooperation between federal, state, and local agencies that is part of soil and water conservation efforts. Give examples.
5. Who in the agencies and programs described here performs agricultural science research? How are the results communicated to the community?

ENRICHMENT ACTIVITIES

1. Visit your local NRCS office and CFSA office. Find out what local projects they have been involved in. Or invite guest speakers from those offices to your class.
2. Tour the content of the USDA and the NRCS on the Internet through their homepages at <http://www.usda.gov> and <http://www.nrcs.usda.gov>. Through the NRCS page you can find information about conservation programs in the 1996 Farm Bill. Also the official criteria for conservation practices by pressing "Technical Resources," then "Conservation Practices." Write a short report on what you find.
3. A history of the CRP is available at <http://www.fsa.usda.gov/dafp/cepd/12crplogo/history.htm>. From this report, describe the cover commonly planted on CRP acres.
4. Information on the Wetland Reserve Program is at <http://www.ncrs.usda.gov/programs/wrp/>. Look at the "fact sheet," and the fact sheet for your state, for basic information about the program. Write a short report about the program, with emphasis on your state.
5. Use an Internet browser to find out information about conservation programs in the 2002 Farm Bill. Did soil conservation funding rise in the 2002 bill?

APPENDIX 1

Some Basic Chemistry

Atoms and Elements

Elements are the building blocks of all matter. Examples of the 109 known elements include oxygen, carbon, and iron. Each element is assigned a symbol made of a letter or letters from the English or Latin word for the element. For example, oxygen is "O," carbon is "C," and iron is "Fe" from the Latin "ferrous."

The smallest unit of an element is the *atom.* While modern models of the atom are more complex, a simple picture of the atom, called the Bohr model, helps us understand how chemical processes occur in the soil. Atoms are made of three particles: a negatively charged *electron,* a positively charged *proton,* and a neutral *neutron.* According to the Bohr model, protons and neutrons inhabit a *nucleus,* and electrons circle the nucleus like planets orbiting the sun, as in figure A-1. These electrons occupy "orbitals," each of which will hold a certain number of electrons.

The simplest element is hydrogen, which is composed of a single proton and electron, occupying the innermost orbital. Elements get successively heavier and more complex as more protons, neutrons, electrons, and electron orbits are added. Oxygen, for instance, has eight of each particle and electrons occupy two orbitals. The total weight of all the protons, neutrons, and electrons in an atom of an element is its *atomic weight.* Electrons are very light and contribute little to atomic weight. Thus, the atomic weight of oxygen, with eight protons and eight neutrons, is about sixteen.

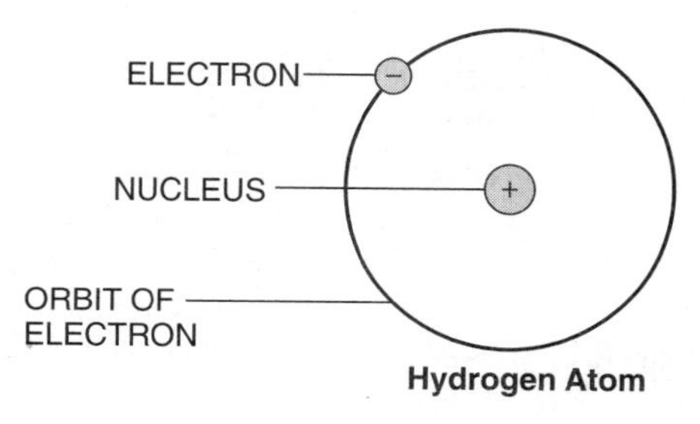

FIGURE A-1 Hydrogen atom.

Compounds

Atoms combine to form *molecules.* A molecule is symbolized by writing the atomic symbols of each element in the molecule; a number in the form of a subscript indicates how many atoms of each element are present in the molecule. Thus, water, or H_2O, is a molecule with two atoms of hydrogen combined with one atom of oxygen. One way for molecules to form is for atoms to join by sharing electrons, as shown for the water molecule in

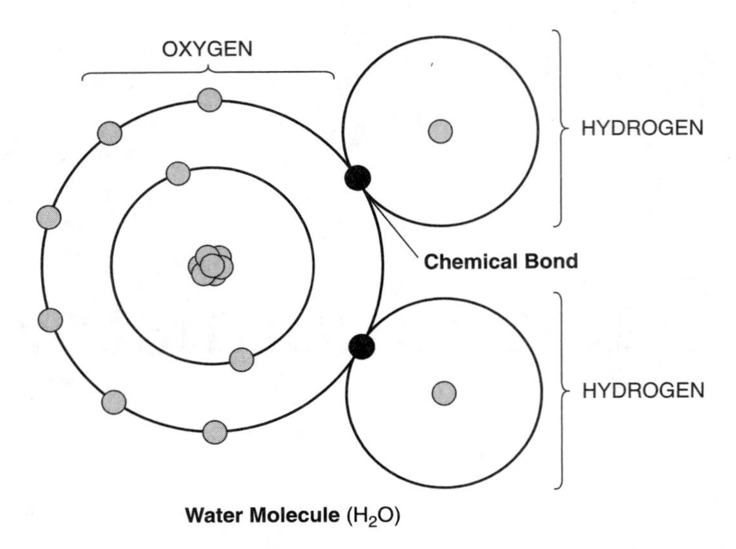

FIGURE A-2 Water molecule (H_2O). In some molecules, chemical bonds result from sharing of electrons.

figure A-2. This sharing fills the outer orbital of the oxygen atom, which can hold eight electrons. A filled outer orbital is more stable than a partially filled one, so oxygen is "eager" to react in ways that will fill that outer orbital. This is one way to form a *chemical bond* between two atoms.

A collection of like molecules is a *compound,* as in the compound water. Silicon dioxide, or SiO_2, is a compound containing two oxygen atoms attached to a silicon atom. Silicon dioxide is the mineral quartz, the major component of sand and one of earth's most common minerals. Pure solid compounds of the earth's crust, like quartz, are called **minerals.** *Rocks* are mixtures of minerals. Granite, for instance, consists of the minerals feldspar, quartz, and others.

Organic Compounds

The common minerals are **inorganic,** distinguishing them from a special class of compounds that are labeled **organic.** Organic compounds all contain carbon and hydrogen, and most have oxygen, sulfur, nitrogen, or other elements. Organic compounds are the stuff of life; all life is made of them. When this text refers to organic matter, it refers to organic compounds in the soil, all of which come from living creatures and plants. However, humans also synthesize a variety of organic compounds.

Ions

A normal atom or molecule has an equal number of electrons and protons; their charges balance and the net charge is zero. Sometimes an imbalance occurs, and the resulting atoms or molecules are called **ions.** For instance, when salt (NaCl) dissolves in water, the molecule breaks apart into sodium ions that are short one electron and chlorine ions with an extra electron. Thus, each sodium ion carries a positive charge, while chlorine ions have a negative charge:

$$NaCl \rightarrow Na^+ + Cl^-$$

salt → sodium cations + chlorine anions

solid salt → salt in solution

Positively charged ions are called **cations,** negatively charged ones **anions.** A **solution** results when compounds disassociate in water with ions dissolved in it.

The hydrogen ion, H^+, is a special case. Having lost its sole electron, the hydrogen ion, in one form, is simply a single proton. Its presence has a potent effect on many biological and nonbiological systems, including the acidity of soil. This is discussed in chapter 11.

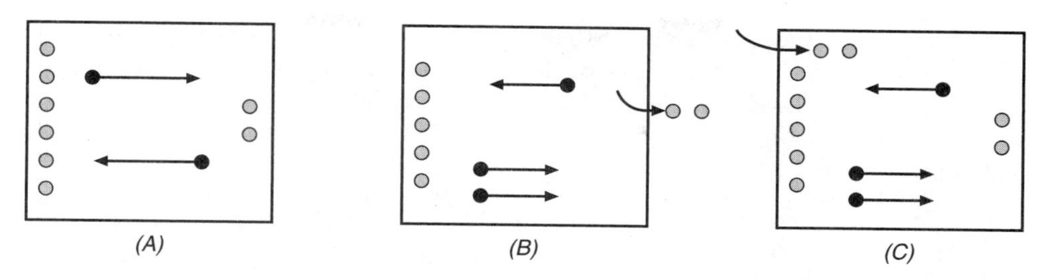

FIGURE A-3 Chemical reaction equilibria as represented by movement of people in a room. (*A*) Reaction in equilibrium; (*B*) reaction in which equilibrium is disturbed by people leaving room to right. Reaction shifts to right to compensate for change. (*C*) Reaction in which equilibrium is disturbed by people entering room from left. Reaction shifts to right to compensate for change.

Chemical and Physical Reactions

Two types of reactions are important in soils: *chemical reactions* and *physical reactions.* Chemical reactions involve the actual rearrangement of atoms to form new molecules and compounds, as in the reaction of hydrogen and oxygen to form water, shown in equation (*a*). The reaction, as written, states that two hydrogen molecules (in the natural environment, hydrogen and oxygen come as two-atom molecules) combine with one oxygen molecule to form two molecules of water.

(*a*) $2H_2 + O_2 \rightarrow 2H_2O$

(*b*) H_2O (liquid) $\rightarrow$ H_2O (solid ice)

In physical reactions, the physical form but not the chemical form changes, as in the freezing of water in reaction (*b*) or the disintegration of rock by physical forces. Both physical and chemical weathering contribute to soil formation.

Reaction (*a*) suggests that a chemical reaction is a one-way process. In actuality, all reactions go in both directions at once but will favor one side of the reaction or the other. One might compare a reaction to a room full of people. Assume that ten people are in a room, and they are asked to go to one side of the room or the other. People that go to the left side are *A* people, those that go to the right are *B* people. In this "reaction," assume seven people go to the left, and three go to the right. We could write this "reaction" as:

$$\underset{\text{(7 people)}}{A} \leftrightarrows \underset{\text{(3 people)}}{B}$$

In a real reaction, the people don't stop; a few continue to walk back and forth across the room, but the numbers of people will continue to balance out to 7:3 (*A* in figure A-3). This balance is called the *equilibrium* of the reaction and is constant for a given reaction at a given condition.

If the conditions of a reaction change, the equilibrium can shift to try to achieve a new balance. Say that in the room full of people, two of the three people on the right side of the room leave altogether. To achieve a new equilibrium, the reaction will shift to the right—that is, more people will walk to the right side of the room than back to the left until a new balance is achieved (see *B* in figure A-3). On the other hand, if a group of people enters the room on the left, the room is "unbalanced" and the "reaction" again shifts to the right to achieve a new equilibrium (see *C* in figure A-3).

This shifting equilibrium that occurs when the amount of a material on one side of a chemical equation changes we will here call "mass action." The equilibrium shift—say moving the reaction to the left—always works in a direction that reduces the original change. This process is important, because it governs chemical processes in the soil, including many related to farming activities such as soil liming or fertilization.

Oxidation-Reduction Reactions

A very important type of reaction in the soil is the *oxidation-reduction reaction.* Technically, an oxidation occurs when an element loses an electron in a reaction (oxidation), and some other element gains that electron (reduction). The electron donor (the "loser") is said to be oxidized and the electron acceptor (the "gainer") is said to be reduced. An important reduction reaction in the soil is the conversion of ferric iron to ferrous iron under low-oxygen conditions:

Fe^{+3}	$\rightarrow$	Fe^{+2}
ferric iron		ferrous iron
oxidized iron		reduced iron

Oxidized iron (ferric), found in aerated soils and colored red to yellow, gains an electron and is reduced to the gray-blue ferrous form. The gray color is a strong indicator of a soil that is usually too wet.

If in the preceding reaction ferric iron acted as an electron acceptor (was reduced), then what gave up that electron or was an electron donor (was oxidized)? The answer was carbon in organic matter, which gave up electrons during the decay process. Ordinarily, in well-aerated soil, oxygen is the most powerful electron acceptor, and aerobic decay is the oxidation of organic matter by microorganisms using oxygen as the electron acceptor. In soils with little or no oxygen, organisms may use other electron acceptors, like ferric iron.

Most oxidation reactions involve oxygen, hence the term oxidation.

Very important oxidized forms of elements in the soil include carbon (CO_2, carbon dioxide or CO_3^-, carbonate ion), nitrogen (NO_3^-, nitrate ion), sulfur (SO_4^-, sulfate ion), and iron (Fe_2O_3, ferric oxide). Reduced forms include reduced nitrogen as the ammonium ion, (NH_4^+), and reduced carbon as methane (CH_4), and organic matter.

Energy

Both chemical and physical reactions in the soil depend on *energy*. Energy is simply the capacity to do work. The greater the energy, the more work that can be done. An understanding of energy is important because everything that happens in the soil—the "work" of soil and plants—is fueled by energy. Energy comes in many forms, such as heat, electricity, light, and motion. The following forms are important to soil:

- *Light energy* is energy contained in sunlight. Plants use that energy for photosynthesis, which creates organic matter for the soil.
- *Chemical energy* is energy contained in a chemical bond. During photosynthesis, plants convert energy in sunlight to energy in chemical bonds. This chemical energy is "stored" for later use by the plant or animals that eat it.
- *Potential energy* is the energy stored in an object that can fall. The energy is released when the object does fall, like rocks falling off a cliff, water falling over a dam, or rain falling from the sky.
- *Kinetic energy* is the energy of a moving object. Soil erosion by wind and water is a result of motion energy. The amount of energy in a moving object is a function of its mass (m) times velocity (v) squared, according to the following formula:

  $$\text{Energy} = 1/2mv^2$$

  Thus, rapidly moving water can do more work than slowly moving water. The equation says, for instance, that if the velocity of running water doubles, it can do four times as much work (two squared) or cause four times as much erosion.

To understand chemical and physical reactions in the soil, it is important to know two rules. First, *energy can change forms*. For instance, sunlight energy can change to heat energy when it strikes pavement—a fact one can become painfully aware of when walking to the beach barefoot on a sunny day. Second, *matter tries to achieve the lowest possible energy state*. An electrical-generating dam is a good example. Water at high elevation behind the dam is in a high state of energy—it has a lot of potential energy. By flowing downhill through or over the dam, that potential energy is released to energy of motion. When the water reaches the lower level, it is now at a lower elevation and in a lower energy state. In the process, the "lost" potential energy was converted to motion. That motion energy can, in turn, be changed to electrical energy when it does the work of turning the turbines in the generating plant.

These two rules of energy control all the physical and chemical reactions in the soil. They are especially important to the behavior of water in the soil (chapter 6) and erosion (chapter 18).

Gradients. One result of the tendency for lower energy states is the concept of *gradient,* which controls the movement of materials in the soil. A gradient is a change of something over distance. The simplest gradient to understand is a slope, or hill, which is an elevation gradient. A cyclist on top of a hill can coast down the hill without pedaling, because he or she is moving to a lower energy state. To go up the hill, however, is to go to a higher energy state. To do that, the cyclist must feed in energy—which is to say, pedal the bike.

The preceding example presents a rule about gradients—that *movement down a gradient occurs without effort, while movement up a gradient requires an input of energy.*

Many gradients exist in nature. A spoon in a hot cup of coffee creates a temperature gradient. Heat will flow naturally from the hot end in the cup to the cool end where you hold it. Water will move in the soil from where it is moist to where it is dry (moisture gradient). Vapors from an opened perfume bottle will scent an entire room as they move from where they are concentrated near the bottle to where they are less concentrated (this is called diffusion along a concentration gradient). A similar diffusion through soil water moves soil nutrients towards plant roots.

Remember that normal movement is down a gradient—from where there is more of something, like heat or water, to where there is less. The opposite movement requires an input of energy—which is the same as saying, it takes work. Imagine trying to get perfume vapors back into the bottle.

APPENDIX 2

Sedimentation Test of Soil Texture

Description

The sedimentation test is an easy way to measure the percentage of sand, silt, and clay in a soil sample. It is based on the fact that large, heavy particles will settle most rapidly in water, while small, light particles will settle most slowly. The Calgon laundry powder is used to "dissolve" the soil aggregates and keep the individual particles separated.

Materials

- Soil sample
- One-quart fruit jar with lid
- Eight percent Calgon solution—mix six tablespoons of Calgon (a laundry powder available in stores) per quart of water
- Metric ruler
- Measuring cup
- Tablespoon

Procedure

1. Place about 1/2 cup of soil in the jar. Add 3 1/2 cups of water and 5 tablespoons of the Calgon solution.
2. Cap the jar and shake for 5 minutes. Leave the jar on the desk and let settle for 24 hours.
3. After 24 hours, measure the depth of settled soil. All soil particles have settled, so this is the TOTAL DEPTH. Write it down and label it.
4. Shake for another 5 minutes. Let stand 40 seconds. This allows sand to settle out. Measure the depth of the settled soil and record as SAND DEPTH.
5. Do not shake again. Let the jar stand for another 30 minutes. Measure the depth, and subtract the sand depth to get the SILT DEPTH.
6. The remaining unsettled particles are clay. Calculate clay by subtracting silt and sand depth from total depth to get CLAY DEPTH.
7. Now calculate the percentage of each soil separate using these formulas:

$$\%\ \text{sand} = \frac{\text{sand depth}}{\text{total depth}} \times 100$$

$$\%\ \text{silt} = \frac{\text{silt depth}}{\text{total depth}} \times 100$$

$$\%\ \text{clay} = \frac{\text{clay depth}}{\text{total depth}} \times 100$$

APPENDIX 3

Soil Orders of the United States

To view the full color detailed version of the map below, go to <http://www.nrcs.usda.gov/technical/land/meta/m4025.html>.

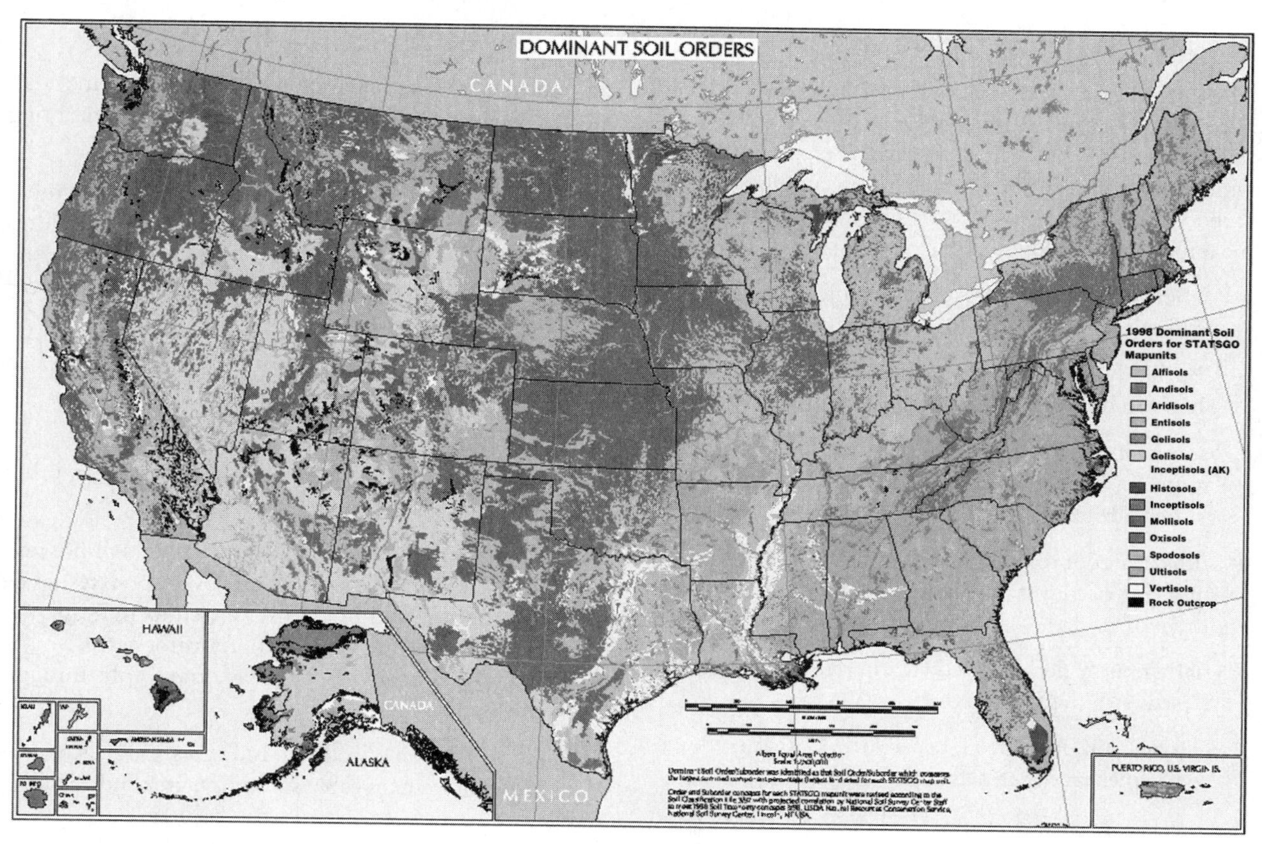

FIGURE A-4 General soils map of the United States. *(Courtesy of USDA)*

APPENDIX 4

Soil Horizon Symbol Suffixes

These lowercase letters are used as suffixes to label certain types of master horizons. One or more suffix may follow a master horizon designation—such as Bky—which indicates a B horizon that has accumulated both carbonates (k) and gypsum (y). The symbols and their meanings are as follows:

a Highly decomposed (sapric) organic material. This suffix is used with the O horizon.

b Buried horizon. Such a soil layer is an old mineral horizon buried by sedimentation or other processes.

c Concretions or hard nodules. A nodule or concretion is a hard "pocket" of soil cemented by iron or other substance.

d Physical root restriction, either man-made or a naturally dense layer that roots can enter only through fractures.

e Moderately decomposed (hemic) organic material. Used with the O horizon.

f Frozen soil. A horizon, usually the C, that contains permanent ice (permafrost).

g Strong gleying. Such a horizon is gray and mottled, the color of reduced (nonoxidized) iron, resulting from saturated conditions.

h Illuvial accumulation of organic matter. The symbol is used with the B horizon to show that complexes of humus and sesquioxides have washed into the horizon. Includes only small quantities of sesquioxides. May show dark staining.

i Slightly decomposed (fibric) organic matter. Used with the O horizon.

k Accumulation of carbonates (CO_3^-). Indicates accumulation of calcium carbonate (lime) or other carbonates in the B or C horizons.

m Cementation. The symbol indicates a soil horizon that has been cemented hard by carbonates, gypsum, or other material. A second suffix indicates the cementing agent, such as "k" for carbonates. This is a hardpan horizon; roots penetrate only through cracks.

n Accumulation of sodium. Indicates a high accumulation of exchangeable sodium, as in a sodic soil.

o Accumulation of sesquioxide clays after intense weathering.

p Plowing or other human disturbance. Horizon was heavily disturbed by plowing, cultivation, pasturing, or other activity. Applies to O and A horizons.

q Accumulation of silica (SiO_2).

r Weathered or soft bedrock. Used with C horizon to indicate bedrock that can be dug with spade that roots can enter through cracks.

s Illuvial accumulation of both sesquioxides and organic matter. Both the organic matter and sesquioxide components of humus-sesquioxide complexes are important.

t Accumulation of silicate clays. Clay may have formed in horizon or moved into it by illuviation, usually as coatings on ped faces. Mostly used in B horizon, sometimes in C.

v Plinthite. An iron-rich, humus-poor material common to tropical soils that hardens when exposed to air. B and C horizons.

w Development of color or structure. The symbol indicates that a B horizon has developed enough to show some color or structure but not enough to show illuvial accumulation of material.

x Fragipan or other noncemented natural hardpans. These are horizons that are firm, brittle, or have high bulk densities from natural processes.

y Accumulation of gypsum ($CaSO_4$) in B or C.

z Accumulation of salts more soluble than gypsum in B or C.

APPENDIX 5

Land Evaluation

Land may be evaluated for a number of uses, such as suitability for row crops, home landscapes, or building sites. Such evaluations might be done by soil scientists, land planners, landscape designers, or even students engaged in a soil-judging contest. While land evaluation purposes and methods vary between regions of the country, this appendix suggests some simple and general soil features to use when evaluating a soil.

Definition of Terms and Criteria

Suitability. Land can be said to be good, fair, or poor for any use. The same terms can also be applied to a particular feature of the land, such as slope or soil texture. Looking at the chart that follows, for each land use there is a series of criteria listed on the same line to the right. Each can be rated, and the "total" used to rate the land:

- Good land is well suited to the use being considered. There are no serious limitations or hazards. Ratings for each criteria are good, or perhaps a couple are fair.
- Fair land is suitable for the use if a few limitations are corrected. Several of the features are rated as fair.
- Poor land is unsuitable for the use, or corrections are too expensive to be practical. Even one or two severe limitations may be enough to gain this rating, or a large number of "fair" descriptions.

Slope. Slope is expressed as a percent. For instance, a slope of 2 percent suggests a two-foot change of elevation over a hundred-foot horizontal distance. Figure 3-6 gives some slope ranges that may help interpret these numbers. Changing slope would involve expensive earth moving.

Soil Texture. Chapter 4 describes how to determine a soil texture. The textural class groupings used in this chart are grouped as follows:

Group	Textural Classes
fine	sandy clay, silty clay, clay
moderately fine	sandy clay loam, silty clay loam, clay loam
medium	silt, silt loam, loam
moderately coarse	sandy loam
coarse	sand, loamy sand

Except for small areas, like a garden, texture cannot be changed practically.

Flooding. Flooding concerns how often or how long the land is actually covered with water during the season. Such flooding could be due to stream flooding, heavy rains, or snowmelt. Flooding may be able to be corrected, in some small areas, by earth moving to change drainage patterns. In many cases, only large-scale projects such as the construction of levees can solve these problems.

Internal Drainage. Internal drainage can be determined from figure 4-20. Many cases of poor drainage can be corrected by installing drainage systems, and excessive drainage may be improved by irrigation.

Depth to Restrictive Layer. Restrictive layers can interfere with rooting depth and plant anchorage, roadbeds, foundations, and home drain fields. Such layers could be bedrock, soilpans, water tables, or others. Some can be corrected, like ripping a soilpan, while bedrock cannot be altered easily.

Available Water for Plants. The available water is here rated as the number of inches of water held in the top five feet of soil. Since most soils will contain layers, one will have to add the capacity of each layer down to five feet. Figure 9-22 rates the water retention of soil textures. Here, a capacity greater than nine inches for five feet is good, six to nine is fair, and less than six is poor. Poor capacity can be alleviated by irrigation, where practical and acceptable.

Erodability. Erodability is related to slope, texture, and other factors discussed in chapter 18. Classes are as follows:

- Slight means that under average conditions there is little chance of excessive erosion. Slopes are gentle and short, internal drainage is good. Wind erosion is unlikely.
- Moderate could occur on medium- or fine-textured soil on gentle slopes over 300 feet long, or shorter moderate slopes. Wind erosion is possible due to texture or lack of cover.
- Severe erodability occurs on slopes over 12 percent, or areas that experience teaches are quite erodable.

The Erosion Index described in chapter 18 could be used to infer erodability.

Other Features. Other features may also limit a soil for certain uses. Examples include stoniness, soil pH, soil fertility, or salinity. Some of these can be improved easily, like liming agricultural soil to raise pH. On the other hand, improving soil fertility is more difficult for forest uses.

Site Inspection

Much of the information needed for this evaluation can be obtained from published soil surveys. As a lab exercise for students, a soil pit should be dug to allow examination of the soil profile.

Select an area of uniform character, preferably a single mapping unit on a soil map. Dig a pit measuring 3 feet × 3 feet at the surface, and about 3 1/2 feet deep. Orient the pit so that the sun shines on the side to be observed. That side should be vertical; the other side can be sloping to save digging.

Some of the information needed can be discovered from examining the pit. Other information, like a slope, can be obtained from the surroundings. The instructor will need to provide such information as flooding.

As the student examines each feature, he or she should circle the best description on the chart included in this appendix. Then consider how important the feature is, or how easily it can be corrected. Finally, come up with an overall rating, as described earlier in this appendix.

This appendix is adapted from the University of Minnesota Ag. Extension Service Environmental Education Activity Sheet #1, "Selecting Suitable Uses for Land," by Clifton F. Halsey.

SELECTING SUITABLE USES FOR LAND

USE	SUITABILITY	% SLOPE	FEATURES OF THE LAND AND SOIL						
			SOIL TEXTURE (Surface soil only unless specified otherwise.)	FLOODING DURING GROWING SEASON or use (frequency-duration)	INTERNAL DRAINAGE	DEPTH TO RESTRICTIVE LAYER (inches)	AVAILABLE WATER FOR PLANTS (in.) Capacity in 5′ deep	ERODIBILITY	OTHER IMPORTANT FEATURES NOT EASILY IMPROVED
Home lawns, shrubs, and gardens	Good—no important limitations	0–6	medium, moderately coarse	none	well drained, moderately well drained	more than 36	more than 9	slight	
	Fair—1 or 2 important limitations	6–12	moderately fine, fine, coarse	once a year, not over 3 days	somewhat poor, poor	20 to 36	6 to 9	moderate	stony, compacted
	Poor—severe limitations	more than 12	organic	more frequent or more than 3 days	excessive	less than 20	less than 6	severe	moderate to strongly alkaline, severely compacted
Cultivated crops	Good	0–6	medium, moderately coarse	none	well or moderately well drained	more than 36	more than 9	slight	
	Fair	6–12	moderately fine, fine, organic	once a year, less than 48 hours	somewhat poor (needs drainage)	20 to 36	6 to 9	moderate	stony
	Poor	more than 12	coarse	more frequent, longer	poor or excessive	less than 20	less than 6	severe	moderately to strongly alkaline
Permanent grass pasture	Good	0–18	medium or moderately fine, fine	up to 4 times annually, less than 48 hours	well or moderately well drained	more than 36	more than 9	slight	
	Fair	18–25	moderately coarse	more than 48 hours	somewhat poor, poor	20 to 36	6 to 9	moderate	
	Poor	more than 25	coarse, organic	5 times or more annually	excessive	less than 20	less than 6	severe	moderately to strongly alkaline
Forests—conifers, for wood products, Christmas trees	Good	0–18	(Entire Root Zone) medium, moderately coarse	never	well drained, moderately well drained	more than 20	more than 6		soil fertility: medium to high, pH: 5 to 6
	Fair	18–45	moderately fine, coarse	never	somewhat poor, excessive		3 to 6		soil fertility: low
	Poor	more than 45	fine, organic	any flooding	poor	less than 20	less than 3		
Forest—deciduous, for wood products	Good	0–18	(Entire Root Zone) medium, moderately fine	once in 10 years	well drained, moderately well drained	more than 20	more than 9		soil fertility: medium to high, pH: 6 to 8
	Fair	18–45	moderately coarse	annual spring flooding	somewhat poor, excessive		6 to 9		soil fertility: low
	Poor	more than 45	coarse, fine, organic	more than once annually	poor	less than 20	less than 6		pH: less than 5, more than 8
Forest—wildlife habitat	Good	0–25	(Entire Root Zone) moderately coarse thru moderately fine	once in 10 years, less than 24 hours	well or moderately well drained	more than 20	more than 9		
	Fair	25–50	coarse or fine, organic	up to 4 times annually, 24 to 48 hours	somewhat poor, excessive		6 to 9		
	Poor	more than 50		more frequent, more than 48 hours	poor	less than 20	less than 6		pH: less than 5, more than 8

SELECTING SUITABLE USES FOR LAND *(continued)*

USE	SUITABILITY	% SLOPE	FEATURES OF THE LAND AND SOIL						
			SOIL TEXTURE (Surface soil only unless specified otherwise.)	FLOODING DURING GROWING SEASON or use (frequency-duration)	INTERNAL DRAINAGE	DEPTH TO RESTRICTIVE LAYER (inches)	AVAILABLE WATER FOR PLANTS (in.) Capacity in 5′ deep	ERODIBILITY	OTHER IMPORTANT FEATURES NOT EASILY IMPROVED
Water fowl habitat	Good	Can have surface water continuously, open water all summer			poor drainage				
	Fair	Can have surface water and open water during spring only			somewhat poorly drained				
	Poor	Can have short periods of open water			moderately well to excessively drained				
A. Athletic, play and picnic grounds B. Public camp-grounds, primitive campsites	A. Good B. Good	0–2 0–6	medium, moderately coarse	none during use	well and moderately well drained	more than 36	more than 9	slight	
	A. Fair B. Fair	2–6 6–12	coarse, moderately fine	once—less than 24 hours	somewhat poor	20 to 36	6 to 9	moderate	
	A. Poor B. Poor	more than 6 more than 12	fine, organic	2–3 times during season	poor, excessive	less than 20	less than 6	severe	
Nursery	Good	0–3	moderately coarse, medium (BR), medium, medium fine (BB)	none	well drained	more than 48	more than 9	slight	
	Fair	3–6	coarse, medium fine (BR), moderately coarse, fine (BB),	once a year, not over 1 day	moderately well drained	36–48	6–9		stony
	Poor	>6	fine (BR) coarse (BB)	more frequent or longer	excessive, some-what poor, poor	less than 36	less than 6	moderate, severe	large stones
Land exposed to erosion during construction or use	Good	0–2	coarse, moderately coarse					none to slight	
	Fair	2–6	medium to fine					moderate	
	Poor	more than 6						severe	
Houses and other low buildings	Good	0–6	To 5 feet deep, coarse, moderately coarse, medium		excessive, well drained, moderately well drained, none	To bedrock, water more than 36			
	Fair	6–12	moderately fine	none	somewhat poor	20 to 36			
	Poor	more than 12	fine, organic	any	poor	less than 20			
Basements and utility excavations	Good	0–6	To 5 feet deep, medium, moderately coarse, coarse	none	well drained, excessive	To bedrock, water more than 5 feet			
	Fair	6–12	moderately fine	none	moderately well drained	3 1/2 to 5 feet			
	Poor	more than 12	fine, organic	any	somewhat poor, poor	less than 3 1/2 feet			

SELECTING SUITABLE USES FOR LAND *(continued)*

			FEATURES OF THE LAND AND SOIL						
USE	**SUITABILITY**	**% SLOPE**	**SOIL TEXTURE (Surface soil only unless specified otherwise.)**	**FLOODING DURING GROWING SEASON or use (frequency-duration)**	**INTERNAL DRAINAGE**	**DEPTH TO RESTRICTIVE LAYER (inches)**	**AVAILABLE WATER FOR PLANTS (in.) Capacity in 5′ deep**	**ERODIBILITY**	**OTHER IMPORTANT FEATURES NOT EASILY IMPROVED**
Home sewage absorption fields	Good	0–6	To 5 feet deep, medium, moderately coarse	none	well drained	6 feet			
	Fair	6–12	moderately fine	none	moderately well drained	6 feet			
	Poor	more than 12	coarse, fine, organic	any	somewhat poor, poor, excessive	less than 6 feet			
Local streets	Good	0–6	To 5 feet deep, moderately coarse, coarse	none	excessive, well drained, moderately well drained	To bedrock, more than 36			
	Fair	6–12	moderately fine, medium	once a year, not over 3 days	somewhat poor	20 to 36			
	Poor	more than 12	fine, organic	more frequent, more than 3 days	poor	less than 20			

APPENDIX 6

Preferred Soil Characteristics for Selected Trees

The following table suggests soil preferences for a number of trees. The information is gathered from a number of sources, including the author's own experience. No information was found for blank entries. These are only general guidelines, and in fact, there is often disagreement between experts. Please keep in mind that a particular plant variety, growing in a particular location, may differ from the suggestions in this table.

Botanical Name	Common Name	Text	pH	Comp	Drain	Salt
Abies balsamea	Balsam fir	CM	4.0–6.5	I	2–4	S
A. concolor	White fir	CM	4.0–6.5	S	3–5	I
Acer ginnala	Amur maple	CMF	4.0–7.5	I	3–5	I
A. campestre	Hedge maple	CMF	5.0–7.5	R	3–5	I
A. negundo	Box elder	MF	5.0–7.5	R	2–5	I
A. palmatum	Japanese maple	CMF	5.0–6.5	S	4–5	I
A. platanoides	Norway maple	MF	5.0–7.5	I	3–5	I
A. rubrum	Red maple	CMF	4.5–6.5	S	2–4	S
A. saccharinum	Silver maple	CMF	5.5–6.5	R	2–5	I
A. saccharum	Sugar maple	MF	6.0–7.0	S	3–4	S
Aesculus glabra	Ohio buckeye	CMF	6.0–7.0	I	3–5	I
Alnus glutinosa	European alder	MF	5.0–7.0	I	1–5	I
Amelanchier sp.	Serviceberries	CM	6.0–7.0	S	4–5	S
Betula nigra	River birch	CMF	4.0–6.5	R	2–4	I
B. papyrifera	Paper birch	CM	5.0–7.0	S	3–4	I
Catalpa speciosa	Eastern catalpa	CMF	6.0–8.0	R	2–5	I
Cedrus atlantica	Blue atlas cedar	CMF	5.0–7.0	S	3–5	I

Botanical Name	Common Name	Text	pH	Comp	Drain	Salt
C. deodora	Deodor cedar	CMF	5.0–7.0	I	3–5	
Celtis occidentalis	Common hackberry	CMF	6.0–8.5	I	2–5	I
Cercidiphyllum japonicum	Katsura tree	MF	5.0–7.0	S	3–4	S
Cercis canadensis	Eastern redbud	CMF	6.0–8.0	I	3–4	S
Cornus alternifolia	Pagoda dogwood	M	6.0–7.5	I	3–4	I
C. florida	Flowering dogwood	CMF	5.0–6.5	S	3–4	S
C. kousa	Kousa dogwood	M	4.5–6.5	S	4–5	S
Crataegus sp.	Hawthorns	CMF	5.5–7.5	I	3–6	I
X Cupressocyparis leylandii	Leyland cypress	CMF	5.0–7.5	I	4–5	R
Diospyros virginiana	Persimmon	CMF	5.5–7.0	I	2–5	I
Eleagnus angustifolia	Russian olive	CMF	5.0–8.0	I	3–6	R
Eucalyptus sp.	Gum trees	CMF	6.0–8.0	I	4–6	I
Fagus grandifolia	American beech	CM	5.0–6.5	S	3–4	I
Fraxinus americana	White ash	CMF	6.0–7.5	I	3–4	R
F. nigra	Black ash	MF	4.0–6.5	R	1–3	R
F. pennsylvanica	Green ash	CMF	5.5–7.0	R	2–5	I
Ginkgo biloba	Ginkgo	CMF	5.5–7.5	R	3–4	I
Gleditsia triacanthos	Honeylocust	MF	6.0–7.5	R	3–4	R
Gymnacladus dioicus	Kentucky coffeetree	MF	6.5–7.5	I	3–4	R
Ilex aquifolium	English holly	CM	4.0–6.0	I	2–4	
I. opaca	American holly	CMF	5.0–7.0	R	3–5	R
Juglans nigra	Black walnut	MF	6.5–8.5	I	3–5	I
Juniperus chinensis	Chinese juniper	CMF	6.0–7.5	S	4–5	I
J. scopulorum	Rocky mountain juniper	CM	6.0–8.0	S	4–6	I
J. virginiana	Eastern red cedar	CMF	6.0–8.0	S	3–6	I
Koelreuteria paniculata	Golden raintree	CMF	5.5–7.5	I	3–5	S
Lagerstroemia indica	Crape myrtle	CMF	5.0–7.5	I	3–4	I
Larix laricina	Tamarac	CMF	4.5–7.5	R	1–4	I
Liquidambar styraciflua	Sweetgum	CMF	5.5–6.5	R	2–5	I
Liriodendron tulipifera	Tulip tree	MF	5.5–6.5	S	4–5	S
Magnolia grandiflora	Southern magnolia	CMF	5.0–7.0	I	2–6	R
M. virginiana	Sweetbay magnolia	CMF	5.0–6.5		1–4	I
Malus spp.	Crabapple, apple	MF	5.0–7.5	R	3–5	I
Morus alba	White mulberry	CMF	5.0–8.0	I	3–5	R
Nyssa sylvatica	Black gum	CM	5.0–6.5	I	1–4	I
Ostrya virginiana	Ironwood	CMF	6.0–8.0	S	4–6	S

Botanical Name	Common Name	Text	pH	Comp	Drain	Salt
Picea glauca densata	Black hills spruce	MF	4.0–6.5	S	3–5	I
P. pungens	Colorado spruce	MF	4.0–7.0	S	3–5	I
Pinus banksiana	Jack pine	CM	4.0–6.5	S	4–6	S
P. flexilis	Limber pine	CM	4.0–7.0	S	4–6	S
P. nigra	Austrian pine	CM	5.0–7.0	S	3–5	I
P. ponderosa	Ponderosa pine	CM	4.0–6.5	S	3–6	I
P. resinosa	Red pine	CMF	4.0–6.5	S	3–6	S
P. strobus	White pine	CMF	4.0–6.5	S	3–5	S
Pistache chinensis	Chinese pistachio	CMF	5.0–7.5	I	3–6	I
Platanus occidentalis	Sycamore	CMF	6.0–8.5	R	2–5	S
Populus deltoides	Cottonwood	CMF	6.5–7.5	R	2–5	I
P. tremuloides	Quaking aspen	CMF	4.5–6.5	S	4–6	I
Prunus sp.	Plums, cherries	CMF	5.0–7.0	S	3–4	I
Pseudotsuga menziesii glauca	Douglas fir	CM	5.5–6.5	S	4–5	S
Pyrus ussuriensis	Ussurian pear	MF	6.0–7.5	I	2–6	
P. calleryana	Callery pear	CMF	5.0–7.0	R	2–5	I
Quercus alba	White oak	CMF	5.5–7.0	S	4–6	I
Q. bicolor	Swamp white oak	MF	5.0–6.5	R	1–5	I
Q. macrocarpa	Burr oak	CMF	4.0–8.0	S	3–6	I
Q. palustris	Pin oak	MF	4.5–6.5	R	2–5	S
Q. phellos	Willow oak	CMF	4.5–6.5	R	2–4	I
Q. rubra	Red oak	CMF	4.5–6.5	S	4–5	I
Q. shumardii	Shumard oak	CMF	6.0–8.0	I	2–5	I
Q. virginiana	Live oak	CM	5.0–8.0	I	3–6	R
Rhus sp.	Sumacs	CMF	5.5–7.5	S	4–6	I
Salix sp.	Willows	CMF	6.0–7.5	I	1–4	I
Sapindus drummondii	Soapberry	CM	6.0–8.0	I	3–6	S
Sassafras albidum	Sassafras	MF	5.5–6.5	S	4–6	I
Sophora japonica	Jap. pagodatree		5.5–7.5		3–5	R
Sorbus sp.	Mountain ash	CMF	5.0–7.0	I	3–5	S
Taxodium distichum	Baldcypress	CMF	6.0–6.5	R	1–4	I
Thuja occidentalis	Arborvitae	MF	6.0–8.5	R	2–4	I
Tilia sp.	Lindens	CMF	5.5–7.5	S	4–5	S
Tsuga canadensis	Eastern hemlock	MF	4.0–6.5	S	3–4	S
Ulmus americana	American elm	M	6.0–8.5	I	2–5	I
Zelkova serrata	Japanese zelkova	CMF	5.0–7.0	R	3–4	I

Key

Texture:

C—coarse texture

M—medium coarse, medium, medium fine

F—fine

Compaction and salt:

S—sensitive

I—intermediate

R—resistant

Drainage:

1—very poorly drained

2—poorly drained

3—moderately poorly drained

4—moderately well drained

5—well drained

6—excessively well drained

Glossary

A

Acid soil. A soil that contains more hydrogen ions than hydroxyl ions; soil pH is less than 7.0.

Actinomycete. An order of microbes common to soil; are bacteria but resemble fungi in having a mycelium. Important decomposers.

Adhesion. Force of attraction between two different substances. In soil, used to define the force that attracts water to soil particles.

Adhesion water. Inner layer of water molecules in a water film around a soil particle, held tightly by adhesion so it cannot move or be absorbed by plants.

Adsorption. Bonding of an ion or compound to a solid surface, usually temporarily. In soil, cations are adsorbed on clay and humus particles.

Aeration pores. Large (noncapillary) pores that normally drain free of liquid water because of gravitational drainage and fill up with air.

Aeration, soil. Process by which air in the soil is replaced by air from the atmosphere. Related to number, size, and continuity of soil pores and to internal drainage.

Aeration, turf. The procedure or removing numerous small cores of soil from turf to relieve compaction and improve soil aeration and water infiltration.

Aerobic. An adjective applied to organisms that grow, or processes that occur, in the presence of oxygen.

Aggregate, soil. A mass of fine soil particles glued together by clay, organic matter, or microbial gums. Aggregates are part of soil structure.

Agricultural lime. Products that neutralize acidity and contain Ca^{+2} and/or Mg^{+2} (carbonates, oxides, and hydroxides).

Air-dry soil. Soil allowed to dry out in open air without heating. Still contains hygroscopic water.

Alfisol. One of twelve soil orders. A mineral soil, usually formed under forest, common to northern and midwestern states. Has leached E horizon with accumulation of bases and clays in Bt horizon.

Algae. Simple chlorophyll-containing plants. Single-celled algae add organic matter to soil by photosynthesis.

Alkaline soil. A soil that contains more hydroxyl ions than hydrogen ions; pH greater than 7.0.

Alkalinity. A measure of the carbonate and bicarbonate content of water.

Allelopathy. Suppression of the growth of plants by substances exuded by the roots of other plants.

Alluvial fan. A fan-shaped alluvial deposit formed where flowing water slows down and spreads out at the base of a slope.

Alluvial soil. A soil developed from mud deposited by running water.

Amendment, soil. A substance mixed into the soil to improve its properties, excluding fertilizer.

Ammonia volatilization. Loss of NH_3 gas from the soil surface.

Ammonification. Process by which certain soil microbes convert organic nitrogen to ammoniacal nitrogen.

Anaerobic. (1) An adjective applied to organisms that grow, or processes that occur, in the absence of oxygen. (2) An adjective applied to soil void of oxygen.

Anchorage. Function of soil to hold plant firmly in place.

Andisol. Soil order with volcanic parent materials.

Anion. An ion with a negative or minus charge.

Anion exchange. Total sum of the number of exchangeable anions a soil can adsorb, expressed as milliequivalents per 100 grams of soil, or as centimoles charge per kilogram soil.

Antagonism. The suppression of the growth of one organism by another organism by the production of toxic or growth-inhibiting (antibiotic) substances.

Antibiotic. A substance produced by one species of organism that will kill or inhibit growth of some other organism.

Antitranspirants. Compounds that slow transpiration by plants to reduce water loss.

Aquifer. An underground formation that holds water. It is porous enough that water can flow through it to a well, so can be a source of groundwater.

Arid climate. A low precipitation region where evapotranspiration greatly exceeds precipitation. Too little rainfall to produce crops without irrigation.

Aridosol. One of twelve soil orders, common to arid regions. Soil is dry and low in organic matter.

Arthropods. A phylum of animals with no backbone, jointed body and legs, and usually a hard shell. Includes such soil animals as insects, spiders, sowbugs, and others.

Association, soil. A soil mapping unit in which two or more taxonomic soil units that occur together are combined.

Autotrophic. Capable of producing one's own food from carbon dioxide and/or carbonates by photosynthesis or oxidation of inorganic compounds.

Available nutrient. An essential element, or nutrient, in the soil in a form that plants can absorb into roots.

Available water. Mostly cohesion water, defined as lying between the field capacity and the wilting point (–1/3 to –15 bars, or –3.3 to –1.5 MPa soil matrix potential).

B

Banding. Placement of fertilizer, at planting, near the seed. May also be applied to surface or subsurface placement of fertilizer in strips.

Bar. An older term used to express water potential; slightly less than one atmosphere pressure.

Base saturation. The ratio of the quantity of exchangeable bases to the cation exchange capacity.

Basic soil. *See* alkaline soil.

Bedrock. Solid, or consolidated, rock lying under the soil. It may be, but is usually not, the parent material of the soil lying above it.

Beneficial element. Chemical element that may play a beneficial role in the development of many plants but that does not meet the criteria for being considered an essential element.

Biopore. Large soil macropore created by life, e.g., earthworth tunnels or channels remaining after root decay.

Bioremediation. The use of biological processes and agents such as plants or microbes to help degrade and thereby clean up environmental contaminants.

Biosolids. *See* sewage sludge.

Bioswale. Swale, or shallow ditch, heavily planted, designed to absorb runoff water.

Blocky structure. Block-like peds commonly found in B horizons. Two kinds are recognized: angular or subangular.

Border strip irrigation. Surface irrigation in which water is run into strips bounded by low earthen borders.

Bradyrhizobia. Genus of symbiotic nitrogen-fixing bacteria hosted by soybeans.

Broadcast. Application of fertilizer by spreading on the soil surface, usually before planting and incorporated by tillage.

Brown fields. Urban land rendered unusable by soil contamination or other problem, improved by remediation or abatement.

Buffer strip. A field border in soil conserving vegetation or mulch; leaving strips of tall vegetation at right angles to the prevailing wind.

Buffer test. Method of measuring lime requirement for an acid soil.

Buffering. The ability of a solution, like the soil solution, to resist changes in pH when acid or alkaline substances are added. Often used when speaking of soil to describe its resistance to pH changes when limed or acidified.

Bulk density. Mass of oven-dry soil per unit volume, usually expressed as grams per cubic centimeter.

Burned lime. CaO formed by roasting $CaCO_3$ in a kiln to drive off CO_2 as a gas; also called quicklime.

C

Calcareous soil. Soil high in calcium carbonate (lime), usually derived from limestone-rich parent materials.

Will bubble or "fizz" if treated with cold, dilute (0.1N) hydrochloric acid.

Calcitic limestone. Limestone rich in calcite ($CaCO_3$).

Calcium carbonate equivalent. The acid neutralizing ability of a lime product compared to that of an equal weight of $CaCO_3$.

Caliche. A zone of soil, near the surface, that is cemented by lime (magnesium or calcium carbonates). Common to arid soils.

Capillary. A very thin tube that water will flow into because of adhesive and cohesive forces. In the soil, small soil pores can act as capillaries.

Capillary fringe. A zone of soil just above a water table that is nearly saturated because of capillary rise.

Capillary rise. Movement of water upward in the soil through soil capillaries. Occurs as soil surface dries, drawing moisture from below.

Capillary water. Old term for water held loosely in capillary pores in the soil, capable of moving in the soil and being absorbed by roots.

Carbon-nitrogen ratio. Ratio of the weight of carbon to weight of nitrogen in an organic material. Obtained by dividing the percent carbon by the percent nitrogen.

Catena. A group of neighboring soils formed of similar parent materials under similar conditions at about the same time. The soils differ because of variations in relief and drainage. Synonym: toposequence.

Cation. An ion with a positive charge.

Cation exchange. Exchange between a cation in solution and one adsorbed on a soil colloid.

Cation exchange capacity. Total number of exchangeable cations a soil can adsorb. A measure of the soil's ability to hold nutrients that are cations in the soil. Expressed as milliequivalents per 100 grams of soil, or centimoles charge per kilogram soil.

Cemented. Hardened because soil particles are "glued" together by substances such as lime or iron oxides. Hardness persists even when wet.

Center-pivot irrigation. Irrigation by a pivoting boom of sprinkler nozzles driven slowly in a circular pattern about the pivot point by the hydraulic pressure of the water in the boom.

Chemical guarantee. The percent Ca, percent Mg and calcium carbonate equivalent for a lime product.

Chemical weathering. Breakdown of rocks and minerals by chemical reactions, mostly with water.

Chiseling. (1) Primary tillage with a chisel plow, which pulls long, curved teeth through the soil to loosen it without turning it over. (2) Using a subsoiling chisel plow to break up deep compacted soil layers. (3) Using chisels to inject fluid fertilizers or ammonia into the soil.

Chlorite. A 2:1 silicate clay in which a fourth sheet holds together the 2:1 layers.

Chlorosis. Common sign of nutrient shortage, showing as a loss of normal green color in a plant. Color loss means a failure to make enough chlorophyll and may result from shortage of nutrients involved in chlorophyll formation (nitrogen, sulfur, iron, and others).

Chroma. One of the three variables in the Munsell color system. Chroma is the purity, or strength, of a color; its opposite is the amount of grayness.

Clay. (1) The class of smallest soil particles, smaller than 0.002 millimeter in diameter. (2) The textural class highest in clay.

Claypan. A dense subsoil layer with a higher clay content than the soil above it. *See* pan.

Clod. A user-made soil aggregate, produced by tillage when a soil is too wet or dry. Clods may vary in size from a quarter inch to ten inches.

Coarse aggregates. For horticultural soil-potting mixes, refers to large, inorganic particles used to create large pores in the mix. These materials include coarse sand, perlite, vermiculite, shredded plastics, and others.

Coarse texture. A soil texture whose traits are largely set by the presence of sand. Includes sands, loamy sands, and most sandy loams.

Cohesion. The force attracting similar substances. In the soil, applied to attraction of water for itself.

Cohesion water. Outer film of water around soil particles, loosely held in place by cohesion to the inner film of adhesion water.

Colloid. A very tiny particle capable of being suspended in water without settling out rapidly. Soil colloids, mostly humus and clay, have a charged surface that attracts cations.

Colluvium. A deposit of rock and soil resulting from materials sliding down a slope under the force of gravity.

Compaction, soil. The squeezing together of soil particles by the weight of farm and construction equipment, vehicles, and animal and foot traffic. Compaction reduces average pore size and total air space in the soil.

Complete fertilizer. A fertilizer containing all three of the primary macronutrients—nitrogen, phosphorus, and potassium.

Composite sample. The soil sample sent to a soil-testing laboratory, resulting from mixing together many individual samples. It should represent the average soil in a field.

Composting. Piling organic materials under conditions that cause rapid decay. Reduces the carbon-nitrogen ratio and destroys many weed seeds and disease organisms.

Conductivity meter. An instrument used to measure the soluble salts concentration in soil or potting mixes. It uses the principle that the more ions dissolved in water the better the water will conduct electricity.

Cone penetrometer. A device with a cone-shaped rod which is pushed into the soil. A dial measures the pressure required to penetrate the soil. Readings are used as an index of compaction.

Conservation tillage. A tillage practice that leaves crop residues on a rough soil surface to reduce erosion.

Consistence, soil. Characteristics of a soil in its response or resistance to pressure, as described at various soil moisture contents. Such characteristics include stickiness, plasticity, hardness, or friability.

Consumptive use. Amount of water transpired from plants, incorporated into plant tissue, and evaporated from soil.

Contour. An imaginary line across a slope that stays at the same elevation.

Contour tillage. Tillage following the contours of a slope, rather than up and down a slope. Helps prevent erosion and runoff.

Conventional tillage. A tillage system which uses primary tillage such as a moldboard plow and secondary tillage such as harrows which produces a fine seedbed.

Cover crop. A crop planted to prevent erosion on a soil. Cover crops can be planted on soils not currently being farmed, between crop rows, or after main crop harvest.

Cropland. Land suited or used for the purpose of raising crops.

Cross-wind ridges. Low ridges in soil at right angles to prevailing wind, creating by tillage, to reduce wind erosion.

Cross-wind trap strips. Bands of vegetation planted at right angles to prevailing wind to trap soil particles blowing across a field.

Cultivation. Tillage to control weeds and loosen soil.

Cyanobacteria. Certain nitrogen-fixing microbes, previously called blue-green algae, now classified as bacteria.

D

Debris basins. Small ponds used to capture runoff and sediment during construction activities.

Decomposers. Microbes that obtain their food from dead organic materials, causing decay and decomposition.

Delta. A usually fan-shaped alluvial deposit created where a stream or river enters a body of quiet water like an ocean or a lake.

Denitrification. Chemical reaction caused by certain microbes in the soil that change nitrate nitrogen to gaseous nitrogen or gaseous nitrogen oxides.

Deposit. Loose material left in a new place after being carried by wind, water, ice, gravity, or man.

Desertification. Conversion of land to desert, often caused by overgrazing, deforestation, or other disturbance.

Detritus. Dissolved or particulate dead, but not decomposed, organic material.

Diagnostic horizon. Any of a series of specific types of soil horizons used to assign a soil to its proper soil order.

Diffusion. Flow of matter through a liquid or gas by the random movement of molecules. In soil science, applied to movement of nutrient ions through soil solution. The movement is caused by a concentration gradient, with the ions moving from more to less concentrated.

Disc plow. A primary tillage tool which inverts the soil by using large (2 to 2.5 feet) disc blades instead of moldboards.

Diversion. Changing the direction of movement. In soil conservation, a special terrace built to divert the flow of running water.

Dolomitic limestone. Limestone rich in the mineral dolomite (calcium-magnesium carbonate).

Double cropping. Harvesting two crops from the same piece of land in one year.

Drainage. (1) The speed and amount of water removal from soil by runoff or downward flow through soil. (2) Amount of time when soil is free of saturation.

Drip irrigation. Irrigation method in which water drips out of a specially designed trickler into the soil surface under a plant.

Drought. A period of soil dryness that seriously harms plant growth.

Dryland farming. Methods of producing crops in low-rainfall areas without irrigation.

Duripan. A soil layer hardened and cemented by silica.

E

Eluviation. Removal of a material, such as clay or nutrients, from a layer of soil by percolating water.

Entisol. One of twelve soil orders. An undeveloped soil with no distinctive subsurface diagnostic horizons within one meter of the soil surface.

Enzymes. Protein molecules synthesized in the cells of living organisms that mediate or accelerate chemical reactions in a catalytic mode.

Eolian soil deposits. Wind-deposited soil material, mostly silt and fine sand.

Ephemeral gullies. Erosion in which running water makes large rills that are not completely filled in when the ground is tilled, leaving a channel for further erosion. Soil carried off fields by ephemeral erosion is not predicted by the Universal Soil Loss Equation.

Erodable. Soil that is easily eroded, due to a variety of factors.

Erosion index (EI) or erosion potential. Measurement of the inherent erodability of a soil used without preventive measures. Using the Universal Soil Loss Equation, $EI = RKLS/T$.

Essential element. An element needed by plants for proper growth and reproduction.

Eutrophication. The rapid increase in growth of aquatic plants and algae caused by pollution of water by phosphates and, to a lesser extent, nitrates.

Evapotranspiration. The sum of water lost from soil by evaporation and transpiration.

Exchangeable base. A cation, excluding hydrogen and aluminum, held on cation exchange sites that can be easily replaced by another cation. Considered to be available for plant growth.

Expanding clays. Clays that expand greatly upon adsorption of interlayer water and shrink upon drying.

F

Fallow. Soil left idle to accumulate water and/or mineral nutrients. Weeds are controlled during fallow chemically or by cultivation.

Family. Level of soil taxonomy just above the soil series.

Fertigation. Fertilizing with soluble fertilizers through an irrigation system, usually sprinkler or trickle.

Fertilizer. A material added to the soil to supply essential elements. State laws may set minimum requirements for materials sold as fertilizers.

Fertilizer analysis. Composition of fertilizer measured by chemical tests. On a bag of fertilizer, would appear as a listing of the percent of each of the nutrients contained in the bag, including primary, secondary, or trace elements.

Fertilizer burn. Damage to plant tissue resulting from overapplication of fertilizer, a form of soluble salt damage.

Fertilizer carrier. A compound mixed into a fertilizer to supply a nutrient.

Fertilizer filler. A non-nutrient material added to fertilizer; for instance, clay, sand, or corncob granules.

Fertilizer grade. The guaranteed minimum analysis in whole numbers of nitrogen, available phosphate, and water-soluble potash, listed as "N-P_2O_5-K_2O."

Fertilizer, inorganic. Fertilizer that contains no carbon. For the purposes of this text, fertilizers that are unaltered minerals are considered separately. Urea, while chemically organic, is often classified inorganic because of rapid hydrolysis in the soil to ammonium ions.

Fertilizer, organic. Fertilizer that contains nutrients plus carbon and hydrogen. Often excludes urea. *See* fertilizer, inorganic.

Fertilizer ratio. Proportion of the primary nutrients in a fertilizer. Obtained by dividing the grade by the lowest common denominator.

Fertilizer, starter. A small amount of fertilizer placed near seeds or transplants to promote early growth, generally high in phosphorus.

Fibric. Adjective applied to only slightly decomposed organic matter.

Field capacity. The percentage of water remaining in the soil after drainage has just stopped.

Filter strip. Strip of heavily vegetated land bordering bodies of water designed to filter sediments and pollutants out of runoff before it enters the body of water.

Fine texture. Soil with a large amount of clay. Usually includes clay, sandy clay, clay loam, silty clay, and silty clay loam.

Finishing harrow. This secondary tillage device completes the job of pulverizing the soil; a drag.

Fixation. (1) A process that changes chemicals from soluble or available forms to insoluble or unavailable forms in the soil. (2) Conversion of gaseous nitrogen to ionic forms.

Flood plain. Land near a stream that is commonly flooded when the stream is high. Soil is built from sediments deposited during flooding.

Fluid fertilizer. Fertilizer used in liquid form, either a solution or a suspension.

Fluid lime. Finely ground lime in a slurry or suspended form in water.

Foliar fertilization. Fertilizing plants by spraying leaves with fluid fertilizers.

Forest soil. Soils developed under forest vegetation.

Fragipan. Naturally occurring hard, brittle subsoil layer high in clay that restricts root growth.

Friable. A consistency term, expressing how easily a moist soil can be crumbled.

Fritted trace element. A slow-release trace element fertilizer in which fertilizer carriers are mixed into glass powder.

Frost wedging. Breakage of rocks caused by pressure created by water freezing in cracks in the rock.

Fungi. Important soil organisms, especially as decomposers. Considered to be either primitive plants or as one of five kingdoms of living organisms.

Furrow diking. Tillage with special tool that creates ridges and furrows with series of small dams and basins in the furrow. The technique retains water in the furrows.

Furrow irrigation. Delivery of irrigation water down furrows from a source ditch.

G

Gelisol. One of twelve soil orders, established in the soil taxonomy in 1998. A soil whose subsoil is permafrost.

Glacial drift. General term for debris deposited by glaciers. Common in northern tier of states.

Glacial outwash. Glacial drift deposited in water flowing away from a melting glacier. Outwash is sorted by the running water.

Glacial till. Glacial drift that deposits in place as glacier melts, unsorted.

Gley. Soil layer that develops under poor soil drainage conditions, has gray color and mottles.

Gleying. A process or condition of reducing conditions (oxygen depletion) indicated by the presence of gray and mottled gray colors in a soil horizon.

Granular structure. Commonly found in A horizons. Each ped is roundish.

Granules. Aggregates of fertilizer particles often all of a uniform guaranteed nutrient content.

Gravitational flow. Water movement driven by gravity in the soil in a downward direction in the earth's gravitational field.

Gravitational potential. That part of the soil total water potential associated with position in a gravitational field. This energy of position (potential energy) could be (+) or (−) in sign depending on vertical position with respect to a reference level, but usually positive.

Gravitational water. Water that moves through the soil under the influence of gravity.

Great group. A taxonomic level of the current soil classification system.

Greenhouse effect. Warming of the earth as energy (long-wave radiation) radiating from the earth is trapped by certain atmospheric gases, including carbon dioxide, methane, and others.

Greenhouse gas. Gases in the atmosphere that block the loss of energy from the earth as long-wave radiation, leading to the greenhouse effect. Includes water vapor, carbon dioxide, methane, and nitrogen oxides; each may be released into the atmosphere by soil processes.

Green manure. A crop grown to be turned under while still green to improve the soil.

Green-tissue test. A qualitative testing or evaluation of nutrient content in freshly collected and macerated green leaf tissue.

Groundwater. Water stored underground in a saturated zone of rock, sand, gravel, or other material.

Gully erosion. A form of water erosion characterized by deep and wide channels. The channels are too wide to cross with ordinary equipment.

H

Hand-move irrigation. Irrigation equipment that is moved by hand from one location of operation to the next.

Hardpan. A hard subsoil layer caused by cementation by carbonates or other chemicals; limits root growth and the infiltration of water.

Heavy metal. Group of high-density metallic elements including lead, cadmium, and others. A few needed as essential elements in trace amounts by plants or animals, otherwise toxic to biological systems. May be naturally occurring in soils or contained in fertilizers, sludges, composts, other soil amendments, or contaminants.

Hemic. Adjective applied to moderately decomposed organic matter.

Heterotroph. Organism not capable of producing own food. Obtains energy for life processes by oxidation of organic compounds.

Histosol. One of twelve soil orders. Histosols are organic soils.

Horizon, soil. A horizontal layer of soil, created by soil-forming processes, that differs in physical or chemical properties from adjacent layers.

Hue. One of the three color variables in the Munsell color system. Refers to the actual color of soil. *See also* chroma and value.

Humification. The process of transforming organic matter to humic substances through biological and non-biological processes.

Humus. Decay-resistant residue of organic matter decomposition. Humus is dark-colored and highly colloidal.

Hydrated lime. Ca $(OH)_2$ produced by adding a controlled amount of water to CaO; also called slaked lime.

Hydraulic conductivity. A trait of soil relating to the ease of water movement in that soil. The finer the soil texture, the lower its hydraulic conductivity.

Hydric soil. A soil that is saturated, flooded, or ponded long enough during the growing season to develop anaerobic conditions in the upper part.

Hydrogen bond. Bond between two molecules in which a hydrogen of one molecule bonds to an oxygen or nitrogen of another molecule.

Hydrologic cycle. The circular route of water from the atmosphere back to the atmosphere after it has undergone precipitation, runoff, percolation, storage, or evapotranspiration.

Hydrology. The science dealing with the properties, distribution, and circulation of water.

Hydrolysis. The reaction of a compound with water to create a new compound.

Hydrophilic gel polymers. Synthetic gel polymers capable of modifying the water retention characteristics of soil or potting soil media.

Hydroponic crops. Crops grown in nutrient solutions rather than soil.

Hydroseeding. Blowing a mixture of seed, water, and mulching materials onto land to establish vegetation.

Hygroscopic water. Water held tightly by adhesion to soil particles. Cannot be used by plants and remains in soil after air-drying. Can be driven off by heating.

Hyphae. Individual strands of the vegetative body of fungi. Hyphae can grow into organic matter to cause decay and can surround a mass of soil particles to make soil aggregates.

Hypoxic. Insufficient oxygen to sustain aerobic respiration. Applied here to severely oxygen-depleted waters.

I

Igneous rock. Rock formed from the cooling of molten rock from deep in the earth.

Illuviation. Deposition in a soil layer of materials transported from a higher soil layer by percolating water.

Immature soil. A young soil that is still changing relatively rapidly in response to its surroundings. Usually has little horizon development.

Immobilization. Absorption of an available nutrient by a soil organism or plant, changing it to an unavailable organic form.

Inceptisol. One of twelve soil orders, usually young soils with weak horizons.

Infiltration. Downward entry of water into the soil. *See* percolation.

Inoculation. Adding microbes to soil, seed, or culture medium. For instance, to treat legume seeds with bacteria from the *Rhizobium* genus.

Inorganic. Any chemical that does not contain both carbon and hydrogen. Inorganic nitrogen, for instance, includes nitrates (NO_3^-) but not urea (NH_2CONH).

Interception. The arrival of nutrients at a root surface simply by displacement of a soil volume by root growth.

Ions. Atoms or molecules that are electrically charged because of gaining or losing electrons.

Irrigation. Artificial application of water to the soil to improve crop growth, or sometimes for other purposes like activating herbicides.

Isomorphous substitution. Replacement of one atom by another of similar size in a crystal lattice. May result in a charge if the atom has a different charge than the one replaced.

L

Labile. Carbon source readily transformed by soil microorganisms, i.e., easily decayed.

Lacustrine. Mineral sediments deposited in fresh water.

Land capability classes. Eight soil classes ranked for their suitability for agriculture according to risk of erosion and other factors.

Land capability subclass. A subclass of a capability class that designates one of four problems: erodability (e), wetness (w), climate (c), or other soil factors (s).

Landscape. (1) The natural features of the earth's surface like hills, trees, and water. Usually the piece of land that can be seen by the eye in a single view. (2) The designed landscape around buildings or other structures of the built environment.

Land-use planning. Developing plans for using land that will best serve the long-term public interest.

Leaching. Removal of soluble material in solution from the soil by percolating water.

Leaching requirement. The fraction of the water entering the soil that must pass through the root zone in order to prevent soil salinity from exceeding a specified value.

Legume. A member of the Fabaceae family of plants, such as soybeans, peas, clover, alfalfa, locust trees, and many other economically important plants. Legumes host the *Rhizobia* bacteria that fix nitrogen.

Levee. Alluvial deposit of a shallow ridge along a river, resulting from coarse deposits during flooding.

Lignins. Complex, decay-resistant organic chemicals that glue together cellulose fibers in a plant to make it rigid.

Lime. Materials used to neutralize acidity, containing calcium or magnesium carbonate, oxide, or hydrate.

Lister plow. A special kind of plow which has two moldboards mounted back to back, resulting in a pattern of ten-inch high ridges and furrows across the field.

Load-bearing capacity. Ability of a soil to carry a load like a roadbed or building without shifting.

Loam. A medium soil texture class, in which sand, silt, and clay contribute almost equally to soil properties.

Lodging. Breaking of plant stems because of stem weakness, often caused by excess nitrogen and low potassium.

Loess. Wind-deposited silt.

Luxury consumption. Absorption by plants of more nutrients, especially potassium, than they need at the time. The nutrients may, however, be used for later stages of growth and can be used by animals feeding on the plant.

M

Macrofauna. Larger, easily visible soil animals.

Macronutrients. An essential element used in large amounts by plants, including nitrogen, phosphorus, potassium, calcium, magnesium, and sulfur.

Macropores. Large, noncapillary pores transmit water rapidly and are important for soils to drain readily. As they drain, water is replaced with air.

Mapping unit, soil. Basis for setting boundaries in a soil map. May be phases of series, families, or other taxonomic units, or may be associations of such units.

Marine sediments. Parent material which settled to the bottom of old oceans and seas.

Marl. A soft limey material deposited in fresh water that can be used as an agricultural lime.

Mass flow. Movement of nutrients by movement of soil water.

Massive soil. A structureless soil in which each soil particle sticks to neighboring particles. Common in C horizons or puddled soils.

Master horizons. O, A, E, B, C, and R horizons.

Matric potential. The soil energy level associated with attraction of water to soil solids.

Mature soil. A soil in equilibrium with its surroundings, usually with well-developed horizons.

Mechanical analysis. The measurement of the amounts of sand, silt, and clay in a soil. This measurement uses the principle that the larger a particle, the faster it sinks in water.

Media. Mixtures which are used to grow potted plants. They have high air- and water-holding capacities.

Medium-textured soil. Soils intermediate between fine- and coarse-textured soils. Includes loam, fine sandy loams, silt loam, and silt.

Mesofauna. Small soil animals slightly larger than microscopic, about 0.2 to 2 mm in size, including large nematodes and small but visible arthropods.

Mesophilic organism. Organisms whose optimum temperature for growth is an intermediate range, between 15 and 35°C. Dominant microorganisms in early and late stages of composting.

Metamorphic rock. Rock that has been changed by heat or pressure in the earth.

Methane. Flammable gas, also known as swamp gas; main component of the fuel natural gas. Produced by biological processes in anaerobic soils, and a potent greenhouse gas.

Micelle. An individual particle of silicate clay.

Microfauna. Microscopic soil animals or animal-like organisms, including protozoa, nematodes, and tiny arthropods.

Microflora. Soil bacteria, fungi, algae, or viruses.

Microirrigation. The frequent application of small quantities of water as drops, tiny streams, or miniature spray through emitters or applicators placed along a water delivery line. Microirrigation encompasses a number of methods or concepts such as drip or microspray.

Micronutrient. Essential element used in small quantities by plants.

Microorganism. An organism too small to be seen without the aid of a microscope.

Micropores. Small, capillary pores which transmit water slowly and hold water against gravity. They are important for the storage of water.

Microspray irrigation. A form of microirrigation that employs small sprayheads to sprinkle a small area, such as a landscape bed of shrubs or a single tree in an orchard.

Mineral. A pure inorganic compound in the earth's crust. Most rocks are mixtures of minerals. Often used in soil science to mean the inorganic solid particles of soil.

Mineral soil. A soil whose traits are determined mainly by its mineral content; mineral soils contain less than 20 percent organic matter.

Mineralization. Conversion of elements in organic forms to inorganic forms by decay. *See* immobilization.

Minimum tillage. Tillage methods that involve fewer tillage operations than conventional tillage. *See* conservation tillage.

Mixed fertilizer. A fertilizer containing more than one primary nutrient.

Moldboard plow. A primary tillage tool which shears off a section of soil, tips it upside down, and fractures it along several planes.

Mollisol. One of twelve soil orders. A soil with a high organic-matter topsoil and high base saturation; usually formed under prairie vegetation.

Mottling. Spots of different colors in a soil, usually indicating poor drainage. *See* gley.

Muck. An organic soil in which the organic matter is mostly decomposed. *See* peat.

Mulch. A material spread on the soil surface, like straw, leaves, plastic, or stones to protect soil from freezing, raindrop impact, evaporation, and heaving.

Munsell color system. A system used to identify soil color by means of three variables: hue (color), value (intensity), and chroma (purity). Identification is done by comparing soil to a set of standard color chips.

Mycelium. A word for the hyphae or filamentous strands of fungi.

Mycorrhizae. Fungi that form a symbiotic relationship with plant roots. The fungi help the plant absorb water and nutrients, while they receive food from the plant.

N

Nematode. Small unsegmented worm; many are parasitic on plant roots.

Nitrification. Microbial conversion of ammonium nitrogen to nitrate nitrogen.

Nitrogen cycle. Series of changes of nitrogen from atmospheric nitrogen, fixation in soil, a series of changes in the soil, and returning to the atmosphere.

Nitrogen depression period. Period of time during decay of organic matter in which nitrogen is being used up by microbes faster than being released by decay (immobilization exceeds mineralization). Resulting temporary nitrogen tie-up can cause nitrogen shortage in crops.

Nitrogen fixation. Microbial conversion of gaseous nitrogen to organic nitrogen in the soil.

Nonexchangeable ions. A term applied to an ion so strongly adsorbed to a soil colloid that it is not normally available to be exchanged with ions in the soil solution.

Nonpoint source. Source of water pollutants that does not come from a single, easily identified point. Example would be runoff from farm fields. *See* point source.

Nonsymbiotic nitrogen fixation. Microbial conversion of gaseous nitrogen to organic nitrogen in the soil by free-living bacteria which do not live on plant roots.

No-tillage. Method of growing crops that involves no tillage. Seeds are planted in slits in soil and chemicals are used to control weeds.

Nutrient carriers. Compounds in fertilizers that are included as useful forms of plant available nutrients.

Nutrients. Elements in the form of ions or molecules used in the metabolism of plants, animals, and microbes.

O

Order, soil. Highest taxonomic level in the USDA soil classification system. There are twelve recognized soil orders.

Organic. A material containing both carbon and hydrogen, and often oxygen, nitrogen, sulfur, or other elements. *See* inorganic.

Organic amendment. A soil amendment that is mostly organic matter.

Organic farming. Farming without using inorganic fertilizers or artificial pesticides.

Organic matter. Material of plant or animal origin that decays in the soil to form humus.

Organic soil. Soil containing more than 20 percent organic matter. Soil properties are dominated by the organic matter.

Osmotic potential. The contribution of dissolved salts or sugars (osmotically active solutes) to the energy of water. The more osmotically active solutes the lower the energy level of the water.

Oven-dry soil. Soil dried at 105°C until it reaches a constant weight.

Oxidation. A chemical reaction in which an element loses electrons to another participant in the reaction, often oxygen. For instance, iron is oxidized by oxygen to form rust.

Oxide clay. *See* sesquioxide clay.

Oxisol. One of twelve soil orders; highly weathered and leached tropical soils, limited to Hawaii and Puerto Rico in the United States.

P

Pan. A dense, hard, or compacted layer in soil that slows water percolation and movement of air and obstructs root growth. Pans may be caused by compaction, clay, or chemical cementation.

Parasite. An organism that lives off another organism (the host). The host is injured by the parasite. *See* symbiosis.

Parent material. The unconsolidated mineral or organic matter from which the solum (A, E, B horizons) has developed.

Particle density. The mass per unit volume of soil particles, excluding pore space. Most mineral soils have a particle density of about 2.65 grams per cubic centimeter.

Peat. Undecayed or slightly decayed organic soil, formed underwater where low oxygen conditions inhibit decay.

Ped. A natural soil aggregate, such as a granule or prism.

Pedology. Study of the formation and classification of soil.

Pedon. The smallest soil body. A section of soil that extends to the root depth and has about ten to ninety square feet of surface area.

Perched water table. A zone of saturated soil that, because of obstructions or other reasons, is maintained above the normal water table. Also applied to the layer of saturated soil in the bottom of a pot of soil, because excess water cannot drain to the normal water table in the soil.

Percolation. Downward movement of water through the soil profile.

Percolation test. A test for soil drainage that involves timing the speed that water will drain out of a hole dug in the ground.

Perforation. A method of fertilizing shade trees by drilling holes in the soil under the tree in a grid pattern and then filling the holes with fertilizer and sand.

Perlite. Large granules of lightweight expanded volcanic glass used in potting mixes.

Permafrost. Continuously frozen material under the solum or a perennially frozen soil horizon. Feature of soils of the order Gelisols.

Permanent wilting point. Water content of soil when a plant wilts and does not recover when placed in a humid chamber. Also called the permanent wilting percentage.

Permeability. Ease with which gases, liquids, and plant roots pass through a specific mass of soil.

pH, soil. Measure of the acidity or alkalinity of a soil. Technically, reciprocal of the logarithm of the hydrogen ion concentration in the soil solution.

Phase. A subdivision of the soil series, designating a difference from the series norm. Includes slope, degree of erosion, stoniness, or surface texture.

Photosynthesis. The reaction, in the presence of chlorophyll, of carbon dioxide and water to form sugar, using light energy.

Physical guarantee. The size guarantee or fineness factor for a lime product.

Physical properties. Traits of a soil caused by physical forces that can be described by physical terms or equations. Examples include soil texture, structure, and bulk density.

Physical weathering. Breakdown of rock particles by physical forces like frost action or wind abrasion.

Plasticity. A consistence term describing how easily a mass of soil can be shaped and molded when wet.

Platy structure. Soil structure in which platelike soil peds lie horizontally in the soil. Common in E horizons or compacted A horizons.

Plinthite. A mixture of various chemicals in the soil that when exposed to cycles of wetting and drying permanently hardens to a bricklike state. Common to tropical soils.

Plow layer. Upper part of a soil profile disturbed by humankind by plowing or other disturbances; *P* horizon suffix.

Plowpan. A tillage pan formed just under the plow layer.

Point source. Source of water pollutants easily traced to a single point, such as a factory pipe or animal feedlot.

Polypedon. A group of similar neighboring pedons that makes up a soil series.

Pop-up fertilizers. Fertilizers placed directly in contact with the crop seeds. If large amounts of water-soluble fertilizers are used, these can cause osmotic injury and not have the desired promotion of early growth.

Pore space. Portion of soil not occupied by solid material but which is filled with air or water.

Porosity. Percentage of soil volume not occupied by solid material.

Potentiometer. Device to measure soil-water potential and particularly soil-matric potential. (Tensiometer is an older term.)

Prairie soil. Soil formed under grassland vegetation.

Precipitation. A form of water falling to earth from the atmosphere; may be rain, snow, mist, or hail.

Precision farming. A system of crop pest and fertility management on very small locations so that very specific (almost prescription) recommendations and applications can be made for fertilizer or pesticides.

Predator. An organism that hunts and eats prey.

Preferential flow. Usually flow of free water through large macropores like biopores, bypassing the general soil matrix.

Pressurized liquid. A fertilizer delivered to an applicator knife through a pressurized delivery system. Anhydrous ammonia is an example.

Prills. Fertilizer pellets that are spherical in shape.

Primary consumer. Organism that feeds on plants; the second level of the food chain.

Primary macronutrient. The three macronutrients (nitrogen, phosphorus, potassium) needing to be added to the soil in the greatest amounts for good crop growth.

Primary producer. Lowest level in the food chain, organisms that produce their own food. Mostly chlorophyllous plants and microorganisms.

Primary tillage. The first step in seedbed preparation in conventional and mulch tillage. Breaks up and loosens soil and buries some or all crop residue.

Prime farmland. Rural land with the best combination of physical and chemical characteristics for producing food, feed, forage, fiber, and oilseed crops, and is available for these uses.

Prismatic structure. Soil structure with large, prism-shaped peds arranged vertically in the soil, usually in the B horizon.

Profile, soil. The vertical section of a soil through all its horizons, ending in the parent material.

Protein. Important nitrogen-containing compounds in living matter, from which comes most of the nitrogen in organic matter.

Protozoan. Single-celled, nonbacterial life form formerly classified mostly as small animals, like amoeba.

Puddling. Dispersal of soil aggregates caused by working soil when wet, creating a massive surface layer.

Pulverized fertilizers. Finely ground or divided fertilizer solids.

R

Rangeland. Land in arid or semiarid areas with permanent plant cover used for grazing.

Reaction, soil. The degree of acidity or alkalinity of a soil, expressed as pH.

Recalcitrant. Carbon source not readily transformed by soil microorganisms, such as high-lignin plant residues. *See* labile.

Recharge basin. Landscape depression in which ponded water percolates through the soil to recharge an underlying aquifer.

Redoxymorphic feature. Soil color patterns indicating alternating reducing and nonreducing conditions (saturated/nonsaturated), such as mottles.

Reduced tillage. *See* minimum tillage.

Relief. Variations in elevation in the landscape.

Residual soil. Soil formed in place from bedrock, rather than from transported parent materials.

Resistance block. A gypsum block with internal electrodes that when in equilibrium with moist soil, allows (with calibration) electrical resistance to be interpreted as soil water content.

Respiration. Biological reaction in which carbohydrates are broken down to carbon dioxide and water with the release of energy. Opposite of photosynthesis.

Retention pond. Man-made depression installed to collect runoff water. May or may not be also a recharge basin.

Revised Universal Loss Equation (RUSLE). A water erosion prediction model incorporating improvements to the Universal Soil Loss Equation (USLE).

Rhizobium. Genus of bacteria that live symbiotically on legume roots and can fix atmospheric nitrogen.

Rhizosphere. A zone of soil immediately surrounding plant roots that supports a high population of microorganisms feeding on organic materials released by the root.

Rill erosion. A form of water erosion characterized by numerous shallow and narrow channels. The channels can usually be filled in by ordinary tillage equipment.

Rip-rap. Large stones laid on top of the soil to protect it from erosion.

River terrace. A former river floodplain now at a higher elevation.

Root interception. *See* interception.

Root wedging. Rocks forced apart by root pressure.

Row crops. Crops planted in wide rows for ease of cultivation or other weed-control practices; commonly corn, soybeans, cotton, peanuts, and others.

Runoff. Water that falls on the soil but fails to be absorbed; flows on the soil surface.

S

Salination. The accumulation of soluble salts in the soil, usually from irrigation.

Saline seep. Small area of saline soil resulting from summer fallow.

Saline soil. A soil high in soluble salts but without too much exchangeable sodium. The pH is between 7.0 and 8.5; electrical conductivity is greater than 4.0 millimhos per centimeter; exchangeable sodium percentage is below 15.0.

Saline-sodic soil. Soil has both high soluble salt and sodium levels. The pH is between 7.0 and 8.5; electrical conductivity is greater than 4.0 mmhos/cm; exchangeable sodium percentage is greater than 15.0.

Saltation. Movement of soil particles in which wind causes fine sand to hop along soil surface.

Sand. (1) Largest of the soil separates, between 0.05 and 2.00 millimeters in diameter. (2) Coarsest textural class.

Sapric. Adjective applied to fully decomposed organic matter.

Saprophyte. Microorganism that feeds on dead organic matter.

Saturated flow. Water movement in the soil when all pores are filled with liquid water.

Saturation. (1) All or most soil pores filled with water. (2) The amount of the cation exchange capacity filled with a certain cation.

Secondary consumer. An organism that feeds on a primary consumer; the third level of a food chain.

Secondary macronutrient. Those macronutrients used less often as fertilizers than the primary elements. Includes calcium, magnesium, and sulfur.

Secondary tillage. Tillage operation following primary tillage that smooths out the soil for a fine seedbed.

Sediment. A layer of loose material deposited by wind or water. Term often applied to eroded soil deposited off the field, such as in lakes or streams.

Sedimentary rock. Rock made of sediments hardened over time by chemicals or pressure.

Seeps. Hillside springs or zones where water flowing laterally within the soil breaks out into the open.

Semiarid climate. Area where rainfall is low enough to limit crop production, but where dryland farming can be used. Evapotranspiration slightly exceeds precipitation.

Series, soil. The basic unit of soil classification, consisting of soils alike in all profile traits except for slight differences in surface texture, slope, erosion, or stoniness. *See* phase.

Sesquioxide clays. Finely divided particles of iron and aluminum oxides, most common in soils of warm, humid regions.

Sewage sludge. Semisolid wastes, collected by sedimentation from sewage, that, when properly treated, can be used as an organic fertilizer.

Sheet erosion. Form of erosion in which soil washes off the land in a thin, uniform layer.

Shifting cultivation. Cropping system common to tropical forest soils. A plot of forest is cut and burned, crops are grown for a few years, then the land is allowed to return to forest to fallow for several years. Also called slash-and-burn agriculture.

Shrink-swell potential. How much a mass of soil swells when wet and shrinks when dry; a function of the amount of swelling clays. An important engineering property of soil.

Sidedressing. Application of fertilizer alongside the rows of an already established crop.

Silicate clay. Clays formed by association of structural units based on silica sheets (silica tetrahedra) and alumina sheets (alumina octahedra).

Silt. Medium-sized soil separate, particles between 0.05 and 0.002 millimeter in diameter.

Silt fences. Tightly woven plastic fabric used to filter sediments from construction sites.

Single-grain soil. A structureless soil in which each soil grain is loose in the soil. Usually in sand.

Site-specific management. Trying to tailor all crop management practices to the unique needs of small sampling units in each field.

Slick spots. Small areas of a field that are slick when wet because of high sodium content.

Slow release. Term applied to fertilizers that, by various means, release their nutrients slowly. Such fertilizers may be only slowly soluble in water or be coated with substances that hinder solution.

Small grains. Crops which are planted in closely spaced rows about seven or eight inches apart or may be broadcast seeded. These crops usually refer to the cereal grains such as wheat, oats, rye, barley, and others.

Sodic soil. Soil high in sodium and low in soluble salts. The pH is between 8.5 and 10.0; electrical conductivity is below 4.0 mmhos/cm; exchangeable sodium percentage is above 15.0.

Sodium adsorption ratio. The balance of Na^{+} to CA^{+2} and Mg^{+2} in soil water extracts or water for irrigation. $[Na^{+2}]/\{([Mg^{+2}]+[CA^{+2}])/2\}^{0.5}$ where [ ] = concentration of the ion.

Soil. Loose mineral and organic material on the earth's surface that serves as a medium for the growth of land plants.

Soil air. Gas phase of soil; space of soil not filled with solid or liquid.

Soil association. *See* association, soil.

Soil classification. The arrangement of soils into classes of several levels.

Soil conservation. Protection of soil from erosion.

Soil degradation. Loss of soil quality through such processes as erosion, salination, contamination, severe compaction, and numerous others.

Soil fertility. Ability of a soil to supply the elements needed for plant growth.

Soil genesis. (i) The mode of origin of the soil with special reference to the processes or soil-forming factors responsible for development of the solum, or true soil, from unconsolidated parent material. (ii) The branch of soil science that deals with soil genesis.

Soil horizon. A layer of soil or soil material approximately parallel to the land surface and differing from adjacent genetically related layers in physical, chemical, and biological properties or characteristics such as color, structure, texture, consistency, kinds and number of organisms present, degree of acidity or alkalinity, etc.

Soil loss tolerance. The average maximum yearly loss of soil in tons per acre that will not lead to loss of productivity over time. This amount, usually between one and five, is characteristic of a soil series.

Soil matrix. The arrangement of solid soil particles and soil pore spaces which is a three-phase system of solid, liquid, and gas.

Soil-moisture tension. An older term used to describe the force attracting water molecules to soil particles; can be thought of as the "suction" required to pull water from the soil particles.

Soil order. *See* order, soil.

Soil sampling. The systematic gathering of soil samples for use in soil testing.

Soil separate. Classes of mineral particles less than 2.0 millimeters in diameter; includes clay, silt, and several sizes of sand.

Soil series. *See* series, soil.

Soil solution. The liquid phase of soil, consisting of water and dissolved ions.

Soil strength. Ability of a soil to resist deformation, related to solid phase cohesion and adhesion, water content, and degree of cementation.

Soil survey. The examination, description, and mapping of soils of an area according to the soil classification system.

Soil taxonomy. *See* taxonomy, soil.

Soil testing. Using various tests to measure properties that affect how well soil will support plant growth.

Soil triangle. Used to place a soil in a soil (textural) class by plotting the percents of sand, silt, and clay.

Soil quality. Capacity of a soil to provide the needed functions for human or natural ecosystems over the long term.

Soil-based potting mix. Potting mixes which have field soil as an important ingredient.

Soil-less potting mix. Potting mixes which contain no soil; they contain mixtures of organic and inorganic ingredients.

Soil-water potential. The amount of work a small amount of water can do in moving from the soil to a pool of pure, free water at the same location. Includes matric, gravitational, and salt potential. Normally, matric potential is the main component, but salt potential is significant in salted soils.

Solarization. A nonchemical method to sterilize soil using the sun to heat the soil under a clear plastic sheet.

Solid-set irrigation. Sprinkler irrigation from a permanently installed delivery system.

Soluble salts. Salts in the soil that are more soluble than gypsum.

Solum. The upper, weathered part of the soil profile; the A, E, and B horizons.

Solution. Water with ions dissolved in it. Applied to soil, the soil solution is the water in the soil with nutrients and other materials dissolved in it.

Splash erosion. The movement of soil particles by splashing from the impact of a drop of water.

Split application. Applying the total desired fertilizers needed in several smaller doses somewhat matching the crop's nutrient uptake needs.

Spodosol. One of twelve soil orders. Mainly acid, coarse-textured forest soils of cool, humid regions. Has subsoil layer with illuvial accumulation of humus, aluminum, and sesquioxides.

Starter fertilizers. A small amount of fertilizer placed near seeds or transplants to promote early growth.

Stem-girdling roots. Tree roots which cross the trunk at or just below the soil surface, inhibiting trunk growth and compressing conducting tissues at that side and leading to poor tree health and stability. Often results from deep planting.

Stomata. Small openings in plant leaves through which oxygen, carbon dioxide, and water vapor are exchanged.

Strip-cropping. Planting different types of crops in alternate strips to prevent wind or water erosion. Strips are usually planted on a slope contour or across the direction of the prevailing wind.

Structure, soil. The arrangement of soil particles into aggregates, or peds.

Structured soil. Man-made growing medium containing gravel, loam, and adhesive installed in urban landscapes to act as base for paving yet permit growth of tree roots.

Structureless soil. Soil with no visible aggregates. *See* massive soil and single-grain soil.

Stubble-mulching. Practice of leaving crop stubble stand between the time of harvest of one crop and beginning stages of growth of the following crop to reduce erosion or capture snow.

Subgroups. In Soil Taxonomy, a category below the great group category and above the family category.

Subirrigation. Method of irrigation in which capillary rise from a saturated zone in the soil carries water to plant roots.

Suborders. In Soil Taxonomy, a category below the order category.

Subsidence. The lowering of a surface. Organic soils subside because organic particles are lost to decay or wind erosion.

Subsoil. Soil below the plow layer, generally the B horizon.

Subsoiling. Breaking up compact subsoils or pans by the use of a chisel or other tool.

Subsurface drainage. A method of artificially draining wet soils by burying a system of tiles or perforated pipes that carry excess subsurface water off the field.

Subsurface tillage. Tillage with special tools that are drawn beneath the soil surface to kill weeds and loosen soil without mixing crop residues into the soil for protection against wind erosion.

Surface creep. Movement of sand particles along soil surface by being rolled in the wind.

Surface drainage. A method of artificially draining wet soils by digging a system of ditches that collect water and carry it off the field.

Surface irrigation. Broad term for generally flooding the soil surface with water released from canals or piping systems.

Surface soil. The top few inches of soil, usually mixed during tillage. Often the same as the plow layer.

Surface water. Water in natural or man-made bodies of water on the earth's surface, such as lakes or reservoirs.

Suspension. (1) A system of tiny particles hanging in a liquid or gas. (2) Movement of clay or silt particles in the wind by being suspended in the air.

Sustainable agriculture. A philosophy and collection of practices that seek to protect resources while ensuring adequate productivity, strives to minimize off-farm inputs like fertilizer and pesticides, and to maximize on-farm resources like livestock manure and nitrogen fixation by legumes. Soil and water management are central components.

Symbiosis. Two organisms living in close association to each other's benefit. *Compare with* parasitism.

Symbiotic nitrogen fixation. Fixation of nitrogen by microbes living in symbiosis with plant roots, usually by *Rhizobium* bacteria and several species of actinomycetes.

Syringing. A light sprinkling of turf or other plants to reduce heat stress and transpiration.

T

Talus. Deposits of dry rock and soil that have slid to the base of a slope under the force of gravity.

Taxonomy, soil. System of classifying soils to show their relationships.

Temporary wilting point. Point of plant water content at which plant wilts but at which plant will recover if watered, cooled, or placed in humid air.

Terrace. a. (Geologic) An older flood plain on a level above the current river channel and modern flood plain. b. (Soil & water conservation) A series of low ridges and shallow channels running across a slope or on the contour to capture water so it can soak into the soil or to control erosion.

Texture, soil The relative proportion of the soil separates in a soil.

Thermophilic organism. Heat-loving organism whose optimum temperature range for growth is above 35°C. Dominates composts during middle stage of decay.

Tillage. Mechanically working the soil to change soil conditions for crop growth, kill weeds, or bury crop residues.

Tilth. Physical condition of the soil in terms of how easily it can be tilled, how good a seedbed can be made, and how easily seedling shoots and roots can penetrate.

Tissue testing. Nutrients in the plant itself, instead of nutrients in the soil, are tested. Can be used as a way to see if something is hindering nutrient uptake.

Topdressing. Applying fertilizer on an established turf, hay meadow, pasture, etc. without incorporation by tillage.

Topography. *See* relief.

Topsoil. The A horizon; or soil stirred during tillage.

Total neutralizing power. An older term for calcium carbonate equivalent.

Total pore space. Portion of soil not occupied by solid material but which is filled with air or water. *See* pore space and porosity.

Trace elements. *See* micronutrients.

Transported soil. Soils formed in parent materials brought to the final location of soil formation by transporting agents such as gravity, flowing water, wind, and glacial ice sheets.

Traveling-gun irrigation. Sprinkler irrigation from a large traveling sprinkler moved by a cable attached to a heavy tractor or winch.

Type, soil. An old term, meaning the same as a textural phase of a soil series.

U

Ultisol. One of twelve soil orders. Leached soils of warm climates.

Universal Soil Loss Equation. Equation developed to predict average soil losses from a soil due to sheet and rill erosion.

Unsaturated flow. Water movement through the soil when the largest pores are not filled with liquid water.

Urban land. Areas altered by urban activities and structures, so that the soil is difficult to identify.

V

Value. One of three variables of the Munsell color system. Refers to the lightness or intensity of color.

Vermiculite. (1) One of the zil expanding clay minerals. (2) Expanded mica that is lightweight and porous. It is used in potting mixes.

Vertical mulching. A band or column of mulching material is placed into a vertical slit or narrow hole in the soil.

Vertisol. One of twelve soil orders. Soils high in swelling clays that have deep wide cracks when dry. Surface soil falls into the cracks, causing soil mixing.

Virgin soil. Soil that has not been disturbed by humans.

Volatilization. Evaporation. Also applies to loss of nitrogen in ammonium fertilizers and urea as soil reactions convert it to ammonia gas which is lost to the atmosphere.

W

Waterlogged soil. Soil whose pores are filled with water and so are low in oxygen. Caused by high water tables, poor drainage, or excess moisture from rain, irrigation, or flooding.

Watershed. The total land area in which runoff water flows into the same stream.

Water content. A measure of the water content of the soil by weight, also, but less accurately, known as moisture percentage. Calculated by subtracting dry weight from moist weight, then dividing by the dry weight.

Water table. Upper surface of a layer of saturated material in the soil.

Water table, perched. *See* perched water table.

Water-use efficiency. Crop production per unit of water reaching the land the crop occupies.

Weathering. Natural process that breaks down rock into parent materials.

Wetlands. Areas that have a predominance of hydric soils and that are inundated or saturated by surface water or groundwater at a frequency and duration sufficient to support, and under normal circumstances do support, a prevalence of hydrophytic vegetation typically adapted for life in saturated soil conditions.

Wheel-move irrigation. A line of sprinklers mounted on wheels which slowly rolls down the field as it sprinkles the land.

Wind Erosion Equation. A wind erosion prediction model which uses the following factors: soil erodability, soil roughness, climatic conditions, field length, and vegetative cover.

X

Xeriscaping. A kind of landscaping adapted to dry climates designed to reduce water use.

Index

C

N

O

T